3rd International Conference on Manufacturing Optimization, Industrial and Material Engineering

(MOIME 2015)

IOP Conference Series: Materials Science
and Engineering Volume 83

Bali, Indonesia
28 – 29 March 2015

Editors:

F. L. Gaol
J. Webb
J. Ding

ISBN: 978-1-5108-0581-1
ISSN: 1757-8981

Printed from e-media with permission by:

Curran Associates, Inc.
57 Morehouse Lane
Red Hook, NY 12571

Some format issues inherent in the e-media version may also appear in this print version.

Printed by Curran Associates, Inc. (2015)

For permission requests, please contact the Institute of Physics
at the address below.

Institute of Physics
Dirac House, Temple Back
Bristol BS1 6BE UK

Phone: 44 1 17 929 7481
Fax: 44 1 17 920 0979

techtracking@iop.org

Additional copies of this publication are available from:

Curran Associates, Inc.
57 Morehouse Lane
Red Hook, NY 12571 USA
Phone: 845-758-0400
Fax: 845-758-2634
Email: curran@proceedings.com
Web: www.proceedings.com

TABLE OF CONTENTS

IOP Conf. Series: Materials Science and Engineering **83** (2015) 011001 doi:10.1088/1757-899X/83/1/011001

The 3rd International Conference on Manufacturing, Optimization, Industrial and Material Engineering (MOIME 2015) was held at the Sheraton Kuta, Bali, Indonesia, from 28 - 29 March 2015.

The MOIME 2015 conference is aimed to bring together researchers, engineers and scientists in the domain of interest from around the world. MOIME 2015 is placed on promoting interaction between the theoretical, experimental, and applied communities, so that a high level exchange is achieved in new and emerging areas within Material Engineering, Industrial Engineering and all areas that relate to Optimization.

We would like to express our sincere gratitude to all in the Technical Program Committee who have reviewed the papers and developed a very interesting Conference Program, as well as the invited and plenary speakers.

This year, we received 99 papers and after rigorous review, 24 papers were accepted.

The participants come from eight countries. There were four parallel sessions and two invited speakers.

It is an honour to present this volume *of IOP Conference Series: Materials Science and Engineering* (MSE) and we deeply thank the authors for their enthusiastic and high-grade contributions.

Finally, we would like to thank the conference chairmen, the members of the steering committee, the organizing committee, the organizing secretariat and the financial support from the conference sponsors that allowed the success of MOIME 2015.

The Editors of the MOIME 2015 Proceedings

Dr, Ford Lumban Gaol
Jeff Webb, Ph.D
Prof. Jun DING, Ph.D

MOIME 2015

The 2015

ID 124:
RELATIONSHIP BETWEEN LEAN PRODUCTION AND OPERATIONAL PERFORMANCE IN THE MANUFACTURING INDUSTRY

IOP Conf. Series: Materials Science and Engineering **83** (2015) 011003 doi:10.1088/1757-899X/83/1/011003

Invited and Plenary Speakers

Keynote Speakers

1. Jeff Webb, Ph.D

Short Bio:

Jeff Webb holds a Bachelor's degree in Applied Physics and a PhD in Photothermal Physics; currently he is an associate professor at the University of Nottingham Malaysia Campus. He has published more than 80 papers in international conferences and journals. He has given keynote speeches in various places including Myanmar, Jakarta and Seoul, Korea. He has a broad range of research interests encompassing quantum cascade lasers, graphene-based devices and structures, and ferroelectric materials. Most of his work is concerned with theory and modelling. He is a member of the Institute of Physics and holds Chartered Physicist status with this institute (CPhys).

Title:

Quantum Cascade Lasers: how to model them with simple, informative two and three level systems.

Abstract:

This talk will discuss how relatively simple models utilising two or three quantum levels can be used to give a good account of the behaviour of quantum cascade lasers. Various parameters useful for characterizing the laser's behaviour and performance will be covered together with some possible applications.

2. Prof. Jun DING, Ph.D

Short Bio:
Dr Jun DING obtained a PhD from Ruhr-University Bochum, Germany in 1990. After working as Research Fellow and Senior Research Fellow at University of Western Australia for 6 years, Dr Ding joined National University of Singapore as a Faculty (NUS). Now, he is Professor at Department of Materials Science & Engineering (NUS) and Visiting Professor at NIMTE, Chinese Academy of Science. Dr Ding has been working in magnetic materials for 25 years. He has published over 300 papers with > 7000 citation and H-index = 47. His current research work focuses on nanomagnetics and magnetic nanoparticles for bio-medical applications.

Title:
Synthesis of Fe_3O_4 nanoparticles for biomedical applications

Abstract:
Magnetic nanoparticles are promising for bio-medical applications, including MRI contrast enhancement, hyperthermia, drug delivery and separation. Superparamagnetic nanoparticles are widely used in order to avoid undesired agglomeration. However, low saturation magnetization and low magnetic susceptibility of superparamagnetic nanoparticles limit their performance. Recently, we have successfully fabricated magnetic nanorings which can be well suspended in water with suitable surfactant. Compared to superparamagnetic nanoparticles, these Fe_3O_4 nanorings exhibited much higher magnetization and susceptibility, so that a relatively low magnetic field is able to achieve a high magnetic flux. High R2* values have been obtained, demonstrating their great potential for T2 MRI agent. More recently, we have tested hyperthermia performance of Fe_3O_4 nanorings and nanodiscs. The SAR values (the measure of hyperthermia performance) obtained are much higher compared to those for reference (Fe_3O_4 nanoparticles). Our calculation for hyperthermia also confirms large SAR values because of their Neel and hysteresis contribution. Our work has shown the great potential of non-superparamagnetic nanorings and nanodiscs.

Peer review statement

All papers published in this volume of *IOP Conference Series: Materials Science and Engineering* have been peer reviewed through processes administered by the proceedings Editors. Reviews were conducted by expert referees to the professional and scientific standards expected of a proceedings journal published by IOP Publishing.

Three hybridization models based on local search scheme for job shop scheduling problem

Tatiana Balbi Fraga

Department of Production Engineering, Centro Acadêmico do Agreste, Universidade Federal de Pernambuco, Caruaru, PE, Brazil.

Email: tatianabf_8@hotmail.com

Abstract. This work presents three different hybridization models based on the general schema of Local Search Heuristics, named Hybrid Successive Application, Hybrid Neighborhood, and Hybrid Improved Neighborhood. Despite similar approaches might have already been presented in the literature in other contexts, in this work these models are applied to analyzes the solution of the job shop scheduling problem, with the heuristics Taboo Search and Particle Swarm Optimization. Besides, we investigate some aspects that must be considered in order to achieve better solutions than those obtained by the original heuristics. The results demonstrate that the algorithms derived from these three hybrid models are more robust than the original algorithms and able to get better results than those found by the single Taboo Search.

1. Introduction

The job shop scheduling problem (JSP) comes from manufacturing environments where jobs usually differ considerably in their processing sequence and times. This problem is classified as NP-Hard in strong sense [1] and it is widely known in the technical literature as one of the most difficult issues in the combinatorial analysis field. As stated by Zhang *et al.* [2], the JSP complexity can be illustrated by the fact that a relatively small instance with 10 jobs and 10 machines, proposed by Fisher and Thompson [3], remained unsolved for more than a quarter of a century and actually it remains difficult to solve optimally benchmarks with more than 20 jobs and 20 machines. Briefly, the JSP can be presented as follows: a set of m machines, $\{M_k\}_{k=1}^{m}$, and a set of n jobs, $\{J_i\}_{i=1}^{n}$, are considered. Each job J_i must be processed exactly once in each machine M_k in a predetermined order, which may vary from one job to another. Therefore, each job can be understood as a sequence of operations, $J_i = \left(O_{ik_s}\right)_{s=1}^{m}$, where O_{ik_s} is the operation of job J_i that must be processed by the machine M_{k_s} for a deterministic processing time, $t_{ik_s} \geq 0$. Additionally: each machine can process only one job at a time; no job can be processed by two or more machines at the same moment; and preemption is not allowed. The objective of this problem is to determine the order in which jobs should be processed on each machine in order to minimize the length of time it takes to complete all operations for all jobs (usually referred to as makespan), without disregarding the restrictions imposed by the problem. Note that the dimensionality of each JSP instance can be specified as $n \times m$ and the total number of solutions is defined by $(n!)^m$.

Due to its practical importance and its complex nature, the JSP has attracted the attention of many researchers, and a wide variety of algorithms has been proposed to solve this problem. According to

Jain and Meeran [4], these algorithms can be classified, in one hand, as optimization algorithms, when they guarantee the generation of optimal solutions (eventually at infinite time and generally involving a high computational cost), and on the other hand, as approximation algorithms, which provide good solutions (but not necessarily optimal) with a reasonable computational cost. Among the optimization algorithms are the family of constructive, or efficient, algorithms for which an optimal solution is built following a simple set of rules defined by the algorithm itself [5, 6, 7], and enumerative algorithms in which successive tentative solutions are generated, one by one, and elimination procedures are used in order to restrict the search space.

The main enumerative algorithms are the Mathematical Programming Techniques, as the Mixed Integer of Manne [8], and the various Branch and Bound algorithms [9, 10]. The approximation algorithms are constituted by a great diversity of families of algorithms ranging from Priority Dispatch Rules [11] and Bottleneck Based Heuristics [12, 13, 14] to artificial intelligence methodologies, such as Constraint Satisfaction Techniques [15], Neural Networks [16] and Ant Colony Optimization algorithms [17], but their relative success is mainly attributed to a class of approximation algorithms named Local Search Heuristics, that includes the well known Simulated Annealing [18, 19], Genetic Algorithms [20, 21], Taboo Search [22, 23, 24] and, more recently, Particle Swarm Optimization [25, 26, 27] heuristics, and until now no algorithm developed was able to find optimal solutions in an acceptable processing time for all the benchmark problems proposed in the literature. An extensive discussion of the family of algorithms applied to solve the job shop scheduling problem is presented by Blazewicz *et al.* [28] and by Jain and Meeran [4].

As each family of algorithms has its own positive and negative aspects, in order to take the best characteristics of the already existing algorithms and overcome their weaknesses, a possible approach consists in the use of hybrid algorithms, and this has been the focus of many authors. These hybrid algorithms can be classified as: *blended*, when generated by the insertion of specific procedures of a family of algorithms, as the taboo list of Taboo Search or the Metropolis criterion of Simulated Annealing, into an algorithm belonging to another family of different nature; *unblended*, when generated by the simultaneous use of algorithms of different nature, *i.e.* deterministic and non-deterministic or population-based and individual-based, so that each algorithm conserves its own characteristics; or *mixed*, when both blended and unblended procedures are applied. As an example of hybrid algorithms, we mention the work by Pezzella and Merelli [29], which based on the idea that the results generated by Taboo Search are tightly dependent on its initial solutions, proposed an unblended hybrid algorithm where the Shifting Bottleneck Procedure of Adams *et al.* [12] is applied in order to generate the initial solutions. Wang and Zheng [30] proposed an unblended hybrid algorithm where initially a Genetic Algorithm is applied in order to generate the initial population, then, for each initial element the Simulated Annealing performs a Metropolis sampling until the equilibrium condition is reached, and then the Genetic Algorithms uses the solution found by Simulated Annealing to continue the parallel evolution. Azizi and Zolfaghari [19] added a taboo list in their algorithm, named Adaptive Simulated Annealing, generating a blended hybrid algorithm that improved its performance significantly. Zhang *et al.* [2] presented an unblended hybrid algorithm where initially the Simulated Annealing is applied in order to find good solutions in a relatively wide search space and then the Taboo Search is applied in order to intensify the search process.

In this work the Local Search Heuristics are designed in a general representation and, based on this representation, three hybridization models are presented and analyzed with two local search algorithms, a deterministic version of the Taboo Search algorithm, with the neighborhood structure proposed by Nowicki and Smutnicki [23] without back-propagation, and the Similar Particle Swarm Optimization algorithm suggested by Lian *et al.* [25]. As the main result, the present work shows that the complementary nature of Taboo Search and Particle Swarm Optimization guarantee the efficacy and robustness of the three models presented providing better results than the separate application of both algorithms. The remainder of the paper is organized as follows: an introduction for the Local Search Heuristics as well as the Taboo Search and Similar Particle Swarm Optimization algorithms used to test the hybridization models are given in sections 2, 3 and 4, respectively. Section 5 gives a brief description of the three hybridization models investigated. Section 6 discusses the experimental results; and, finally, in section 7 conclusions are drawn, and some proposals for future research are suggested.

2. Local Search Heuristics

The Local Search Heuristics - also referred to as neighborhood techniques - are characterized by the fact that, at each iteration, a set of new solutions is generated in the neighborhood of a set of "parent" ones by introducing small perturbations (usually called "moves") of the latter (figure 1). Then a selection procedure eliminates a part of the available solutions in order to get the new set of "parent" solutions which will be used for the neighborhood generation of the next iteration. The process continues until a given stopping criterion is reached.

According to the choice of the neighborhood generation techniques and selection criterion, different heuristics are obtained. As an example, in Taboo Search heuristic, at each cycle, the neighborhood is generated from a single parent solution, usually swapping the positions of two of its operations, and then the new parent solution is formed by the best solution within the neighborhood that does not belong to the taboo list. The Simulated Annealing heuristic usually applies the same neighborhood generation techniques that in Taboo Search but, at each cycle, just one neighbor is generated from a single parent solution and, if its makespan is smaller than the parent solution's makespan, this solution replaces the parent one, otherwise, a Metropolis criterion is applied in order to define if the generated solution replaces or not the parent one. In the case of the Genetic Algorithms and Particle Swarm Optimization heuristics, at each cycle, the neighborhood is generated from a set of parent solutions (population), usually applying crossover and mutation operations on the pairs of these solutions. The basic difference of both methods is the way in which these pairs of solutions are selected. While the Genetic Algorithm usually applies a selection criterion based on a probability related to an efficiency index of each solution, the selection criterion applied by the Particle Swarm Optimization is always based on the best solution ever generated and the best solution visited by each individual.

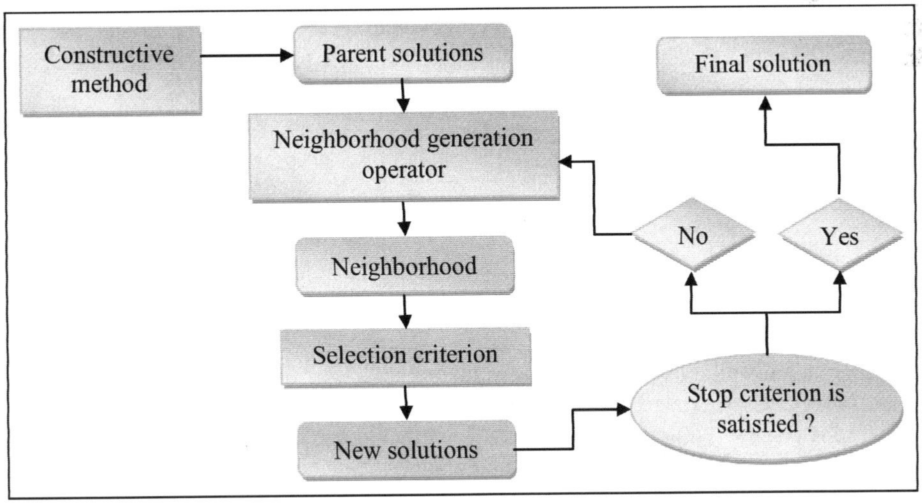

Figure 1. Scheme of Local Search Heuristics.

Additionally, the choice of the specific method that generates the initial parent solutions, the specific operator that must be applied to generate the neighborhood and the specific selection and stop criteria, allow generating a wide variety of algorithms for each family. As stated previously, in this work two algorithms were chosen to be applied in the hybrid models, a deterministic version of the

Taboo Search algorithm, with the neighborhood structure proposed by Nowicki and Smutnicki [23] without back-propagation, and the Similar Particle Swarm Optimization algorithm suggested by Lian *et al.*[25]. In the following sections, both of them will be presented.

3. Taboo Search algorithm

The Taboo Search (TS) heuristic was originally developed by Glover [31] and its first application to JSP is attributed to Taillard [22]. In its classical form the TS algorithm can be described as follows: at first an initial solution is generated randomly or by a constructive method, which is stored as the current and the best solutions. Then, at each iteration, a neighborhood is generated by applying "moves" (*i.e.*, small perturbations) to the current solution and so this solution is replaced by a non taboo neighbor, *i.e.* a solution in the neighborhood generated (usually the best one), under the restriction that its generating movement doesn't belong to the taboo list. This last one is a vector containing the inverse movements that generated the last t current solutions, and is introduced as a way to prevent the search process of getting trapped in a locally optimal solution – where t is the size of taboo list and is always referred as "tenure". An aspiration criterion can be defined in order to allow the replacement of the current solution by a solution that belongs to the taboo list when it is useful for the search (*i.e.* when this solution is better than the best solution found). The best solution is replaced by the current solution if the latter presents a smaller makespan than the first one. The process continues until a given stopping criterion is satisfied.

generate an initial solution
store the initial solution as the current and the best solutions
while stopping criterion is not satisfied
    generate a neighborhood
    replace the current solution by the not taboo best neighbor
    update the taboo list adding the inverse move that generated the new current solution and removing the oldest one (when taboo list size is bigger than t)
    if the makespan of the current solution is smaller than the makespan of the best solution
        replace the best solution by the current solution
    end if
end while
return the best solution

Figure 2. Taboo Search algorithm.

Figure 2 presents a general TS algorithm. In this scheme, the neighborhood generation process has a direct impact on the efficiency of the method and several neighborhood structures have been presented in the literature (*i.e.*, see [22, 23, 24]). Among them, the one usually referred as to N5 introduces the real breakthrough in both efficiency and effectiveness for the job shop problem. This method generates a neighborhood substantially smaller than the others and was chosen to be applied in this work. The N5 method, the best solution selecting process, the taboo list, and the stopping criterion of the TS algorithm applied in this work will be described in more detail in the following sections.

3.1. N5 neighborhood generation method

Given a solution, a critical path can be defined as a sequence of operations wherein:

- the last element in this sequence is the last operation completed. So its final processing time is the makespan of the respective solution;

- for each operation in this sequence, except for the first one, its immediately preceding operation is either: the operation that precedes it in the machine order or in the task order. Between the two, it is the one which presents the higher finishing time that is chosen;
- the start time of the first operation of this sequence is zero.

It is important to note that a solution can contain more than one critical path, as happens when there are two or more operations that are finalized last, or when operations preceding an operation on the machine and task orders are completed at the same moment. In their article, Nowicki and Smutnicki [23] argue that the choice of the critical path does not have a strong influence on the final result. The authors suggest that it should be chosen randomly. In this study, the critical path is selected in accordance with the following priorities:

- if two operations are finalized last, the operation selected will be the one belonging to the lowest machine index;
- if the operations, which precede an operation in machine and task orders, finish at the same moment, the operation selected will be the one above in the task order.

The critical path can also be subdivided into blocks of operations where each block consists of a subsequence of successive operations that must be processed by the same machine. By the N5 method, a neighborhood is formed by swapping the first two (and the last two) operations of each block on a critical path (where each swap will generate a new neighbor). For the first block, only the last two operations are swapped and for the last block, only the first two operations are swapped. If any block is formed by only one operation, no exchange will be made and if it is formed by two operations only one exchange must be performed. (for more details see Nowicki and Smutnicki [23]).

3.2. Selecting the best non-taboo solution

In the TS operator applied in this work, a taboo list consists of a vector containing the moves inverse to those used to generate the latest t current solutions (where t is a parameter, the value of which can be adjusted). Thus, the best neighbor (the solution chosen as the new current solution) is the solution with the smallest makespan, whose motion generator is not on the taboo list. Additionally, an aspiration criterion is considered which allows the current solution to be replaced by a solution whose motion generator belongs to the taboo list if the makespan of this solution is less than that of the best solution found so far. Should there be a tie for the best solution, the current solution remains the same.

3.3. Stopping criterion

In this algorithm the iterative process stops when after $Nite$ iterations the best solution found is not updated, *i.e.* no new current solution has a lower makespan than that of the best solution found. $Nite$ is a parameter, the value of which can be adjusted.

4. Similar Particle Swarm Optimization

Particle Swarm Optimization (PSO) is a population-based heuristic developed by Kennedy and Eberhart [32]. In this method, each solution is interpreted as the position of a particle (individual of the swarm), which is represented by a D-dimensional vector, where D is the number of variables that must be determined. The trajectories of the particles look for the position corresponding to an optimal solution. According to the representation given in figure 3, the implementation of PSO can be described as follows: the initial position of the swarm is randomly generated and then the individuals or potential solutions, named particles, search for an optimum by updating their own positions. At each iteration the position of each particle is adjusted according to its velocity which is randomly generated towards the best position visited by the particle (p_{best}) and the best position visited by the swarm (g_{best}). At iteration k, for each particle i ($1 \leq i \leq NP$, where NP represents the total number of particles), the velocity v_i and the position x_i can be updated by the following equations:

$$v_i(k+1) = w \times v_i(k) + c_1 r_1 \left(p_{\text{best}\,i}(k) - x_i(k)\right) + c_2 r_2 \left(g_{\text{best}}(k) - x_i(k)\right) \quad (1)$$

$$x_i(k+1) = x_i(k) + v_i(k+1) \quad (2)$$

The inertia weight w, first proposed by Shi and Eberhart [33], is used to control exploration and exploitation. A larger w can prevent particles to becoming trapped in local optima, and a smaller w encourages particle exploitation of the same search space area. The constants c_1 and c_2 are learning factors used to decide whether particles prefer moving toward a p_{best} or g_{best} position. Usually $c_1 = c_2 = 2$. The r_1 and r_2 are random variables between 0 and 1.

initialize a population of particles with random initial positions and velocities on *D* dimensional space
while stopping criterion is not satisfied
    update the velocity of each particle, according to Eq. (1)
    update the position of each particle, according to Eq. (2)
    evaluate the fitness value of each position according to the desired optimization fitness function and at the same time, update p_{best} and g_{best} position if necessary
end while
return the best global solution (g_{best})

Figure 3. Particle Swarm Optimization algorithm for problems in continuous space.

The original PSO algorithm was developed to solve continuous optimization problems. When working with combinatorial optimization problems, we have to modify the representation of the positions and the way how velocity and movement are adjusted (some examples of adaptations of the original PSO applied for JSP can be found in [25, 26, 27]). Lian *et al.* [25] proposed a Similar Particle Swarm Optimization (SPSO) algorithm for the JSP where the position of each particle is mapped by the work procedure code (see below), which considers only feasible solutions, and its velocity and position are adjusted according to the following equations:

$$v_i(k+1) = p_{\text{best}\,i}(k) \ \Theta \ g_{\text{best}}(k) \quad (3)$$

$$v_i(k+1) = \begin{Bmatrix} v_i(k+1) \ \ if \ mut_i = 0 \\ M\left(v_i(k+1)\right) \ \ if \ mut_i = 1 \end{Bmatrix} \quad (4)$$

$$x_i(k+1) = x_i(k) \ \Theta \ v_i(k) \quad (5)$$

$$x_i(k+1) = \begin{Bmatrix} x_i(k+1) \ \ if \ mut_i = 0 \\ M\left(x_i(k+1)\right) \ \ if \ mut_i = 1 \end{Bmatrix} \quad (6)$$

where Θ and $M(x)$ represents, respectively, the crossover and mutation operators applied in Genetic Algorithms and the boolean variable mut_i is a flag destined to indicate if the mutation operation is ($mut_i = 1$) or not ($mut_i = 0$) applied on particle i. At each iteration, N_{mut} particles are randomly chosen to suffer a mutation operation:

$$\sum_{i=1}^{NP} mut_i = N_{\text{mut}} \tag{7}$$

The mutation probability (p_{mut}), *i.e.* the percentage of particles that suffer mutation at each iteration, is defined by:

$$p_{\text{mut}} = \frac{N_{\text{mut}}}{NP} \times 100 \tag{8}$$

The authors tested four crossover (C1 – C4) and ten mutation (M1 – M10) operators on three benchmark problems proposed by Fisher and Thompson [32] named FT6, FT10 and FT20, and the SPSO algorithm has shown to be more efficient than the Genetic Algorithms in the solution of job shop scheduling problems. Despite the SPSO presenting better results than the Genetic Algorithms, these results have not yet reached the desired purpose and, additionally, like the TS heuristics, the SPSO is quite sensitive to the value of its control parameters, as, for example, the percentage of particles that suffer a mutation operation by iteration in which small values lead the algorithm to falling in a cyclic process, and the large values impaired its convergence.

From all four crossover and ten mutation operators we chose to show in this work the results of the application of the M7 (single job moving-inserting) mutation operator and a new crossover operator, described as C1 with 1 floating point. The working procedure code and the reported operators are described in the following sections.

4.1. Work procedure code

In the work procedure code (WPC), a schedule is represented by a sequence of the indexes of jobs in which each job index is repeated by the number of its corresponding operations. A corresponding operation can be known by the job index and the position that it appears, *i.e.* the first time that a job index appears in the sequence represents the first operation of this job, the second time that the same job index appears in the sequence represents the second operation of this job and so on. For example: in a job shop problem with 3 jobs and 3 machines a possible solution can be represented as follows:

WPC (1 3 2 1 2 3 3 2 1)

In WPC, the position of each operation determines the order of precedence in which it should be processed. So the sequence, in which tasks should be processed on each machine, is determined by taking operations for their respective machines in the same order as that in which they appear in the solution given by WPC. One inconvenience of this form of representation is that two apparently different solutions, when plotted by WPC, can portray the same solution (when represented by the sequence of tasks on each machine). Observe that the N5 neighborhood generating method is based on the critical path. Therefore, for the application of the TS operator, the solution should be represented by the Task Sequence Code (TSC) (*i.e.*, the sequence of tasks on each machine). During the iterative process, the representation of the solutions is converted to WPC or TSC, depending on the algorithm applied.

4.2. C1 with one floating point

A crossing point is randomly selected along the length of the first chromosome (sequence of operations). The sub-section of jobs from the first position to the crossing point is copied into the offspring. The remaining places of the offspring are filled up by taking in order each legitimate gene of the second chromosome.

4.3. Single job moving-inserting (M7)

One moving and one inserting point are randomly selected along the length of the chromosome then the job at moving point are moved and placed in the inserting point.

5. Hybridization models

The hybridization idea presented in this work is based on the schema of Local Search Heuristics (figure 1) and consists basically on generating a hybrid neighborhood by applying two or more neighborhood generation operators. Table 1 presents some of the possible neighborhood generation operators, classified according to their nature. These operators can be the same neighborhood generation operators applied in the Local Search Heuristics, as previously mentioned, or any other algorithm capable of transforming a group of parent solutions in a new group of solutions, through the perturbation of the first, as in the case of the own Local Search Heuristics when using as their respective initial solutions, the set of parent solutions. The basic difference between both kinds of operators is that, in the first case, the hybrid algorithm generated will be blended and, in the last case, unblended.

Table 1. Neighborhood generation operators.

				Local Search Heuristics	
	Critical path swap operators	Crossover operators	Mutation operators	New solutions are generated by introducing small perturbations in just one solution	New solutions are generated by crossing two solutions
Deterministic	N1, N2, N3, N4, N5, N6 – Zhang *et al.* [24]			Taboo Search	
Nondeterministic		C1, C2, C3, C4 – Lian *et al.* [25]	M1, M2, M3, M4, M5, M6, M7, M8, M9, M10 – Lian *et al.* [25]	Taboo Search, Simulated Annealing	Genetic Algorithms, Similar Particle Swarm Optimization

Usually the hybridization of neighborhood generation operators enables the generation of algorithms more robust, *i.e.* less sensitive to the adjustment of parameters and able to produce better results than each single algorithm separately, provided that the operators used in the hybridization are of different nature. In this study the TS algorithm (as presented in section 3) and the SPSO algorithm (as presented in section 4) were chosen as operators for generating the hybrid neighborhood because they present different aspects in different directions. This choice was based on the following considerations:

- The TS heuristic improves every single parent solution in order to bring it to a minimum point that can be local or global. The SPSO heuristic generates an iteration between the parent solutions, converging the solutions to an optimal point (local or global) common to all solutions;
- The TS heuristic can be easily presented as deterministic while the random aspect is a fundamental and essential part of the SPSO heuristic;
- The TS algorithm with the neighborhood generation method N5 proposed by Nowicki and Smutnicki [23] is still one of the most effective and efficient algorithms for solving the JSP;
- Despite the SPSO and Genetic Algorithms heuristics presenting similar characteristics, the SPSO algorithm proposed by Lian *et al.* [25] has achieved better results than the Genetic Algorithms proposed in the literature;

- Despite another PSO algorithms were developed to solve JSP, in the SPSO algorithm proposed by Lian *et al.* [25] it is not necessary to set the parameters of the original PSO (*i.e.*, the inertia weight and the learning factors).

This work analyzes some of these possibilities of combining the operators in order to select which are the main aspects to be considered to achieve an improvement in the results of the original algorithms. The following sections present the hybridization schemes analyzed in this paper, respectively defined as Hybrid Successive Application (HSA), Hybrid Neighborhood (HN) and Hybrid Improved Neighborhood (HIN).

5.1. *Hybrid Successive Application*

In the Hybrid Successive Application scheme (figure 4) a set of initial parent solutions is randomly generated. Then, at each cycle, the first neighborhood generation operator (NGO) is applied to the parent solutions and the other NGO are successively applied on the set of solutions furnished by the preceding one. The final neighborhood provided in each cycle corresponds to the neighborhood generated by the last NGO.

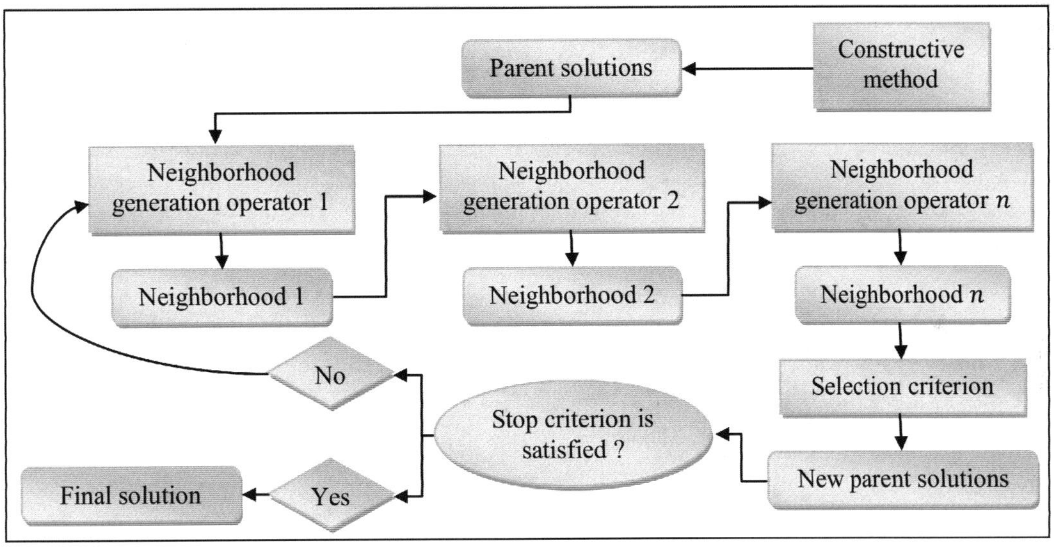

Figure 4. Scheme of the Hybrid Successive Application.

5.2. *Hybrid Neighborhood*

In the Hybrid Neighborhood scheme (figure 5) in each cycle all the NGO are applied on the same set of parent solutions (set of initial solutions of the cycle) and then the set of parent solutions of the next cycle is randomly selected from the group of solutions furnished by all the neighborhood generation operators.

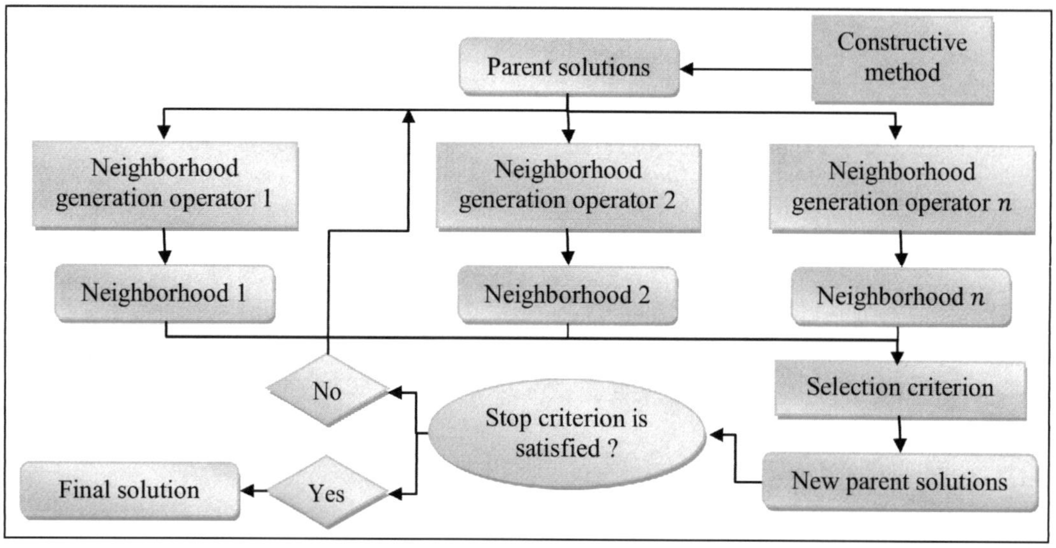

Figure 5. Scheme of the Hybrid Neighborhood.

5.3. Hybrid Improved Neighborhood

In the Hybrid Improved Neighborhood scheme (figure 6), the same procedure of the Hybrid Successive Application is used, but the set of new parent solutions of a cycle derives from a random selection on the group of all solutions obtained by the different neighborhood generation operators.

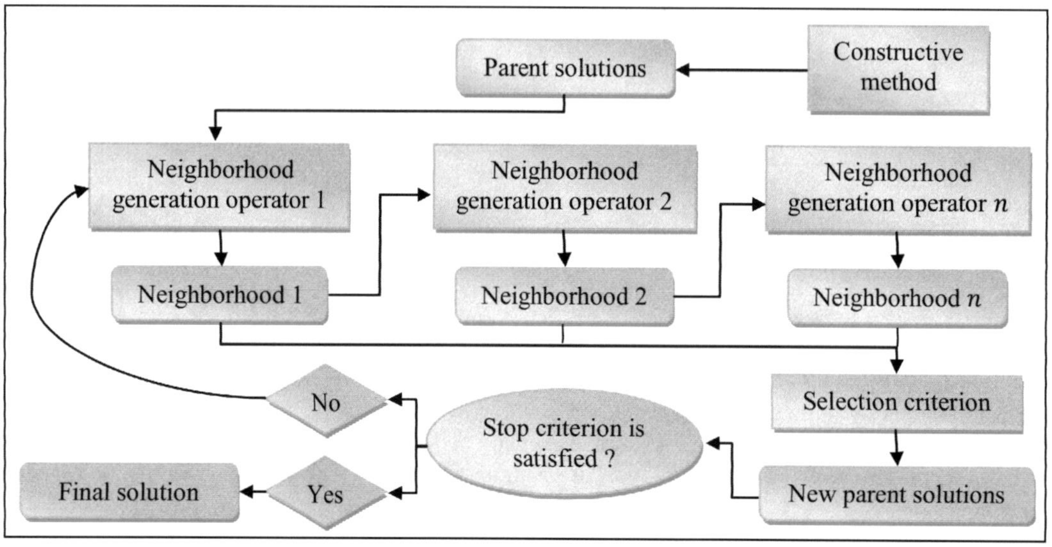

Figure 6. Scheme of the Hybrid Improved Neighborhood.

6. Computational results

To analyze the robustness and efficiency of the three hybridization models proposed in this work, the schemes presented in Figures 4, 5 and 6 were tested on the benchmark problems FT6 (with 6 jobs and 6 machines), FT10 (with 10 jobs and 10 machines) and FT20 (with 20 jobs and 5 machines) proposed by Fisher and Thompson [34] and problems ABZ5 and ABZ6 (with 10 jobs and 10 machines), and ABZ7, ABZ8 and ABZ9 (with 20 jobs and 15 machines) proposed by Adams *et al.* [12] with the algorithms TS, as presented in section 3, and SPSO, as presented in section 4, respectively, and inverting their orders. For each benchmark problem twelve initial solutions (initial configurations) were randomly generated and they were used as initial solutions of all algorithms for all the tests applied. As SPSO is a population based algorithm and TS is an individual based one, the first was applied on the set of solutions and the last was applied on every single solution. As the SPSO is non-deterministic, in order to minimize the distortion generated by the random appearance, each test was repeated five times.

Initially, a sensitivity analysis of the parameters "maximum number of iterations without upgrade" of the TS algorithm and "number of iterations" of the SPSO algorithm was performed. The results obtained demonstrated that when the SPSO algorithm starts with random positions (solutions), the method searches into the space delimited by the initial positions a local optimum and the percentage of particles that suffer mutation (p_{mut}) holds a fundamental role in the convergence process. Small values lead the algorithm to a cyclic process and large values impaired its convergence. When applying successively the TS and the SPSO algorithms, as TS walks through local minima, the SPSO tends to converge to the best point found by the previous algorithm and, unless p_{mut} has a large value, the SPSO is not able to improve the results of TS. When setting large values for p_{mut}, the SPSO allows TS to escape from local minima and the fast convergence of TS leads to a global optimal point. However, this method is quite sensitive to the p_{mut} parameter. Otherwise, when SPSO is applied with no more than one iteration, the hybrid algorithms tend to combine the global convergence property of SPSO with the local convergence property of TS, and all results tends to converge to a global optimum. When applying the HN and the HIN schemes we can observe the same behavior except that the SPSO tends to concentrate on the best available solution.

Based on this observation a sensitivity analysis of the parameters "maximum number of iterations without upgrade" and "tenure" of TS algorithm, and "percentage of particles that suffer mutation" of the SPSO algorithm, as well as a convergence analysis based on the parameter "cycles number" of the hybrid algorithms presented here, were applied fixing the parameter "number of iterations" of SPSO as one. In the next sections we present the results of these analyses - all results generated and an extensive discussion of these analyses are presented in Fraga [35].

6.1. Sensitivity analysis with respect to the maximum number of iterations without upgrade of the makespan

First, a sensitivity analysis was applied on the single TS on each one of the set of initial solutions defined, fixing the value of the parameter "tenure" to eight and setting the parameter "maximum number of iterations without upgrade" to the values 10, 10^2, 10^3, 10^4, 10^5 and 10^6. For all benchmarks analyzed, excepted FT6 for which the method found optimal solutions with no more than 10 iterations, the results demonstrated that the TS method is quite sensible to the initial solution, as it generated different results for each one. For all initial solutions of the defined set, it was identified a saturation point (ST) for the value of the parameter analyzed in which the method falls into cyclic processes and the result cannot be upgraded (figure 7). It was observed that, for most of the initial solutions, ST is about 10.000 for the problems with 10 jobs and 10 machines (FT10, ABZ5 and ABZ6) and about 100.000 for the problems with 20 jobs and 15 machines (ABZ7, ABZ8 and ABZ9), but some solutions presented smaller or larger ST.

Afterwards, the same sensitivity analysis was applied to the three hybrid models proposed here with ten cycles, fixing the value of parameters "tenure" of TS and "percentage of particles that suffer mutation" of SPSO, respectively at eight and 20%, and setting the parameter "maximum number of iterations without upgrade" of TS to the values 10^2, 10^3, 10^4 and 10^5. Despite the hybrid algorithms not being able to find optimal solutions for all benchmark problems, when applied with ten cycles and the defined small set of initial solutions, they demonstrated to be able to significantly improve the results found by Taboo Search (figure 8), even when the order of their neighborhood generation operators was inverted (figure 9). For the 20×15 problems it was also observed that these improvements are smaller when the value of the parameter analyzed is increased. For the majority of the analyses, the

Hybrid Successive Application was the model that generated the best results, especially when considering small values for the parameter analyzed.

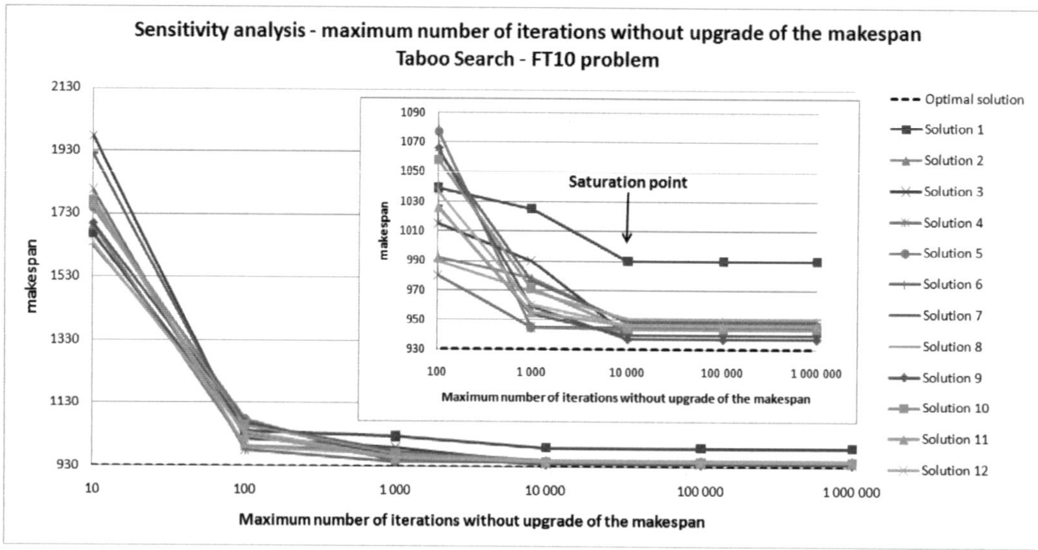

Figure 7. Sensitivity analysis with respect to the "maximum number of iterations without upgrade" of the Taboo Search algorithm for the FT10 problem with the following fixed parameter: tenure = 8.

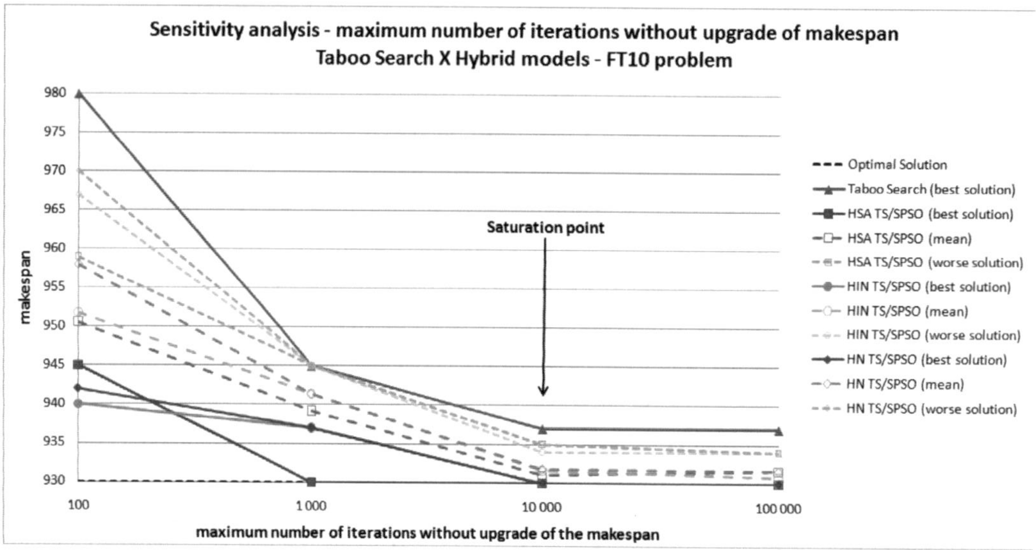

Figure 8. Sensitivity analysis with respect to the "maximum number of iterations without upgrade" of the Hybrid Models (HSA, HIN and HN) for the FT10 problem with the following fixed parameters: tenure = 8; percentage of particles that suffer mutation = 20%; cycles number = 10. Comparison between the best, the mean and the worse results generated by the three hybrid models and the best result generated by the Taboo Search algorithm.

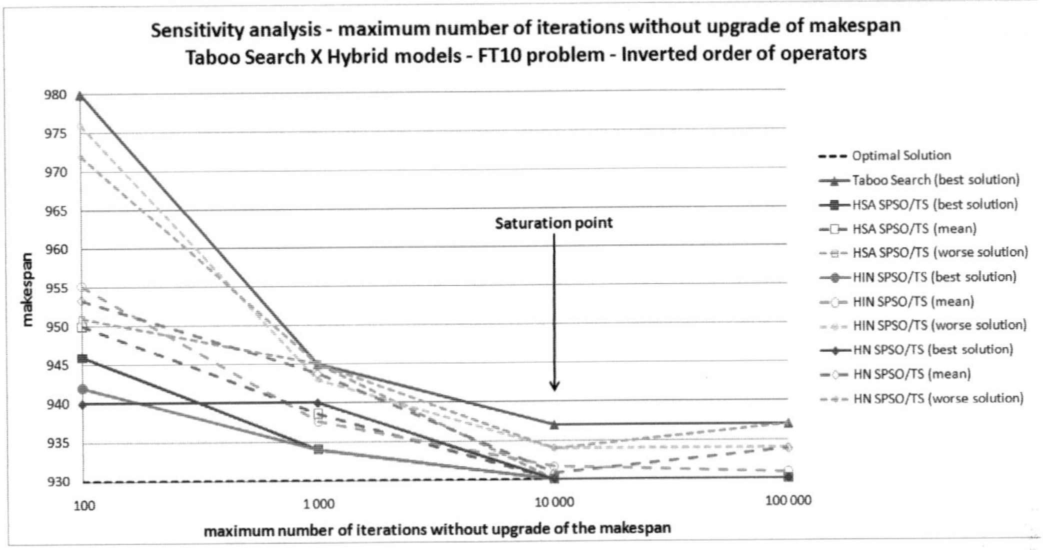

Figure 9. Sensitivity analysis with respect to the "maximum number of iterations without upgrade" of the Hybrid Models (HSA, HIN and HN), with inverted order of the neighborhood generation operators, for the FT10 problem with the following fixed parameters: tenure = 8; percentage of particles that suffer mutation = 20%; cycles number = 10. Comparison between the best, the mean and the worse results generated by the three hybrid models and the best result generated by the Taboo Search algorithm.

6.2. Sensitivity analysis with respect to the Taboo Search tenure

The sensitivity analysis with respect to the value of the "tenure" parameter was also first applied to the single TS algorithm on each solution of the set of initial solutions generated for each different benchmark, fixing the value of the parameter "maximum number of iterations without upgrade" of TS at 10^3 for the problem FT20, at 10^4 for the problems FT10, ABZ5 and ABZ6, and at 10^5 for the problems ABZ7, ABZ8 and ABZ9. Also the value of the parameter "tenure" was set from 6 to 15. The results confirmed the high sensitivity of the algorithm to the initial solution and also demonstrated that it is quite sensitive to the tuning of the tenure parameter. Different initial solutions presented different sensitivity for the tuning of this parameter, as can be seen in solution 6 and in solution 11 generated for the FT10 problem (figure 10). After the same analysis was applied to the hybrid models fixing the value of the parameter "maximum number of iterations without upgrade" of TS at 10^3 for the problem FT20, at 10^4 for the problems FT10, ABZ5 and ABZ6 and at 10^5 for the problems ABZ7, ABZ8 and ABZ9, the value of parameter "percentage of particles that suffer mutation" of SPSO at 20%, and setting the value of the parameter "tenure" from 6 to 15. The results demonstrated that, despite the hybrid algorithms also presenting sensitivity to the parameter analyzed, this sensitivity is smaller and, when the neighborhood generation operators TS and SPSO were applied, in this order, even the worst result from the five runs for which each test was applied was better or, at least, equal to the best result provided by the single TS, when applied to each component of the same set of initial solutions and with the same parameters. Additionally, the results of best solutions, in most cases, significantly improved the best result provided for the single TS. As the value of the parameter "maximum number of iteration without upgrade" of TS was fixed as 10^4 (for 10×10 problems) or 10^5 (for 20×15 problems), both models the HSA and the HN presented the best results.

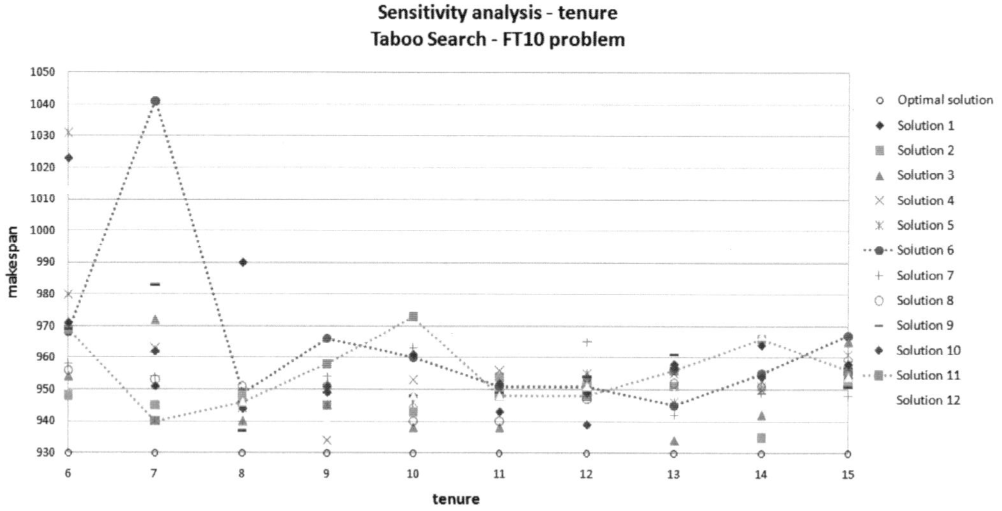

Figure 10. Sensitivity analysis with respect to tenure of the Taboo Search algorithm for the FT10 problem with the following fixed parameter: maximum number of iterations without upgrade = 10^4.

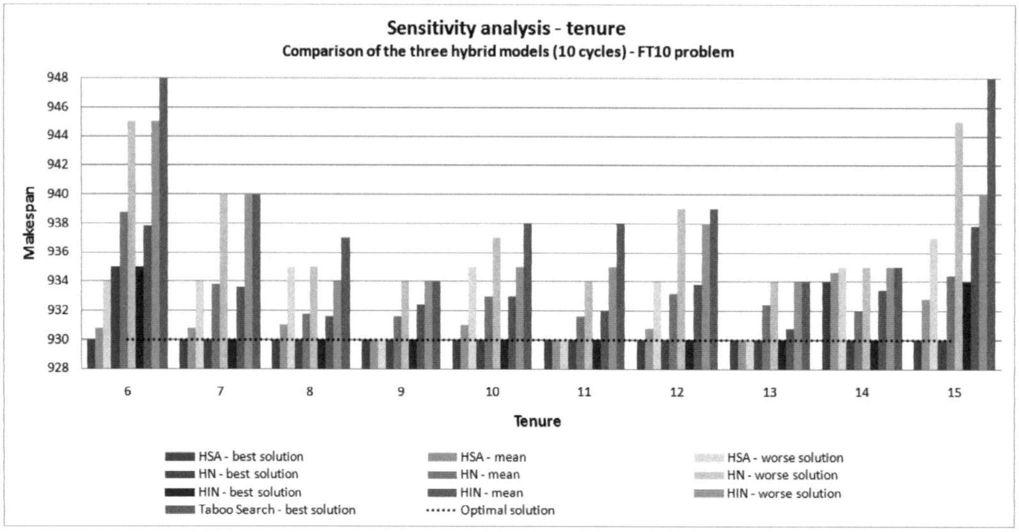

Figure 11. Sensitivity analysis with respect to the tenure of the Hybrid Models (HSA, HIN and HN) for the FT10 problem, with the following fixed parameters: maximum number of iterations without upgrade = 10^4; percentage of particles that suffer mutation = 20%; cycles number = 10. Order of neighborhood generation operators = TS – SPSO. Comparison between the best, the mean and the worse results generated by the three hybrid models and the best result generated by the Taboo Search algorithm.

6.3. Sensitivity analysis with respect to the SPSO mutation probability

The mutation probability sensitivity analysis was applied on the hybrid models fixing the value of the parameter "maximum number of iteration without upgrade" of TS at $1O^3$ for the problem FT20, at

10^4 for the problems FT10, ABZ5 and ABZ6 and at 10^5 for the problems ABZ7, ABZ8 and ABZ9, the value of parameter "tenure" of TS at eight, and setting the value of the parameter "percentage of particles that suffer mutation" of SPSO from 0 to 100%. In this analysis it was identified a small variation of the results generated for each value of the parameter analyzed. As the hybrid models were applied with no more than one iteration of the algorithm SPSO by cycle, this variation must be main attributed to the random nature of the hybrid models more than to the sensitivity to the parameter analyzed.

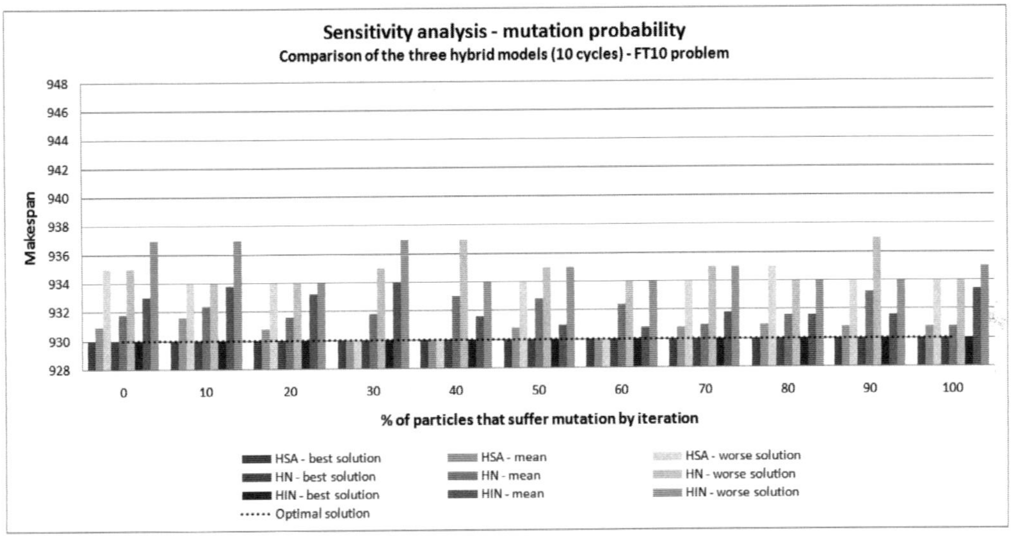

Figure 12. Sensitivity analysis with respect to the "percentage of particles that suffer mutation" of the Hybrid Models (HSA, HIN and HN) for the FT10 problem, with the following fixed parameters: maximum number of iterations without upgrade = 10^4; tenure = 8; cycles number = 10. Order of neighborhood generation operators = TS – SPSO.

6.4. Convergence analysis

For the convergence analysis, the three hybrid algorithms were tested fixing the value of parameters "tenure" of TS and "percentage of particles that suffer mutation" of SPSO, respectively, at eight and 20%, the value of the parameter "maximum number of iterations without upgrade" of TS at 10^2 for FT20 problem, at 10^3 for FT10, ABZ5 and ABZ6 problems and at 10^4 for ABZ7, ABZ8 and ABZ9 problems and fixing the maximum number of cycles for FT20, FT10, ABZ6 and ABZ7 problems at 10^2 and at 10^3 for ABZ7, ABZ8 and ABZ9. For the FT20 and the 10×10 problems (FT10, ABZ5 and ABZ6) all the solutions converged to the optimal solutions with a smaller number of cycles. In the case of the 20×15 problems (ABZ7, ABZ8 and ABZ9), despite the hybrid algorithms demonstrated to be able to improve the results found by the single Taboo Search, the computation time has exceeded the maximum time we allowed for our numerical experiments (one week) without finding the optimal solution and without completing the limit defined by the number of cycles. This failure may be connected to the fact that all experiments involved a small set of initial solutions (12).

7. Conclusions

In this work, we presented three possible schemes of hybridization based on Local Search Heuristics, defined as Hybrid Successive Application, Hybrid Neighborhood, and Hybrid Improved Neighborhood. These models were tested applying a Taboo Search (TS) and a Similar Particle Swarm Optimization (SPSO) algorithms as their neighborhood generation operators. Many sensitivity analyzes have been performed on the single TS and SPSO as well as on the hybrid algorithms derived from the three hybrid approaches presented. The results confirmed the following conclusions. The single TS and SPSO algorithms are very sensitive to the tuning of their respective parameters. Also, the results of the single TS algorithm strongly depends on the initial solutions. The TS and SPSO algorithms have complementary nature, however, when they are applied in the hybrid schemes, we constated that the SPSO must be used with no more than one iteration.

In addition, the hybrid algorithms have shown good properties of robustness and accuracy. The results demonstrated the ability of the hybrid algorithms to improve the original techniques significantly, generating better results at acceptable computing times. The convergence to an optimal solution has been obtained in all the experiments, except for the 20×15 problems, where the computational time has exceeded the previously fixed limit (one week). Since all the tests involved a small set of initial solutions, increasing the size of this set may significantly improve the results. This will be the subject of further research.

Another important remark is that, as previously observed, the hybridization is more efficient when the algorithms involved have complementary characteristics. We suggest that the algorithms must be classified into classes according to some general criterions and the hybridization has to be performed by using elements of different classes. This point will also be matter of future work.

Acknowledgements

The author are grateful for the financial support provided by: MCTI, Ministério da Ciência, Tecnologia e Inovação; CNPq, Conselho Nacional de Desenvolvimento Científico e Tecnológico; and CAPES, Coordenação de Aperfeiçoamento de Pessoal de Nível Superior.

References

[1] Garey M R, and Johnson D S 1979 *Computers and Intractability: A Guide to the Theory of NP-Completeness* (San Francisco, CA: Freeman and Company)
[2] Zhang C Y, Li PG, Rao YQ and Guan ZL 2008 A very fast TS/SA algorithm for the job shop scheduling problem *Computers & Operations Research* **35** 282–94
[3] Muth J F and Thompson G L 1963 *Industrial Scheduling* (Englewood Cliffs, NF: Prentice-Hall)
[4] Jain A S and Meeran S 1999 Deterministic job-shop scheduling: past, present and future *Journal of Operational Research* **113** 390–434
[5] Johnson S M 1954 Optimal two- and three-stage production schedules with set-up times included *Naval Research Logistics Quarterly* **1** 61–8
[6] Akers S B 1956 A graphical approach to production scheduling problems *Operations Research* **4** 244–5
[7] Jackson J R 1956 An extension of Johnson's result on job lot scheduling *Naval Research Logistics Quarterly,* **3(3)** 201–3
[8] Manne A S 1960 On the job-shop scheduling problem *Operations Research* **8** 219–23
[9] Carlier J and Pinson E 1989 An algorithm for solving the job shop problem *Management Science* **35(29)** 164–76
[10] Brucker P, Jurisch B and Siever B 1994 A branch and bound algorithm for the job-shop scheduling problem *Discrete Applied Mathematics* **49** 105–27
[11] Panwalkar S S and Iskander W 1977 A survey of scheduling rules *Operations Research* **25(1)** 45–61
[12] Adams J, Balas E and Zawack D 1988 The shifting bottleneck procedure for job shop scheduling *Management Science* **34** 391–401
[13] Dauzere-Peres S and Lasserre J -B 1993 A modified shifting bottleneck procedure for job-shop scheduling *International Journal of Production Research* **31** 923–32
[14] Wenqi H and Aihua Y 2004 An improved shifting bottleneck procedure for the job shop scheduling problem *Computers & Operations Research* **31** 2093–110
[15] Erschler J, Roubellat F and Vernhes J P 1976 Finding some essential characteristics of the feasible solutions for a scheduling problem *Operations Research* **24(4)** 774–83

[16] Sabuncuoglu I and Gurgun B 1996 A neural network model for scheduling problems *European Journal of Operational Research* **93** 288–99
[17] Blum C and Sampels M 2004 An ant colony optimization algorithm for shop scheduling problems *Journal of Mathematical Modelling and Algorithms* **3** 285–308
[18] Van Laarhoven P J M, Aarts E H L and Lenstra J K 1992 Job shop scheduling by simulated annealing *Operations Research* **40(1)** 113–25
[19] Azizi N and Zolfaghari S 2004 Adaptive temperature control for Simulated Annealing: a comparative study *Computers & Operations Reasearch* **31** 2439–51
[20] Cheng R, Gen M and Tsujimura Y 1996 A tutorial survey of job-shop scheduling problems using genetic algorithms - I. representation *Computers & Industrial Engineering* **30(4)** 983–97
[21] Cheng R, Gen M and Tsujimura Y 1999 A tutorial survey of job-shop scheduling problems using genetic algorithms: Part II. Hybrid genetic search strategies *Computers & Industrial Engineering* **36(2)** 343–364
[22] Taillard É D 1994 Parallel taboo search techniques for the job shop scheduling problem *ORSA Journal on Computing* **6(2)** 108–117
[23] Nowicki E and Smutnicki C 1996 A fast taboo search algorithm for the job shop problem *Management Science* **42(6)** 797–813
[24] Zhang C Y, Li P, Guan Z and Rao Y 2007 A taboo search algorithm with a new neighborhood structure for the job shop scheduling problem *Computers & Operations Research* **34** 3229–42
[25] Lian Z, Jiao B and Gu X 2006 A similar particle swarm optimization algorithm for job-shop scheduling to minimize makespan *Applied Mathematics and Computation* **183** 1008–17
[26] Sha D Y and Hsu C-Y 2006 A hybrid particle swarm optimization for job shop scheduling problem *Computers & Industrial Engineering* **51** 791–808
[27] Lin T-L, Horng S-J, Kao T-W, Chen Y-H, Run R-S, Chen R-J, Lai J-L and Kuo I-H 2010 An efficient job-shop scheduling algorithm based on particle swarm optimization *Expert Systems with Applications* **37** 2629–36
[28] Blazewicz J, Domschke W and Pesch E 1996 The job shop scheduling problem: conventional and new solution techniques *European Journal of Operational Research* **93(1)** 1–33
[29] Pezzella F and Merelli E 2000 A taboo search method guided by shifting bottleneck for the job shop scheduling problem *European Journal of Operational Research* **120** 297–310
[30] Wang L and Zheng D-Z 2001 An effective hybrid optimization strategy for job-shop scheduling problems *Computers & Operations Research* **28** 585–96
[31] Glover F 1989 Taboo search–Part I *ORSA Journal on Computing* **1(3)** 190–206
Glover F 1990 Taboo search–Part II *ORSA Journal on Computing* **2(1)** 4–32
[32] Kennedy J and Eberhart R 1995 Particle Swarm Optimization *Int. Conf. on Neural Networks (Perth, Australia)*
[33] Shi Y and Eberhart R C 1998 Parameter selection in particle swarm optimization *Proceedings of the 7th international conference on evolutionary programming (New York).*
[34] Fisher H and Thompson G L 1963 Probabilistic learning combinations of local job-shop scheduling rules *Industrial Scheduling*, ed J Muth and G Thompson (Englewood Cliffs, New Jersey: Prentice Hall) pp 225–51.
[35] Fraga T B 2010 *Proposition and analysis of hybrid models for the job shop scheduling problem* (*Ph.D. Thesis*) (Nova Friburgo, RJ, BR: Universidade do Estado do Rio de Janeiro - Instituto Politécnico)

Structural Properties of ZnO/SnO_2-Composite-Nanorod Deposited Using Thermal Chemical Vapour Deposition

N.D. Md Sin[1,3]*, A.K Shafura[1], M.F. Malek[1], M.H. Mamat[1], M. Rusop[1,2]
[1]NANO-ElecTronic Centre, Faculty of Electrical Engineering,
Universiti Teknologi MARA (UiTM), 40450 Shah Alam, Selangor, Malaysia
[2]NANO- SciTech Centre (NST), Institute of Science (IOS),
Universiti Teknologi MARA (UiTM), 40450 Shah Alam, Selangor, Malaysia
[3]Fakulti Kejuruteraan Elektrik, Universiti Teknologi MARA Cawangan Johor,
Kampus Pasir Gudang, 81750 Masai, Johor.

*E-mail: nordiyana86@yahoo.com, diyana0366@hohor.uitm.edu.my

Abstract. In this work, we report on the effect of substrate temperatures on ZnO/SnO_2 composite nanorods deposited by thermal chemical vapour deposition (CVD) onto a ZnO template layer. The substrate temperature varied from 200 ~ 600°C. The FESEM image reveals that the size of the thin film created by the ZnO/SnO_2 composite nanorods decreased as the substrate temperature increased.

1. Introduction

SnO_2 is a promising candidate for nanotechnology applications. SnO_2 is an n-type semiconductor possessing a wide band gap (3.6 eV). The electronic conductivity of SnO_2 is significantly influenced by the effects of molecular adsorption on its surface [1]. It is considered to be a good candidate sensor material because of the wide applications inherent to its defect structure and the variable oxidation state of Sn [2]. SnO_2 nanostructures can be prepared by many methods, such as chemical vapour deposition [3], radio frequency (RF) magnetron sputtering [4], the sol-gel method [5], spray-pyrolysis [6], and hydrothermally [7]. The advantageous properties of SnO_2 include good optical, chemical, electrical and thermal stabilities that enhance the abilities of this material in various device applications, such as solar cells [8], gas sensors [9] and light emitting diodes (LEDs) [10].
ZnO is naturally an n-type semiconductor material with direct band gap energy of 3.37 eV at room temperature and has a free-exciton binding energy of 60 meV [11, 12]. ZnO has been proposed for many technological applications, such as optical devices, solar cells, piezoelectric devices, varistors, surface acoustic wave (SAW) devices and gas sensors [12]. The properties of ZnO, including the electrical and optical properties, heavily depend on its crystallinity, crystallographic orientation, crystal size and morphology [12]. It has been proposed for numerous functions as a semiconductor material including as a catalyst, conduction layer and transport layer for device fabrication.
Many researchers have been exploring using composites to improve performance in many applications. ZnO/SnO_2 composites have been proposed for various applications such as OLEDs, gas

sensors and solar cells. The advantage of these materials is their high electron mobility and improvement in surface reactions as a result of surface modification. A substrate temperature study is important in order to produce high quality of ZnO/SnO_2 composite nanorods. The issue of sensor problems such as low sensitivity and high defect are generally caused by low temperature processing during the growth phase.
This is the first work that reports on ZnO/SnO_2 composite nanorods deposited using different substrate temperatures via thermal CVD methods. In this paper, we study the effect of substrate temperature on the properties of ZnO/SnO_2 composite nanorod growth when deposited using thermal CVD on a ZnO template.

2. Methodology

2.1 ZnO template preparation

The ZnO template was prepared using sputtering system. A glass substrate was used to prepare the ZnO template. The glass substrate was cleaned with acetone, methanol and deionised water via several steps before experimentation. A ZnO target (high purity (99.999%)) was used to deposit the ZnO template. The sputtering system was set at 200 watts with a pressure maintained at 7 mTorr. A molecular pump was used to evacuate the sputtering chamber to $5x10^{-4}$ Pa. Argon gas was injected into the remaining oxygen at a ratio of 45:5 sccm. The substrate temperature was set to 500°C and the deposit time was 60 min.

2.2 ZnO/SnO_2 composite nanorods preparation

The ZnO templates were used for the ZnO/SnO_2 composite nanorod deposition. Two furnaces were used to grow doped ZnO/SnO_2 composite nanorods. A quartz tube was inserted in the furnace. Zinc nitrate hexahydrate ($Zn(NO_3)_2.6H_2O$, 98%, Systerm, Laboratory Chemicals And Reagents, Classic chemicals Sdn. Bhd.) and tin(IV) chloride pentahydrate ($SnCl_4$.H2O, 98%, Sigma Aldrich) were used as the precursor and were placed inside furnace 1. The ZnO template was placed in furnace 2. The argon (Ar) and oxygen (O_2) were injected at 40 sccm and 5 sccm, respectively. The mass of both precursors was set at 3 g. The substrate temperature varied from 200 ~ 600°C over a deposition time of 1 hour.

2.3 ZnO/SnO_2 composite nanorods preparation

The surface morphology was characterised using FESEM (JEOL JSM 6701F) and the crystalline properties of the ZnO/SnO_2 composite nanorods were characterised using x-ray diffraction (XRD, Rigaku Ultima IV).

3. Results and discussions

Figure 1 shows the FESEM images of the ZnO/SnO_2 composite nanorods synthesised at different substrate temperatures: (a) 200ºC, (b) 300 ºC, (c) 400ºC, (d) 500ºC and (e) 600ºC for a working voltage of 5 kV and magnifications of 30,000x and 100,000x. The insets in Figures 1 (a), (b), (c), (d) and (e) show the measured sizes of all of the samples. These surface images show that all of the thin films were uniformly deposited onto the ZnO template layer. At a substrate temperature of 200˚C, the size of the nanostructures ranged from approximately 128 to 165 nm, as shown in Figures 1 (a). Figures 1(b) and (c) show the ZnO/SnO_2 composite nanorods with lengths of approximately 185 to 550 nm and 163 to 279 nm that were fabricated using substrate temperatures of 300˚C and 400˚C, respectively. Figures 1 (g-h) show that as the substrate temperature increases to 500˚C, the ZnO/SnO_2 composite nanorods grew in alignment on the ZnO template layer with lengths of approximately 19 to 30 nm. However, the ZnO/SnO_2 composite nanorods with lengths of approximately 43 to 58 nm began to bend downwards as the substrate temperature increased to 600˚C, as shown in Figure 1 (i-j). This behaviour may have been caused by the substrate starting to melt; the melting temperature of glass is 565˚C [13]. This result also shows that the ZnO/SnO_2 composite nanorods became closely packed at high substrate temperatures. The density of the ZnO and SnO_2 crystal nuclei within the nucleated phase is expected to increase with the decomposition temperature. Thus, the collision frequency of the particles increases. Increasing the substrate temperature can facilitate better atomic packing and enhance the crystalline quality of the film [14, 15]. The substrate temperature influences the growth of the ZnO/SnO_2 composite nanorods formed by thermal CVD [16, 17]. The thickness of the ZnO/SnO_2 composite nanorods formed at substrate temperatures of 200, 300, 400, 500 and 600°C were approximately 476 nm, 580 nm, 748 nm, 850 nm and 568 nm, respectively. The different morphology sizes resulted from changing the ratios of the Zn, Sn and O_2 vapour pressures at high temperatures [18, 19].

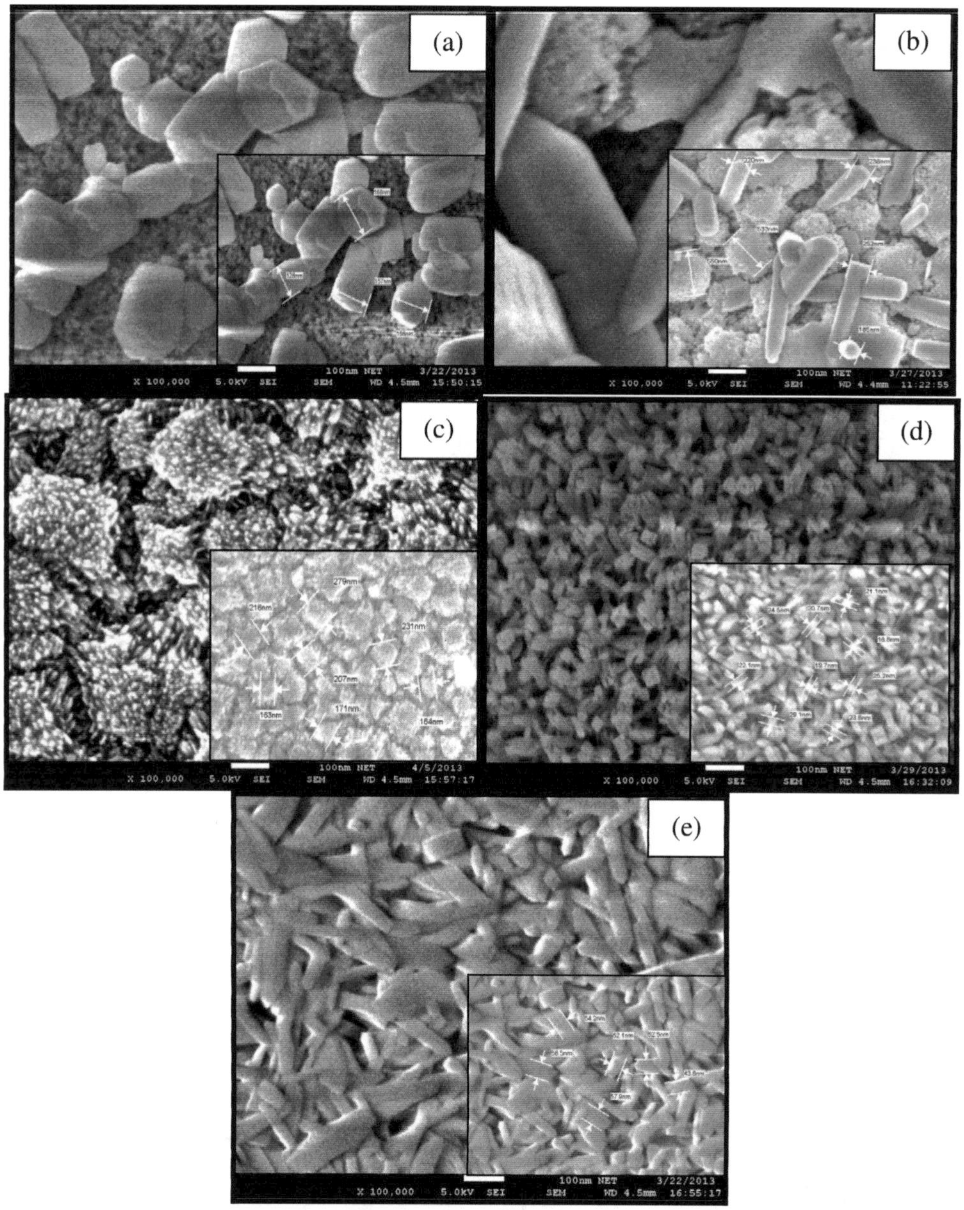

Figure 1. Surface Images of the ZnO/SnO_2 Composite Nanorods at Different Substrate Temperatures: (a) 200°C, (b) 300 °C, (c) 400°C, (d) 500°C and (e) 600°C.

In our experiments, the Zn vaporised first because the Zn source had a lower melting point than the Sn source. In the evaporation process, vaporised Zn and Sn atoms impinged on the substrate and were adsorbed by the ZnO template layer. Then, the atoms reacted with the ZnO template layer and formed a bond. Two conditions arose during the growth process at low and high substrate temperatures. At low temperatures, the surface diffusion energy was insufficient to form nanorods, and the deposited atoms on the substrate had limited surface diffusion energy. This decreased the diffusion length and mobility of the atoms, thereby affecting the accumulation of the constituent atoms on the substrate [20, 21]. Atoms that were deposited at low temperatures were quenched from the high source temperature to the low temperature of the substrate. Decreasing the temperature from the evaporation temperature to the substrate temperature affected factors such as the temperature and the partial pressures, which may have affected the direction of the growth of the nanostructures [22]. At high temperatures, the crystallinity of the ZnO/SnO_2 composite nanorods improved due to the enhancement in the surface diffusion and the mobility of the adatoms on the substrate surface. Surface diffusion facilitated the transport of these adatoms and their nucleation into nanorod structures as the Zn particles condensed to form liquid droplets on the ZnO template layer. This nucleation form is preferred for the nanorods on the substrate with face grains [21]. Besides, the ZnO template was important to act as the holder or based for ZnO/SnO_2 composite nanorods. Moreover, the growth of ZnO/SnO_2 composite nanorods on the ZnO template was used to increase the adherent between the substrate and the ZnO/SnO_2 composite nanorods. In sensor application, the ZnO template can act as the transport layer to the ZnO/SnO_2 composite nanorods.

Figure 2. EDS images for ZnO/SnO_2 Composite Nanorods at Different Substrate Temperatures: (a) 300 °C, (b) 500°C and (c) 600°C

IOP Conf. Series: Materials Science and Engineering **83** (2015) 012002 doi:10.1088/1757-899X/83/1/012002

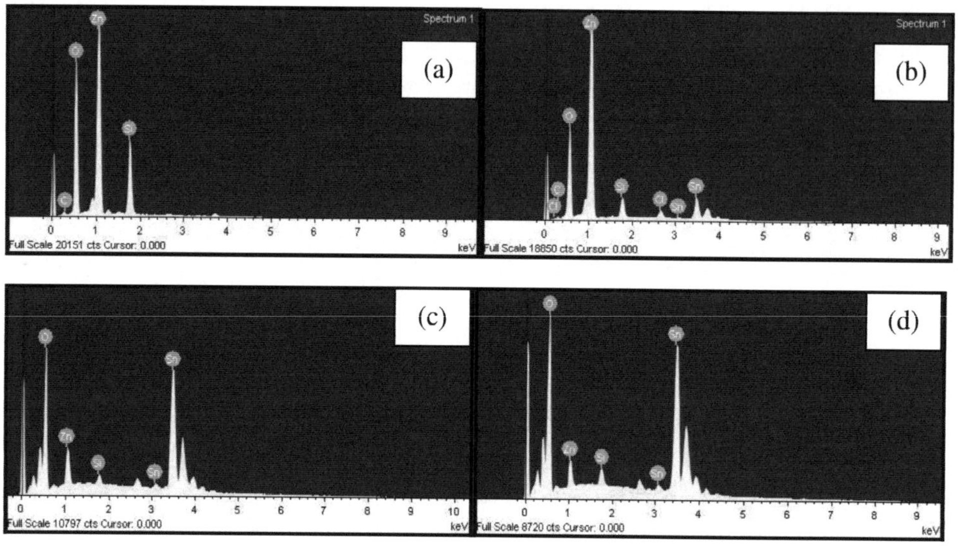

Figure 3. EDS spectrum images for ZnO/SnO_2 Composite Nanorods at Different Substrate Temperatures: (a) 200°C, (b) 300 °C, (c) 500°C and (d) 600°C.

Table 1: Atomic percent of ZnO nanoparticle, SnO_2 nanorod and ZnO/SnO_2 nanoflower like structures.

	Zn	Sn	Si	O	C	Cl
200°C	20.17	0	14.11	59.94	5.77	0
300°C	26.99	7.00	3.91	56.39	3.79	1.91
500°C	4.88	28.81	1.27	65.04	0	0
600°C	2.84	29.36	2.08	65.72	0	0

Table 1 above show the atomic percent of ZnO/SnO_2 composite nanorods for (a) 200°C, (b) 300 °C, (c) 500°C and (d) 600°C as refer to the EDS spectrum images in Figure 2. Table 1 and Figure 3 show the possible corresponding atomic percent(%). It reveals that the element at substrate temperature 200°C the zinc (Zn), oxygen (O) silicon (Si) and carbon (C). As the substrate temperature increase to 300°C, the elements consist of tin (Sn), zinc (Zn), oxygen (O) silicon (Si) and carbon (C).

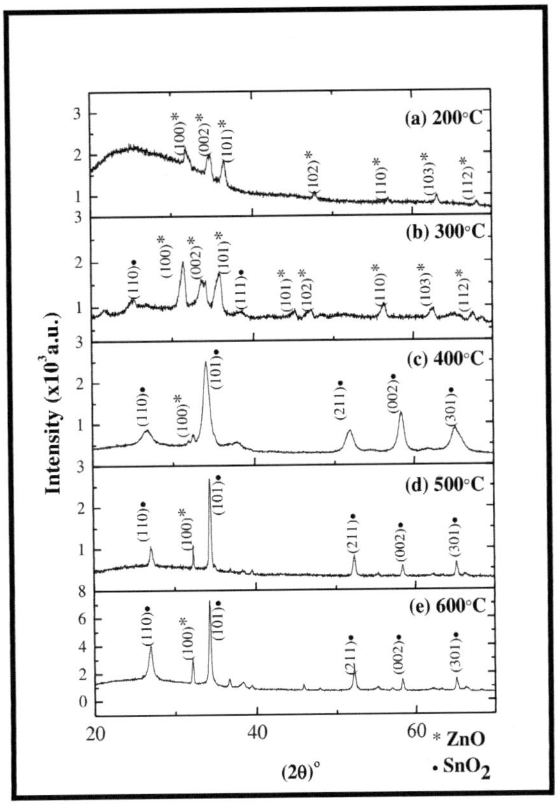

Figure 4. XRD Patterns for the ZnO/SnO_2 Composite Nanorods at Different Substrate Temperatures: (a) 200°C, (b) 300 °C, (c) 400°C, (d) 500°C and (e) 600°C.

Figure 4 shows the XRD patterns for the ZnO/SnO_2 composite nanorods at different substrate temperatures: (a) 200°C, (b) 300 °C, (c) 400°C, (d) 500°C and (e) 600°C. The XRD patterns of these samples at different substrate temperatures exhibited diffraction peaks corresponding to hexagonal wurzite ZnO (JCPDS card no. 36-1451). The diffraction peaks for ZnO that were formed at a substrate temperature of 200°C appeared at (100), (002),(101), (102), (110), (103) and (112). All of the diffraction peaks could be perfectly indexed to the hexagonal wurtzite structure of ZnO and the tetragonal anatase-rutile crystal of SnO_2 (JCPDS card no. 41-1445), as shown in Figure 4 (b) for substrate temperatures of 300 °C, (c) 400°C, (d) 500°C and (e) 600°C. For the diffraction peaks obtained for samples that were synthesised at substrate temperatures of 300°C, the major peaks corresponded to the hexagonal wurtzite structure of ZnO and were indexed to (100), (002), (101), (102), (110), (103) and (112), whereas the minor peaks were assigned to SnO_2 and were indexed to (110) and (111). For the substrate temperatures of 400°C, 500°C and 600°C, minor ZnO peaks appeared at (110). However, the peaks for the SnO_2 phase were assigned to (110), (101), (211), (002) and (301). The reaction of the ZnO and SnO_2 began at a substrate temperature of 300°C, as shown in Figure 4 (b), with the formation of the SnO_2 orientation first. The crystallinity of the ZnO/SnO_2 composite nanorods improved because of the high deposition rate at the high substrate temperatures, which increased the electron mobility in the composite nanorods [20]. The enhanced crystallinity improved the performance of a sensor application. As the substrate temperature increased from 400°C to 600°C, a small amount of the ZnO phase with a peak at (100) was formed, and the amount of the SnO_2 phase increased. Al-Ghaashi et al. [18] studied the thermal decomposition of tin (II) chloride dehydrate at different substrate temperatures (over a range of 300 to 900 °C) and found that the growth of SnO_2 crystal nuclei began at approximately 400°C. This result supports the XRD results shown in

Figure 4, where the SnO_2 major crystal structure formation begins at 400°C. The SnO_2 phase with a preferential (101) orientation exhibited the highest intensity because of its low surface energy [23, 24].

4. Conclusions

In summary, ZnO/SnO_2 composite nanorods were successfully prepared on a ZnO template using thermal CVD. The size of ZnO/SnO_2 composite nanorods decreased as the substrate temperature increased. Zinc nitrate and tin tetrachloride were used as precursors. Observation of the initial growth phase of ZnO crystal began to form at 200 °C and for SnO_2 crystal nuclei began to form at approximately 300 °C. Enhancing the crystallinity of the ZnO/SnO_2 composite nanorods improved the sensing characteristics.

5. Acknowledgement

The authors would like to express their thanks to Institute of Science and Faculty of Mechanical UiTM for providing the laboratory facilities. Thanks also to Ministry of Higher Education (MOHE) for financial support. Thanks also Universiti Tun Hussein Onn Malaysia (UTHM) for providing the XRD measurement facilities.

6. References

[1] Wang G X, Park J S, Park M S and Gou X L 2008 Synthesis and high gas sensitivity of tin oxide nanotubes *Sensor. Actuat. B-Chem.* **131** 313-7
[2] Kim H, Sathaye S D, Hwang Y K, Jhung S H, Hwang J, Kwon S H, Park S and Chang J 2005 Humidity Sensing Properties of Nanoporous TiO~ 2-SnO~ 2 Ceramic Sensors *B. Korean Chem. Soc.* **26** 1881
[3] Jeong J, Choi S-P, Chang C I, Shin D C, Park J S, Lee B T, Park Y-J and Song H-J 2003 Photoluminescence properties of SnO2 thin films grown by thermal CVD *Solid State Commun.* **127** 595-7
[4] Wang Y, Ma J, Ji F, Yu X and Ma H 2005 Structural and photoluminescence characters of SnO2:Sb films deposited by RF magnetron sputtering *J. Lumin.* **114** 71-6
[5] Rani S, Roy S C and Bhatnagar M C 2007 Effect of Fe doping on the gas sensing properties of nano-crystalline SnO2 thin films *Sensor. Actuat. B-Chem.* **122** 204-10
[6] Paraguaydelgado F, Mikiyoshida M, Antunez W, Gonzalezhernandez J, Vorobiev Y and Prokhorov E 2008 Corrigendum to``Morphology and microstructure of textured SnO2 thin films obtained by spray pyrolysis and their effect on electrical and optical properties"[Thin Solid Films 516 (2008) 1104 1111] *Thin Solid Films* **516** 4293-4
[7] Wang Y, Guo M, Zhang M and Wang X 2009 Hydrothermal synthesis of SnO2 nanoflower arrays and their optical properties *Scripta Mater.* **61** 234-6
[8] Chen Z, Tian Y, Li S, Zheng H and Zhang W 2011 Electrodeposition of arborous structure nanocrystalline SnO_2 and application in flexible dye-sensitized solar cells *J. Alloy. Compd.* **515** 57-62
[9] Iizuka K, Kambara M and Yoshida T 2012 Highly sensitive SnO2 porous film gas sensors fabricated by plasma spray physical vapor deposition *Sensor. Actuat. B-Chem.* **173** 455-61
[10] Wen Z, Zheng F, Yu H, Jiang Z and Liu K 2013 Hydrothermal synthesis of flowerlike SnO2 nanorod bundles and their application for lithium ion battery *Mater. Charact.* **76** 1-5
[11] Schmidt-Mende L and MacManus-Driscoll J L 2007 ZnO – nanostructures, defects, and devices *Mater. Today* **10** 40-8
[12] Wang Z L 2004 Nanostructures of zinc oxide *Mater. Today* **7** 26-33
[13] Rosenbaum S and Clasen R 1999 Electrospraying of glasses—Preparation of glass coating on glass *J. Aerosol Sci.* **30** 975-6

[14] Junwei M, Feng R, Meihua X and Huijied J 2011 Influence of substrate temperature on the performance of zinc oxide thin film transistor *Journal of Semiconductors* **32** 044002

[15] Zadsar M, Fallah H R, Haji Mahmoodzadeh M, Hassanzadeh A and Ghasemi Varnamkhasti M 2012 Substrate temperature effect on structural, optical and electrical properties of vacuum evaporated SnO2 thin films *Mat. Sci. Semicon. Proc.* **15** 432-7

[16] Shaalan N M, Yamazaki T and Kikuta T 2011 Synthesis of metal and metal oxide nanostructures and their application for gas sensing *Mater. Chem. Phys.* **127** 143-50

[17] Dai Z R, Pan Z W and Wang Z L 2003 Novel nanostructures of functional oxides synthesized by thermal evaporation *Adv. Funct. Mater.* **13** 9-24

[18] Al-Gaashani R, Radiman S, Tabet N and Daud A R 2012 Optical properties of SnO2 nanostructures prepared via one-step thermal decomposition of tin (II) chloride dihydrate *Mater. Sci. Eng. B* **177** 462-70

[19] Lu J G, Chang P and Fan Z 2006 Quasi-one-dimensional metal oxide materials—Synthesis, properties and applications *Mater. Sci. Eng. B* **52** 49-91

[20] Mageshwari K and Sathyamoorthy R 2012 Influence of substrate temperature on the physical properties of thermally evaporated nanocrystalline bismuth sulfide thin films *Vacuum* **86** 2029-34

[21] Yadav J B, Patil R B, Puri R K and Puri V 2007 Studies on undoped SnO2 thin film deposited by chemical reactive evaporation method *Mater. Sci. Eng. B* **139** 69-73

[22] Wang Q, Yu K and Xu F 2007 Synthesis and field emission of two kinds of hierarchical SnO2 nanostructures *Solid State Commun.* **143** 260-3

[23] Karslıoğlu R, Uysal M and Akbulut H 2011 The effect of substrate temperature on the electrical and optic properties of nanocrystalline tin oxide coatings produced by APCVD *J. Cryst. Growth* **327** 22-6

[24] Batzill M and Diebold U 2005 The surface and materials science of tin oxide *Prog. Surf. Sci.* **79** 47-154

An Overview of Okra Fibre Reinforced Polymer Composites

Nadendla Srinivasababu[1]
Department of Mechanical Engineering, Vignan's Lara Institute of Technology and Science, Vadlamudi–522 213, Andhra Pradesh, India.

E-mail: cnjlms22@yahoo.co.in

Abstract. Increasing attention towards "sustainable environment" invited the development of new materials to satisfy the needs of the public with less/no damage on surroundings. In this regard a specific attempt is made by the author to do review and understand the performance of the variable vegetable fibres of okra which is botanically called 'Abelmoschus Esculentus' and their composites under various conditions of load. Further the results exist in the literature are also reviewed for the purpose of comparison.

1. Introduction

Natural fibres may be classified based on the origin i.e. plant, animals or minerals of birth. A typical bio fibre consists of (i) helically arranged crystalline micro fibrils of cellulose, (ii) amorphous region mainly consists of lignin and hemicellulose, (iii) disorderly arranged crystalline cellulose micro-fibrils network, (iv) primary wall, (v) secondary wall S1, (vi) secondary wall S2, (vii) secondary wall S3, (viii) Lumen. The overall properties of the fibre will dependent on the property of each constituent whereas the reinforcing ability of the natural fibre is dependent on the nature of cellulose and its crystallinity. Cellulose fibrils are aligned along the length of the fibre which renders maximum tensile, flexural strengths along with rigidity [**1**].

In order to achieve strong interfacial adhesion an intimate molecular contact at the fiber-matrix is required. Based on adhesion characteristics an interface can be classified as follows [**2**].

1. Sharp interface with weak molecular force, as in the case of non-polar polymer with a polymer fibre: in such a case the interface will be mechanically weak because of lack of chemical affinity and wettability leading to interfacial slippage.
2. Sharp interface with strong intermolecular, such as the dispersion force between a non-polar polymer and a high energy material or intermolecular forces involving specific interactions i.e. primary or secondary. In this case interface slippage will not occur and the interface will be mechanically strong.
3. Diffuse interface with any intermolecular: when there is sufficient molecular diffusion, entanglement and, and/or anchoring sites, interfacial slippage will not occur. Therefore both primary chemical bonds and secondary bonds are important to attain adhesion depending on interfacial structure.

[1] Corresponding Author Address: Dr. Nadendla Srinivasababu, Professor & Head, Department of Mechanical Engineering, Vignan's Lara Institute of Technology and Science, Vadlamudi – 522 213. Andhra Pradesh, India. E-mail: cnjlms22@yahoo.co.in. Ph: 0863– 2118507, Fax: 0863-2118456.

The researchers have struggled in order to reinforce variety of natural fibres (as is condition or modified physically/chemically) into diversified matrices to achieve target properties. In this paper an attempt is made to do review on the mechanical and dielectric performance of okra fibre (Individual form and Oval chain structured form) reinforced polyester composites at various fibre contents. Further analysis is carried out by to compare the results with the values exist in the literature.

2. Fibre, Matrix and Composites Manufacturing

The present section describes fibres, matrices used for the fabrication of the composites and the testing procedures adopted.

2.1. Fibre

Okra is botanically called Abelmoschus Esculentus and this crop is extensively cultivated in few parts of the India. An attempt was made by Nadendla srinivasababu in the year 2007 to manufacture the composites reinforced with okra individual fibres of Srikakulam origin and the results were published elsewhere [**3**]. Then he extracted the good and fine okra fibre in 2008 from the crop grown in the village of Gorigapudi, Bhattiprolu Mandalum in the state of Andhra Pradesh, India. The crop is very neatly grown because the seeds (2405133 variety) are purchased from Syngenta Ltd., Shivaji Nagar, Pune, India. Another variety of okra fibre was extracted from the crop grown from the seeds of Indra–32 variety purchased from Prabhakar Seeds, Gandhi Nagar, Bangalore, India. The two varieties of the fibre are extracted by "Retting and Manual Extraction Method". Hybrid okra variety (2405133) exhibited two kinds of fibres i.e. (i) Fibres are in individual form/filaments; (ii) Fibres are in Oval Chain Structured Layered forms and are shown in Figure 1, 2 respectively.

Figure 1 Extracted Okra hybrid Variety 2405133 individual fibre

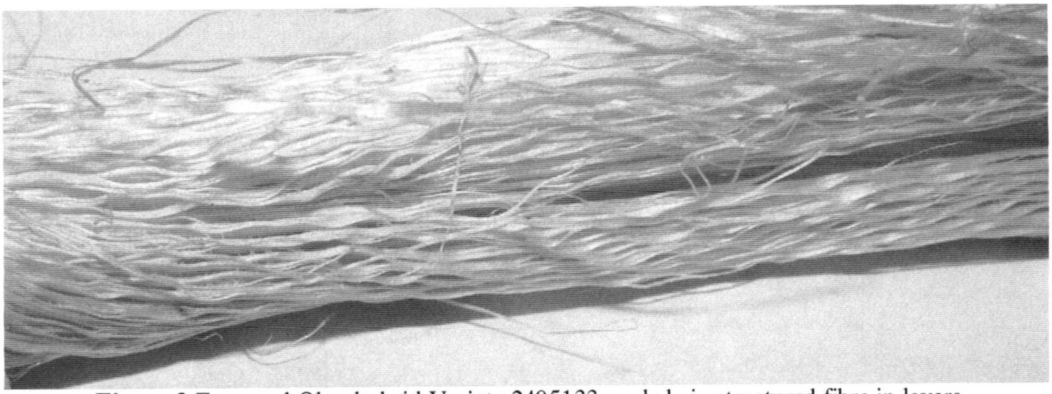

Figure 2 Extracted Okra hybrid Variety 2405133 oval chain structured fibre in layers

Now the focus is made on other okra fibres reported by the authors available in the literature. Okra Bast Fibre (OBF) was collected from Jhenaidah regions in Bangladesh. In order to remove the dirty substances, OBF was scoured in a solution containing 5 mg Na_2CO_3 and 5 gm detergent per liter of water at 70 ^{0}C in a large beaker [**4**]. Here also the authors collected okra plants from the same region and were retted into water for 13 – 15 days and the fibre was collected [**5**]. Retting time to extract the okra fibre was 20 days [**6**]. The plant had been collected from in Kushtia District (Bangladesh) and was kept under water (15 – 20 days) to allow microbial degradation and the extracted fibre was washed with deionised water [**7, 8, 9, 10 and 11**].

2.2. Matrix and Method of Manufacturing
An Ecmalon 4413 Unstaurated polyester resin was used to fabricate the OIFV–1, OIFV–2, OOCSFV-1 and OOCSFV-2 composites by wet lay-up technique i.e. according to the regular procedure. In hand lay-up technique the fibre is placed, the resin is poured in the mold as per the predetermined proportions and is allowed to cure for 24 h.

Cut okra fibres 2 – 3 mm size was reinforced into bakalite by hand lay-up technique and tried the possible dispersion during the preparation of the composites [**7**]. Poly(lactic) acid was used as a matrix to fabricate composites by injection molding [**11**]. In injection molding technique the predetermined quantity of fibre is blended thoroughly with matrix and is processed as per the normal procedure for making the composites.

2.3. Chemical Treatment
Various chemicals and treatment procedures were adopted in order to improve the mechanical behavior of the composites by enhancing their bonding i.e. reinforcement and matrix. The details of the chemical treatments followed by the author and the procedures adopted in the literature are given in Table 1. The author has also made a specific effort to manufacture the composites reinforced with the fibres which were treated under accelerated environmental conditions i.e. affect of acid rain. The author has considered the acid rain condition because the environment is exposed to gigantic pollutants, hence it is essential to understand this affect on degradable materials i.e. fibre and its composites.

Table 1. Chemical Treatment of Okra Fibre.

Fibre	Form	Treatment	Procedure (Concentration, Soaking time)	Reference
OIFV–1 CT–1	Individual long	NaOH $KMnO_4$ H_2SO4	0.125 M – 6h 0.03164 – 14 1.87602×10^{-5}	Corresponding authors study
OIFV–1 CT–2	,,	NaOH $KMnO_4$ H_2SO4	0.125 M – 45 min 0.006328 – 5 min 3.752×10^{-6} – 5 min	,,
OIFV–1 CT–3	,,	NaOH	0.1875 M – 1 h	,,
OIFV–2 CT–1	,,	NaOH $KMnO_4$ H_2SO4	0.25 M – 5h 55 min 0.01266 – 2 min 3.752×10^{-6} – 2 min	,,
OOCSFV-1 CT–1	Oval Chain structured layered	NaOH $KMnO_4$ H_2SO4	0.125 M – 6h 0.03164 – 14 h 1.876×10^{-5} – 14 min	,,
OOCSFV-1 CT–2	,,	NaOH $KMnO_4$	0.125 M – 45 min 0.006328 – 5 min	,,

IOP Conf. Series: Materials Science and Engineering **83** (2015) 012003 doi:10.1088/1757-899X/83/1/012003

		H_2SO4	3.752×10^{-6} – 5 min	
OOCSFV-1 CT–3	Oval Chain structured layered	NaOH	0.1875 M – 1 h	”
OOCSFV-2 CT–1	”	NaOH $KMnO_4$ H_2SO4	0.25 M – 5h 55 min 0.01266 – 2 min 3.752×10^{-6} – 2 min	”
OBF	Individual	Solution (3.5 gm Na_2CO_3 and 6.5 gm detergent per litre of water) Ratio of fibre-to-solution	60 ^{0}C – 30 min 1:50	5
”	”	Bleaching	Digesting the scoured fibre with 0.7 % sodium chlorite solution buffered at pH 4–2 h 90 – 95 ^{0}C	”
”	”	Grafting	Stating – 30 ^{0}C increased up to 90 ^{0}C – 30 min Reaction continued for 210 min.	”
Okra	50 cm length	Bleached	10 % NaOH + 10 % Na_2SO_4 at 30 ^{0}C	7
Okra	Individual	SOF	Scouring with a solution of 3.2 % sodium carbonate and 6.3 % soap flake	9
”	”	AAROF	Acetylation with 10 % acetic acid	”
”	”	AABOF	Bleaching with 10 % sodium chlorite	”
”	”	BOF	Bleaching with 10 % sodium chlorite	”
”	”	SAKPOF	Treatment with 1 % sulphuric acid followed by 0.055 % potassium permanganate	”
”	”	SSROF	Treatment with 10 % sodium dodecyl sulphate	”
”	”	SHBOF	Bleaching with 10 % sodium chlorite followed by 1 % sodium hydroxide alkalisation	”
Okra	Chopped 5-10 mm	Boiling in toluene/ethanol mixture	6 h	10

3. Test, Results and Discussion

The present section describes fibres, matrices used for the fabrication of the composites and the testing procedures adopted along with the procedures adopted in literature. Then a specific care is taken to understand the physical, mechanical properties of the fibre and its composites which were tested under tensile, flexural and impact load at different levels of fibre reinforcement. The author has done the manual modification of the existing fibre to form as a woven fibre and experimentation on okra woven FRP composites and the results were published elsewhere.

3.1. Physical and tensile properties of fibre

The chemical composition of the okra bast muticellular fibre was determined. The values reported by Md. Shamsul Alam et al. are [**4**] 67.5 % α-cellulose, 15.4 % hemicelluloses, 7.1 % lignin, 3.4 % pectin matter, 3.9 % fatty and waxy matter and 2.7 % aqueous extract. The density of the okra fibre was determined by ASTM D 3800-99 and its average value is 881, 817 kg/m^3 for OIFV-1, OOCSF-1 respectively.

Initially the fibre was tensile tested on Electronic Tensometer and its tensile strength, modulus were calculated. Out of all the fibres investigated in the present work OI FV1 CT-2 had shown highest tensile strength, modulus of 151.46 Mpa, 2.66 GPa respectively. In case of OOCSFV-1 CT-3, CT-1 fibre had exhibited more tensile strength (36.19 Mpa), modulus (1.85 Gpa) respectively. One may understand that the chemical treatment was responsible for the enhancement in bonding between the fibre and matrix. Tensile strength, modulus of untreated and treated OI, OOCSFV1, 2 against percentage strain of all the tested samples is graphically shown in Figure 3 a, b, c and d. Out of raw, bleached, alkali treated, AN-grafted 7.38%, 11.43% okra bast fibres investigated, grafting of bleached fibre with acrylonitrile brought a substantial increase in tensile stress and its value is nearly 80 N/mm [**6**].

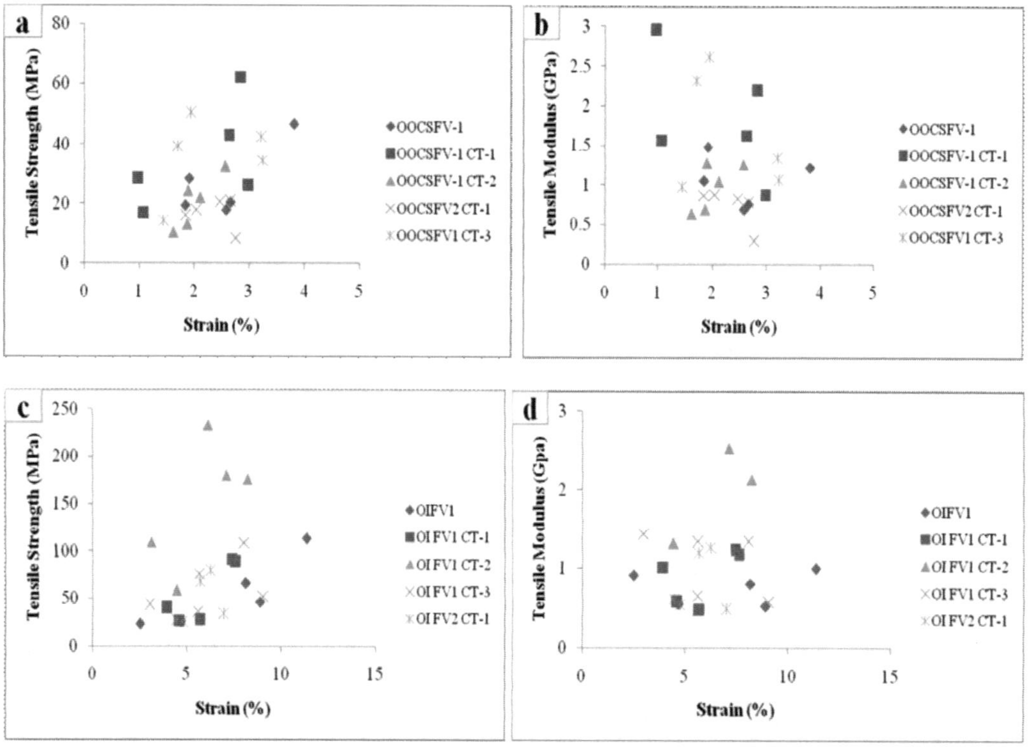

Figure 3. Tensile strength, modulus of okra individual, oval chain structured fibre

The tensile strength, modulus were decreased with increase in diameter of okra fibre. Khustia Dt. Okra fibres (more than eighty) were tested (10 mm gauge length) according to ASTM D 3379-75 on Lloyd Dynamometer LR30K with 20N load cell at a crosshead speed of 1mm/min [**8**].

The mechanical properties of okra fibre (raw and treated forms) were reported by I. M. De Rosa et al. [**9**]. The tensile strength, modulus of okra fibres had shown variability with respect to diameter and their values were in the following range: diameter 88.3 – 218.9 μm, tensile strength 52.6 – 233.8 MPa and modulus 3.2 – 12.8 GPa [**9**].

3.2. Scanning Electron Micrograph Analysis of fibre

The surface morphology of the OIFV-1, 2 treated and untreated fibres were shown elsewhere in the paper published by corresponding author. Now the SEM images of untreated and treated OOCSF fibres are given below along with the individual fibres for comparison, **Figure 4–13**.

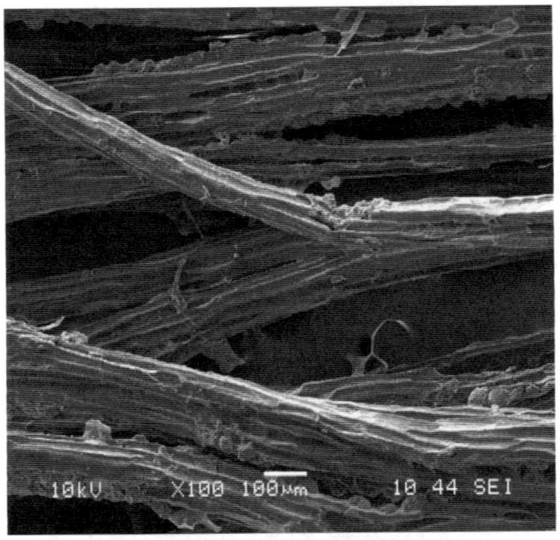

Figure 4. SEM image of OOCSFV-1

Figure 5. SEM image of OOCSFV-1 CT-1

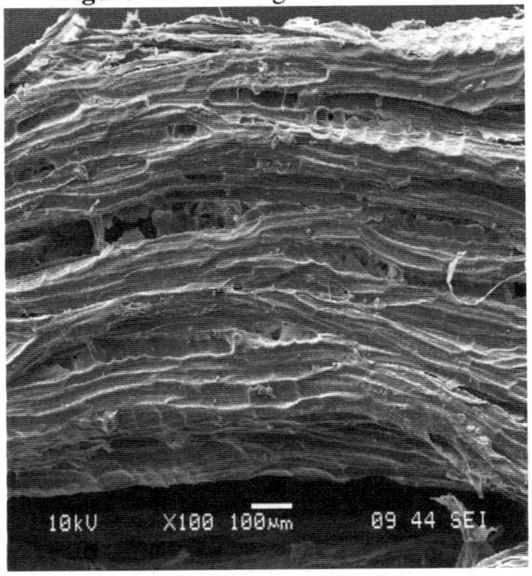

Figure 6. SEM image of OOCSFV-1 CT-2

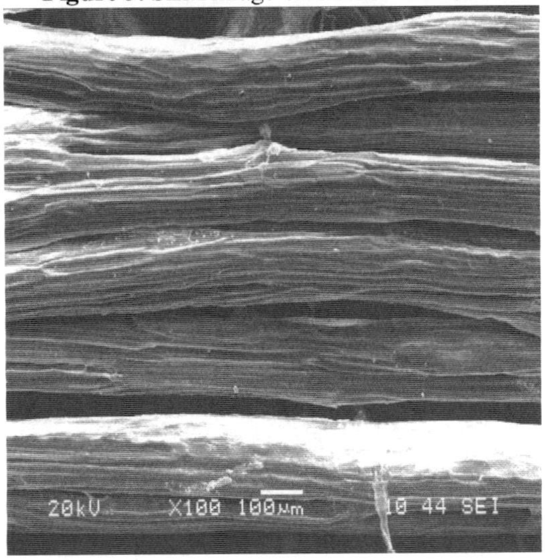

Figure 7. SEM image of OOCSFV-1 CT-3

IOP Conf. Series: Materials Science and Engineering **83** (2015) 012003 doi:10.1088/1757-899X/83/1/012003

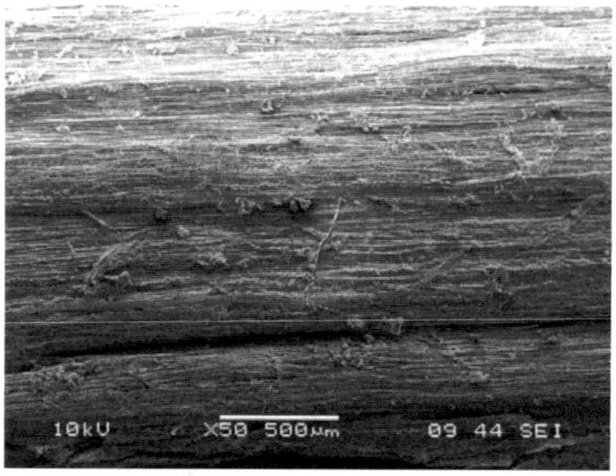

Figure 8. SEM image of OOCSFV-2 CT-1

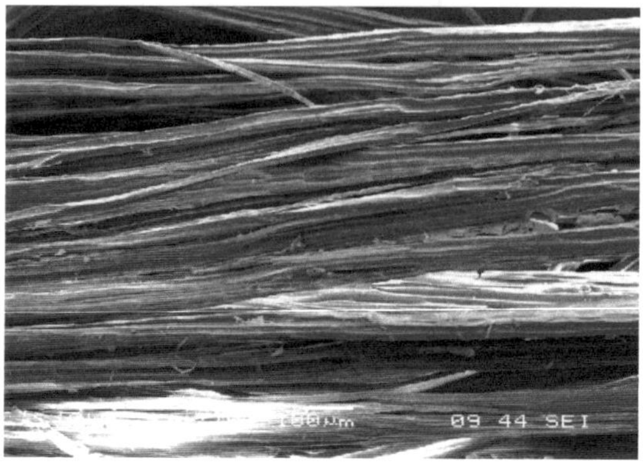

Figure 9. SEM image of OIFV1

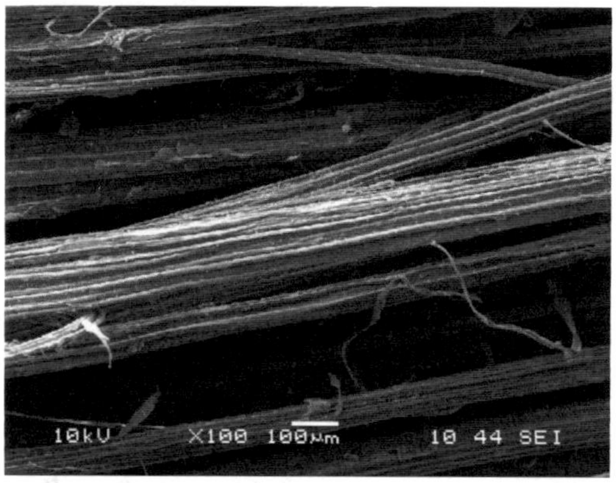

Figure 10. SEM image of OIFV1 CT-1

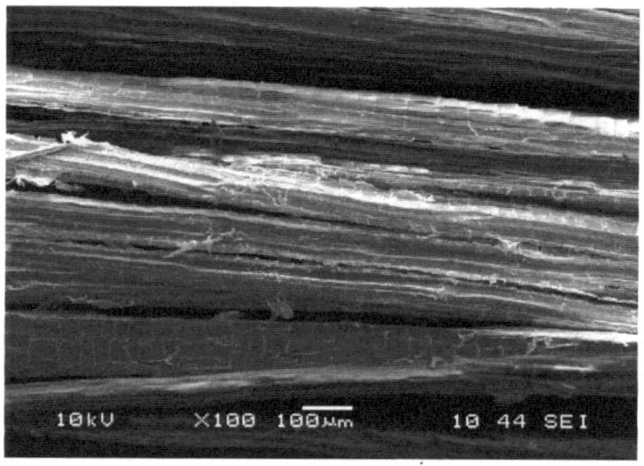

Figure 11. SEM image of OIFV1 CT-2

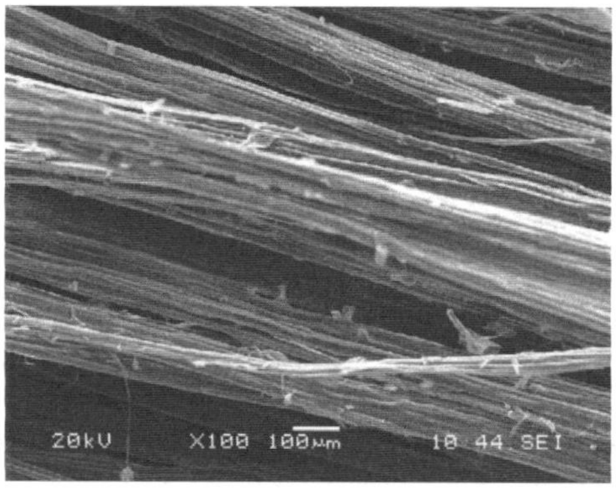

Figure 12. SEM image of OIFV1 CT-3

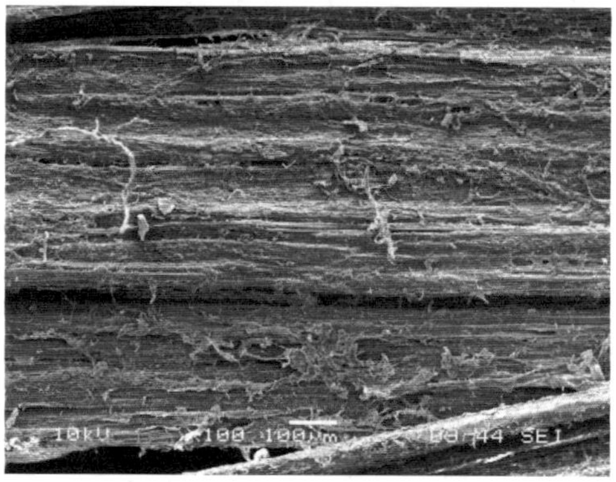

Figure 13. SEM image of OIFV2 CT-1

Okra oval chain structured fibre with channels and other matters are visible from Figure 4 whereas a clear flattened rough surface obtained after CT-1 is observed from Figure 5. The composition CT-2 resulted in much rougher surface with loss of fibrils is noticed from Figure 6 whereas alkali treatment resulted in the exploiting of neat cylindrical filaments, Figure 7. The flat filaments bonded together are understood from Figure 8. Very fine filaments are visualised from Figure 9 in case of OI FV1. Some of the useful cellulosic materials lost in CT-1 of OIFV1 and was resulted in channel formation, Figure

10. The formation of near square boxes on circumference of fibre surface is due to the CT-2 on OI FV1, Figure 11. The Figure 12, 13 represents the formation of filaments and deteriorated surfaces respectively. A considerable amount of polyacrylonitrile was grafted onto okra bast fibre surface and intercellular gaps were also reduced due to the decomposition of PAN grafts which was identified when bleached and AN-grafted okra bast fibre micrographs were observed [5]. When raw, bleached and alkali treated okra fibres were observed at different magnifications the multicellular nature of okra fibre is more clearly visualized [6]. The presence of reasonable number of voids along okra fibre length was observed. The non selective nature of treatment on okra fibre resulted in increase of voids in the fibre and was also observed in the image [7].

The cross section, longitudinal surface of okra fibre was microscopically examined by I. M. De Rosa et al [8]. The structure of an okra fibre consists of several elementary fibres or ultimate fibres or cells overlapped along the length of the fibres and bonded firmly together by pectin and other non cellulosic compounds. The region at the interface of the two cells is called middle lamella and the bundles of elementary fibres are referred to as technical or single fibres. They have also observed a polygon shape that varies from irregular to circular. Brittle behaviour with the absence of significant fibril splitting along with the fibre pull-out in the outer part of the bundle was observed. The presence of cell wall defects along the fibre length may create stress intensities and were responsible for different fibre cells fractured at various stress levels [8].

I. M. De Rosa et al. [9] have used SEM micrographs to measure cross sectional area of fibre more accurately and the results were used to recalculate the mechanical properties. Fracture surfaces present a variable level of twisting as an effect of chemical modification. Chemical modification resulted in variable level of twisting and was observed on fracture surfaces. The surface both along the length and in section appear smoother in case of acetylated and bleached okra fibre, which may be ascribed to the regeneration effect of the fibres due to acetylation. The presence of small holes around the mid section of the fibre(s) was observed due to permanganate treatment on okra fibres which may be ascribed to some degradation of the structural part of the fibre owed to the treatment.

The increase in porosity with 90.02% permanganate treatment in alkaline solution has also been observed in sisal fibre [9]. E. Fortunati et al. have described that the dimensions of unmodified okra fibre is in the order of 90 μm. Chemical treatment (dewaxing, bleaching, and alkali treatment) of fibre split the fibres and increased their surface area [10]. Pristine okra fibres (untreated) had shown polygonal shape with a diameter of 88.3 μm (which was calculated in a series of fifty fibres). After chemical treatment the fibre diameter was 9.8 μm. The morphology of the okra derived cellulose lateral surface shows a quite rough surface, and is oriented in the longitudinal direction [11].

3.3. Tensile Properties of Composites

The specific tensile strength and modulus of the composites was determined and is shown in Figure 14, 15. The composites reinforced with OI FV1 CT-2 fibre have exhibited highest specific tensile strength (60.75 Mpa/kg.$m^{-3}\times10^{-3}$), modulus (978.75 MPa/kg.$m^{-3}\times10^{-3}$) than all other okra fibre reinforced polyester composites investigated by Nadendla Srinivasababu. Under the composites reinforced with okra oval chain structured fibres in layered form CT-2 to the fibre resulted in good interfacial bonding between fibre and matrix which is evidenced from the experimental results i.e. specific tensile strength of 56.83 MPa/kg.$m^{-3}\times10^{-3}$ and modulus of 834.99 MPa/kg.$m^{-3}\times10^{-3}$ was observed at maximum (27.85 %) fibre volume fraction. All the composites were failed due to tensile load only and no fibre was pulled out from the matrix.

Lady finger fibre (untreated, bleached) reinforced Bakelite composites had shown decrease in tensile strength with increase in fibre content. The maximum value achieved in case of untreated, bleached fibre composites was nearly 18.2, 16.5 MPa respectively at its lowest fibre content [7]. E. Fortunati et al. [10] reported that PVA/10CNC composites have showed the highest value of elastic modulus of 1.9 GPa.

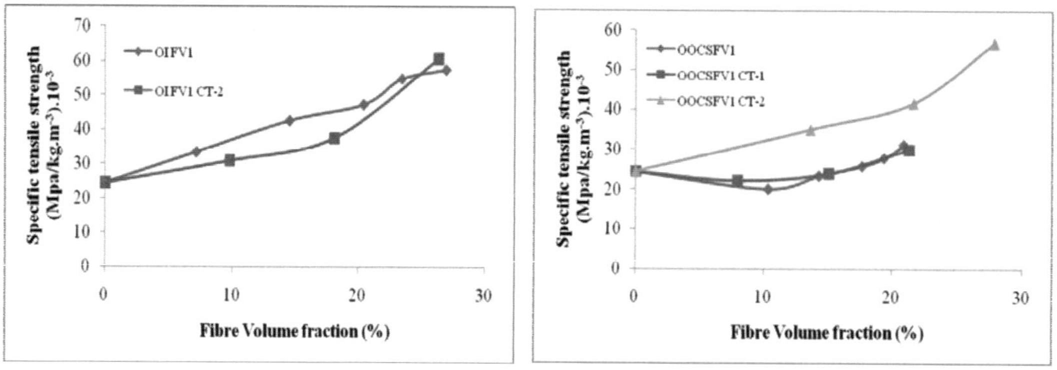

Figure 14 Specific tensile strength of OI and OOCS fibre reinforced polyester composites

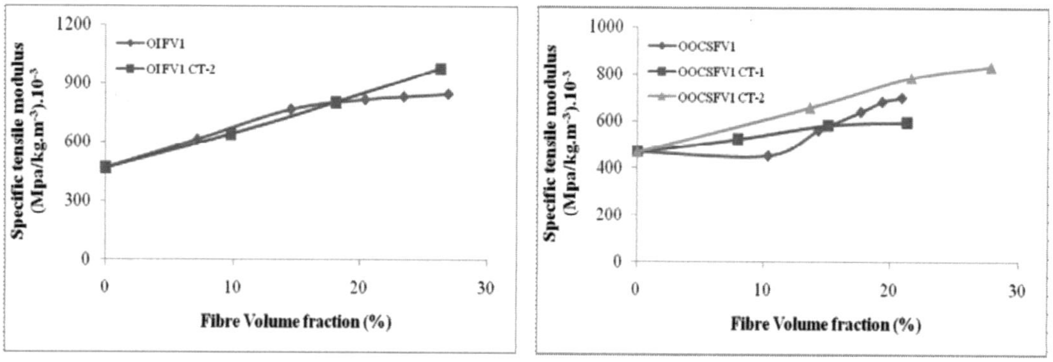

Figure 15 Specific tensile modulus of OI and OOCS fibre reinforced polyester composites

3.4. Flexural Properties of Composites

The specific flexural strength and modulus of the composites (reinforced with OOCS, OI fibre) against fibre volume fraction is graphically represented in Figure 16, 17 respectively. All the okra fibre reinforced polyester composites (except OI FV1) have shown increasing trend of flexural strength with increase in fibre volume fraction. In these composites the composites reinforced with individual fibres at 35.05 %, 31.21 % fibre volume fraction have shown highest flexural strength, modulus of 108.7 MPa/kg.m^{-3}x10^{-3} and 9.52 MPa/kg.m^{-3}. Very good interfacial locking is shown by the OOCS FV2 CT-1 FRP composites and its flexural modulus was 5.39 MPa/kg.m^{-3}, which was highest in okra oval chain structured fibre reinforced polyester composites. All the composites were failed due to bending only and failed in its outer most layers, which is highly desired.

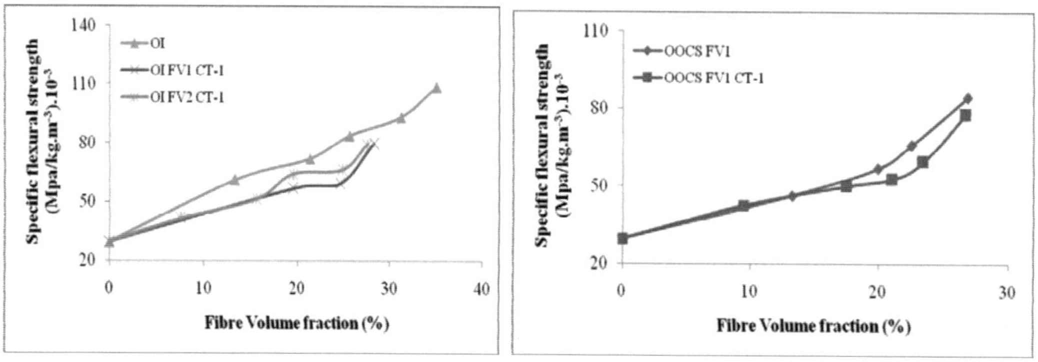

Figure 16 Specific flexural strength of OI and OOCSF reinforced polyester composites

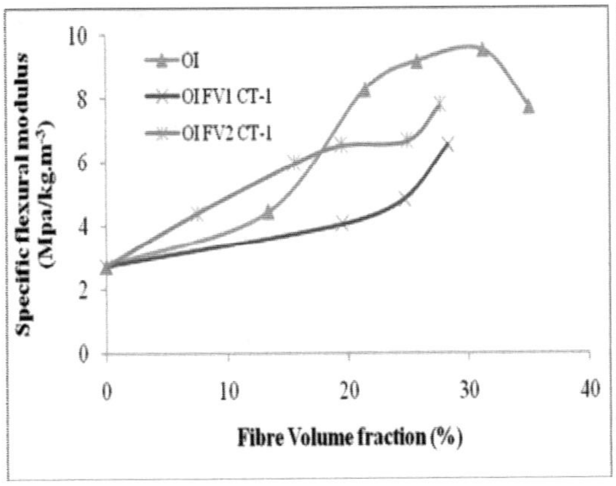

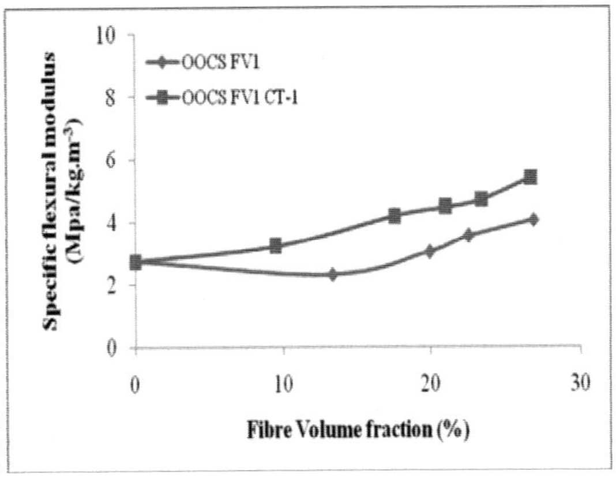

Figure 17 Specific flexural modulus of OI and OOCSF reinforced polyester composites

Bleached okra fibre reinforced Bakelite composites exhibited haphazard flexural strength and is visualized from the graph and a maximum flexural strength of 42 MPa [10].

3.5. Impact Properties of Composites

OI FV1 FRP composites with 12.7 mm width have exhibited impact strength of 52.45 kJ/m^2 which is higher than all other OI and OOCS fibre composites investigated. OI, OOCS FV1 reinforced polyester composites had shown increase in impact strength with fibre content. The okra FRP composites with 10 mm width specimens have shown an increase in impact strength with increase in fibre volume content which is recognizable from Figure 18. All the composite specimens had shown clear break of the specimens and is designated as 'C' as per ASTM D 6110 – 08.

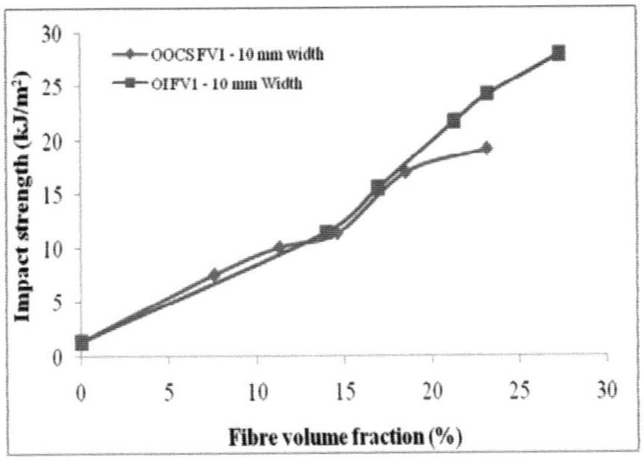

Figure 18 Impact strength of OOCSF and OI fibre reinforced polyester composites

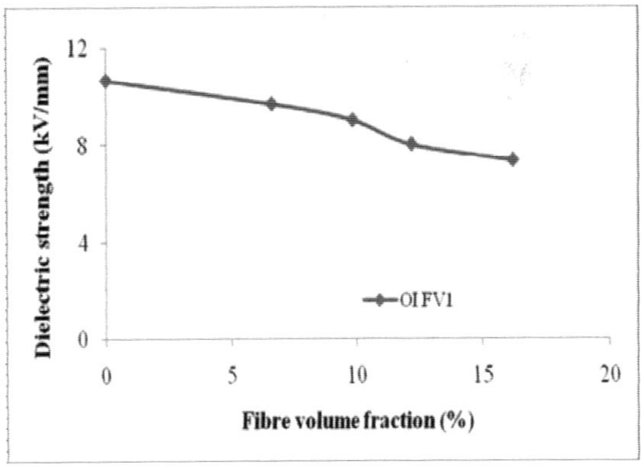

Figure 19 Dielectric strength of OOCSF and OI fibre reinforced polyester composites

3.6. Dielectric properties

The insulation property of the composites is assessed for okra individual fibre reinforced polyester composites. The dielectric strength of the composites is determined and is shown against fibre volume fraction, Figure 19. The highest value of dielectric strength (9.67 kV/mm) is achieved at 6.59 % fibre volume fraction. Hence the designer will get an opportunity to select the light weight insulating materials based on the data available.

3.7. Environmental Impact Assessment

In order to understand the present section one has an idea of conditions of formation of acid rain. The pure rain water possesses a pH of about 5.6 and is acidic due to the equilibrium established between water and carbon dioxide in the atmosphere which is resulted in the formation of carbonic acid as per the following reaction.

$$CO_2 + H_2O \leftrightarrow H_2CO_3 \leftrightarrow H^+ + HCO^-_3$$

Acid rain or precipitation refers to rain, snow, fog, or gaseous particles that have a pH significantly below 5.6. In general precipitation below 5 can be considered as acidic. Acid rain occurring as rain or snow is referred to as wet deposition whereas dry deposition occurs when gaseous acid particles are directly deposited on the surfaces. Robert Angus Smith a Scottish chemist was identified acid rain and studied it in depth in 1852.

The principal cause of acid rain is the combustion of fossil fuels that produce sulphur and nitrogen emissions and the primary sources are electrical power plants, automobiles and smelters. Power plants, automobiles produce most of the sulphur emissions and nitrogen emissions respectively. The reactions involved in the formation of acid rain are understood from the following reactions.

$$S_8 + 8\,O_2 \rightarrow 8\,SO_2$$

$$SO_2 + H_2O \rightarrow H_2SO_3$$

$$2\,H_2SO_3 + O_2 \rightarrow 2H_2SO_4$$

Nitric acid, HNO_3 and nitrous acid HNO_2 form when nitrogen dioxide reacts with water.

$$NO_2 + H_2O \rightarrow HNO_3 + HNO_2$$

Many equations contribute to the overall formation of acid rain and the reactions often involved a complicated number of steps that depend on the atmospheric conditions. The above reactions shows the formation of sulphuric acid and nitric acid and do not show the numeric reactions that actually occur. Acid rain has number of negative effects on humans, *plants*, building materials, and entire eco system [**12**].

Hence in order to understand the impact of such kind of rain, the accelerated environmental conditions to the fibre were created through the chemical treatment which contains very minute levels of concentration of H_2SO_4. A clear decrease in the mechanical performance is visualized from the experimental results, which indicates that the fibre was degraded through conditions of exposure viz. in case of individual fibres chemical treatment 1, 2 for OI FV1, 1 for FV2 whereas for oval chain structured fibre 1, 2 for FV1, 1 for FV2.

4. Conclusion and Applications

The present review has given the following impressions. Till today there is a little utilization of the okra fibre and its composites and the literature available is quite small which is evidenced from the list given at the end. Further the OOCSFV1 is a new finding and was identified by the author in 2007 after the fibre was extracted. This kind of fibre with enormous potential properties can be utilised in the preparation of various automobile, structural parts. Acid rain will have a clear affect on degradation of the performance of the composites. So the performance of natural fibres is clearly dependent on the environment where they will grow. Okra FRP composites have shown reasonable insulation ability i.e. dielectric strength.

Acknowledgments

The author firstly thankful to GOD, created diversified natural fibres in a single okra plant. He is further thankful to Dr. Katta Murali Mohan Rao, who has given an encouragement by providing Srikakulam okra fibre during the intial stage of authors work. The results of the work are explored to the world only with the kind help of the authors' grandmother Smt. Veeramachineni Jhansi Lakshmi and mother Smt. N. V. L. R. Mani help to manufacture the composites and in all the stages of the research work.

References

[1] Maya John J and Sabu Thomas 2012 *Natural Polymer Volume 1: Composites* (Cambridge: Royal Society of Chemistry)

[2] Sabu Thomas and Laly Pothan A 2009 *Natural Fibre Reinforced Polymer Composites: From Macro to Nanoscale* (USA: Old City Publishing, Inc.)

[3] Srinivasababu N and Sravanthi N 2011 Tensile and dielectric properties characterization of srikakulam abelmoschus esculentus fiber reinforced polymer composites, *National Conference on Processing and Characterization of Materials, NIT Rourkela*, pp. 39.

[4] Shamsul Alam Md, Arifuzzaman Khan G M, 2007 Chemical analysis of okra bast fiber (Abelmoschus esculentus) and its physic-chemical properties *J. Text. Appar. Tech. And Mgt.* Vol. 5, pp. **1-9**

[5] Arifuzzaman Khan G M, Saheruzzaman Md, Abdur Razzaque S M, Sakinul Islam Md, and Shamsul Alam Md 2009 Grafting of acrylonitrile monomer onto bleached okra bast fibre and its textile composites Indian *J. Fibre. and Text. Res.* Vol. 34, pp. **321-327**

[6] Arifuzzaman Khan G M, Saheruzzaman Md, Rahman M H, Abdur Razzaque S M, Sakinul Islam Md, and Shamsul Alam Md 2009 Grafting of acrylonitrile monomer onto bleached okra bast fibre and its textile composites *Fibers and Polymers* Vol. 10, pp. **65-70**

[7] Moniruzzaman Md, Maniruzzaman Mohd, Gafur M A, and Santulli C 2009 Lady's Finger Fibres for Possible Use as a Reinforcement in Composite Materials *J Biobased Mat. And Bioenergy* Vol. 3, pp. **1-5**

[8] Igor Maria De Rosa, Jose Maria Kenny, Debora Puglia, Carlo Santulli, and Fabrizo Sarasini 2010 Morphological, thermal and mechanical characterization of okra (Abelmoschus esculentus) fibres as potential reinforcement in polymer composites *Comp. Sci. and Tech.* Vol. 70, pp. **116-122**

[9] Igor Maria De Rosa, Jose Kenny M, Maniruzzaman Mohd, Moniruzzman Md, Macro Monti, Debora Puglia, Carlo Santulli and Fabrizio Sarasini 2011 Effect of chemical treatments on the mechanical and thermal behaviour of okra (Abelmoschus esculentus) fibres *Comp. Sci. and Tech.* Vol. 71, pp. **246-254**

[10] Fortunati E, Puglia D, Monti M, Santulli C, Maniruzzaman M, and Kenny J M 2012 Cellulose nanocrystals extracted from okra fibers in PVA nanocomposites *J App. Polym. Sci.* Vol. 122, pp. **3220-3230**

[11] Fortunati E, Puglia D, Monti M, Santulli C, Maniruzzaman M, Foresti M L, Vazquez A, and Kenny J M 2013 Okra (Abelmoschus esculentus) fibre based PLA Composites: mechanical behaviour and biodegradation *J Polym. Environ.* Vol. 21, pp. **726-737**

[12] Richard Myers 2003 *The Basics of Chemistry* (Greenwood Publishing Group)

Appendix A.

OIFV–1: Okra Individual Fibre Variety–1

OIFV–1 CT–1: Okra Individual Fibre Variety–1 Chemical treatment–1

OIFV–1 CT–2: Okra Individual Fibre Variety–1 Chemical treatment–2

OIFV–1 CT–3: Okra Individual Fibre Variety–1 Chemical treatment–3

OIFV–2 CT–1: Okra Individual Fibre Variety–2 Chemical treatment–1

OOCSFV–1: Okra Oval Chain Structured Fibre Variety–1

OOCSFV–1 CT–1: Okra Oval Chain Structured Fibre Variety–1 Chemical Treatment–1

OOCSFV–1 CT–2: Okra Oval Chain Structured Fibre Variety–1 Chemical Treatment–2

OOCSFV–1 CT–3: Okra Oval Chain Structured Fibre Variety–1 Chemical Treatment–3

OOCSFV–2 CT–1: Okra Oval Chain Structured Fibre Variety–2 Chemical Treatment–1

Electrochemical Impedance Spectroscopy Study on Corrosion Protection of Acrylate Nanocomposite on Mild Steel Doped Carbon Nanotubes

M R Mahmud [1,2], M M Akhir [1,2], M S Shamsudin [1,2], A N Afaah [1,2], A Aadila [1,2], N A M Asib [1,2], Salman A H Alrokayan[4], Haseeb A Khan[4], M K Harun [1], M Rusop [2,3] and S Abdullah [1]
[1]Faculty of Applied Sciences, Universiti Teknologi MARA (UiTM), 40450 Shah Alam, Selangor, Malaysia
[2]NANO-SciTech Centre, Institute of Science, Universiti Teknologi MARA (UiTM), 40450 Shah Alam, Selangor, Malaysia
[3]NANO-ElecTronic Centre, Faculty of Electrical Engineering, Universiti Teknologi MARA (UiTM), 40450 Shah Alam, Selangor, Malaysia
[4]Research Chair of Targeting and Treatment of Cancer Using Nanoparticles, Department of Biochemistry, College of Science, King Saud University (KSU), Riyadh, Saudi Arabia

E-mail: mrashidmahmud@yahoo.com

Abstract. Acrylate:carbon nanotubes (A:CNTs) nanocomposite thin film was prepared by sol-gel technique. The corrosion coating protection of acrylate:carbon nanotubes (CNTs) nanocomposite thin film has been coated on mild steel characterised by electrochemical impedance spectrometer (EIS) measurement and equivalent circuit model are employed to analyse coating impedance for corrosion protection. In this study, 3.5 w/v % sodium chloride (NaCl) solution was immersed the acrylate:carbon nanotubes nanocomposite thin film. As the results, the surface morphology were found that there formation of carbon nanotubes with good distribution on acrylate-based coating. From EIS measurement, A:CNTs nanocomposite thin film with 0.4 w/v % contain of CNTs was exhibited the highest coating impedance from Nyquist graph after immersed in sodium chloride solution and may provide the excellent corrosion protection. The Bode plots have shown the impedance is high at the beginning from the time at high frequency and slightly decreases with value of frequency become smaller.

1. Introduction

The coatings corrosion protection has been used widely in industries due to possess protection from any defects or corrosion activity [1, 2]. Corrosion occurs in metal due to the surface exposures to the environment changes and moistures such as solution. Polymers anti-corrosion protection have been used to avoid corrosion activity from penetrate to the coated materials. This study using acrylate as polymer materials to coat surface of mild steel [3]. Polymeric materials known as acrylate based coating possess excellent mechanical strength and adhesion for several substrates. Acrylate provides excellent coatings protection for home appliances, aircraft, marine instrument, ships, oil planter and others metal materials avoid from corrosion activity [4-6]. In this research, mild steel substrate was used as coated materials for thin films of acrylate. Carbon nanotubes discovered by Sumio Iijima in 1991 which has invented carbon nanotubes (CNTs) using arc-discharge method in powder form [7-11]. CNTs possess versatile nanofillers for blend with polymer-based such as acrylate due to tremendous properties in mechanical, electrical, thermal and structure [12, 13].

Acrylate:carbon nanotubes (A:CNTs) nanocomposite thin film used for coating protection on coated mild steel by provides coating barrier between the metal coated and exposure medium such as sodium chloride solution used in this characterisation. Furthermore, Sol-gel technique was prepared of A:CNTs nanocomposite thin film provide a cost effective and easy way to incorporating the polymers composite and nanomaterials in order to obtain well dispersion and homogenously distributed [4, 14].

Electrochemical impedance spectrometer (EIS) has used for evaluation corrosion performance on coating metal materials that exposure to the aqueous solution and non-destructive testing for corrosion of the coating protection [15]. In term of organic coatings, possess tremendous and effective protection of metals coated and as corrosion inhibitors to the materials. Furthermore, metals substrate and intermediate layer of coating protection naturally can be determine the protection characteristics of the coated materials. All coating protection influence the response of impedance to the metal coated that extracted by analysis used of EIS spectra. Polymeric materials exposed to the environment changes and given protection to the coated metal but it is influence to the polymeric are potentially permeable to corrosion activity such as oxygen, water and ions [16]. Macedo et al. has discussed anomalous impedance variations on the samples coated as a function of time exposure in a sodium chloride solution with approach used the five-electrode system [15, 17]. However, most of the popular method for measured the coating corrosion protection was used electrolyte as an solution to exposure the coated metal by using composition of the sodium chloride (NaCl) around 3.5 w/v % of NaCl [18-20], the results from EIS measurement are typically fluctuated from different samples through the coating defects or pinholes in solution. In this present work, were prepared acrylate:carbon nanotubes nanocomposite thin film by doped with varies amount of weight percentages that increment of 0.2 w/v % of CNTs into acrylate such as 0.2 w/v %, 0.4 w/v %, 0.6 w/v %, 0.8 w/v % and 1.0 w/v % to determine the coating corrosion protection of the A:CNTs nanocomposite thin film.

2. Experimental

The chemicals of sodium chloride (NaCl) powder was used in order to produce 3.5 w/v % of NaCl electrolyte. The contact area of mild steel was then polished with high graded sand paper and cut into small size with 2 cm x 2 cm dimension. In other hand, the carbon nanotubes (CNTs) was synthesised by chemical vapour deposition (CVD) technique. Camphor oil and ferrocene as a precursor and catalyst were used in this research, respectively [7]. CVD were employed with 180 °C for the pre-cursor temperature (first furnace) and 800 °C for the deposition temperature (second furnace) takes around 60 minutes duration for deposition time. The experimental set up of CVD can be found in our previous paper [21-24]. The acrylate:carbon nanotubes (A:CNTs) nanocomposite was prepared by sol-gel technique. The acrylate was dissolved in organic chloroform solvent with ratio of 1:5 (acrylate:chloroform). Then, the CNTs were introduced by 0.2 to 1.0 w/v % with 0.2 w/v % increment in each acrylate solutions. The sonochemistry technique was employed in order to ensure CNTs mix well in acrylate solution by using ultrasonic probe processor (Sonotrode H3UP 400S, Hielscher) with parameter sonication time is 60 minutes. After that, spin-coater with speed around 5000 rpm for deposition time is 60 second for 10 drops were used for A:CNTs nanocomposite deposited on mild steel substrate. Then, the sample have been dried for 10 minutes with 100 °C to remove the residual from substrate.

Characterisation parameter on immersed in NaCl solution by using EIS analysis with initial frequency (50 kHz) and final frequency (0.015 Hz). Samples exposure area to the solution is 2.85 cm^2 with AC voltage 10 mV_{rms}. EIS measurement were performed by Gambry Instrument for this corrosion behavior and used Gambry Echem Analyst software for corrosion analysis in this study. Hence, field emission scanning electron microscope (FESEM) were observed for surface morphology studies of acrylate/carbon nanotubes nanocomposite thin film with 5 kX magnification and working distance is 5.4 mm.

3. Results & Discussion

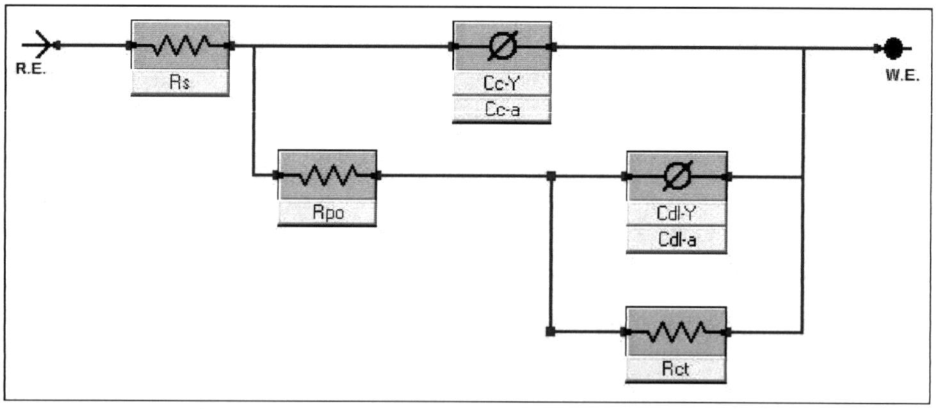

Figure 1. Equivalent circuit model used for EIS measurement of the samples immersed in 3.5 w/v % NaCl solution.

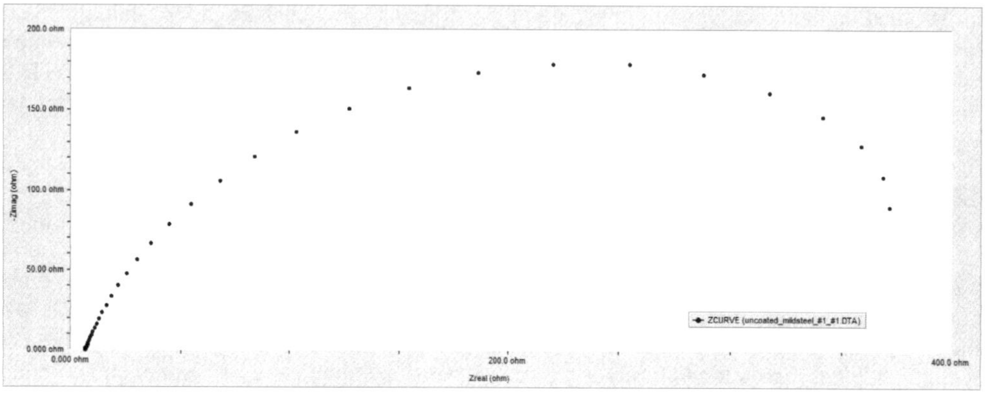

Figure 2. The Nyquist graph has shown EIS measurement of the uncoated mild steel.

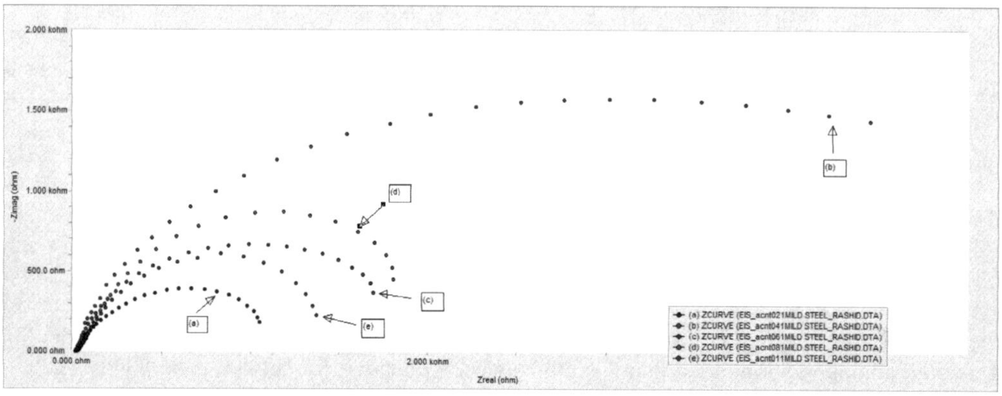

Figure 3. The Nyquist graph on EIS measurement for acrylate:carbon nanotubes nanocomposites for several weight percentage between 0.2 w/v % - 1.0 w/v % of CNTs contain with increment of 0.2 w/v %.

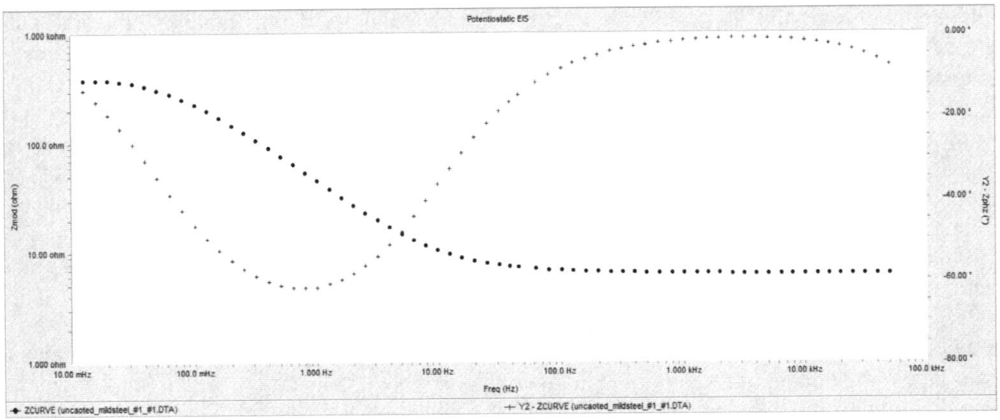

Figure 4. The Bode plot for the uncoated mild steel.

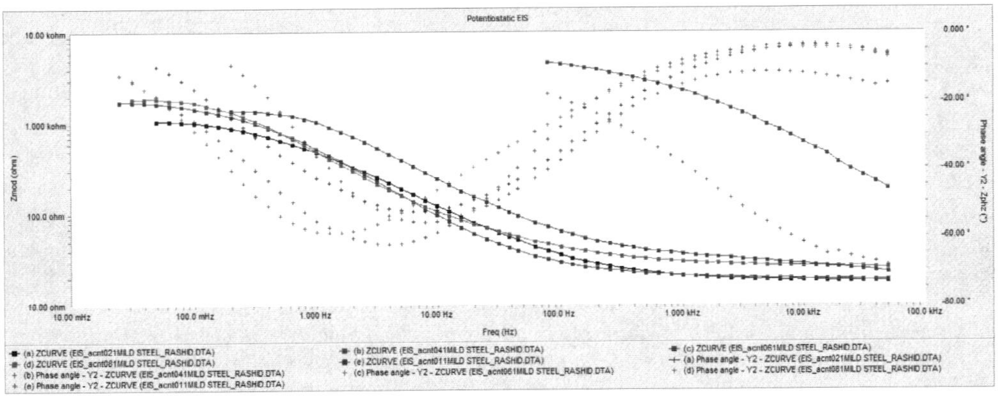

Figure 5. The Bode plots of acrylate:carbon nanotubes nanocomposites thin film on mild steel with increment 0.2 w/v % of CNTs contains (a) 0.2 w/v %, (b) 0.4 w/v %, (c) 0.6 w/v %, (d) 0.8 w/v % and (e) 1.0 w/v %.

Figure 1 has shown the equivalent circuit model are employed to analyze the corrosion protection of the coatings under percolation of sodium chloride, (NaCl) solution. Basically, The electrochemical impedance spectroscopy (EIS) were performed in this research well-known for investigated the coating corrosion protection for the acrylate:carbon nanotubes (A:CNTs) nanocomposite thin films [25]. Hence, R_s is solution resistance; C_c is coating capacitance; R_{po} is pore resistance; C_{dl} is double layer capacitance; R_{ct} is charge transfer resistance. Where, ω is angular frequency; α is a dimensionless parameter; Y is a parameter with dimensions Ω^{-1} s^{α} representative for constant phase element (CPE). Commonly, $0 < \alpha < 1$; when $\alpha=1$ and Y=C, CPE function like ideal capacitor with capacitance, C; when $\alpha=0$, a resistance is presented. It also resemblance equivalent circuit model founded to stimulate the degraded behavior of organic coating in existence publication [17, 26].

$$Z_{CPE} = \frac{(j\omega)^{-\alpha}}{Y} \tag{1}$$

Employing EIS results, the impedance spectroscopy were analyzed using Gambry software (Gambry Echem Analyst) which is used initial frequency 50 kHz and final frequency 0.015 Hz under continuous immersion in 3.5 w/v % NaCl solution probably half an hour. In figure 2, obtained the Nyquist graph of uncoated mild steel corresponding to the natural corrosive activity apparently happen due to there is no coating protection on the metal surface. In addition, the graph shown single capacitive loop at high frequency at the starting measurement. The protection of the coating being attributed to the solution protection in the conductive paths of the sol-gel coating and corrosion activity only occuring in the layer of free coating protection [14].

In other side, figure 3 shows that the Nyquist graph of acrylate:carbon nanotubes nanocomposite for several samples with different contains of CNTs such as (a) 0.2 w/v %, (b) 0.4 w/v %, (c) 0.6 w/v %, (d) 0.8 w/v % and (e) 1.0 w/v % with increment of weight percentage 0.2 w/v % of CNTs. From the table 1 obtained that the sample with 0.2 w/v % of CNTs shows the highest impedance value compared to the rest has possess Z_{real} = 4.454 kΩ and $Z_{imaginary}$ = 1.439 kΩ, respectively. Then, the coating impedance slightly decreases interm of Z_{real} and $Z_{imaginary}$ from 0.8 w/v % (1.804 kΩ and 450 Ω), 0.6 w/v % (1.689 kΩ and 369.2 Ω), 1.0 w/v % (1.374 kΩ and 226.1 Ω), 0.2 w/v % (1.051 kΩ and 182.1 Ω) of CNTs contain and the lowest impedance is uncoated mild steel (371.8 Ω and 89.59 Ω). The most excellent corrosion protection of acrylate with contain 0.4 w/v % of CNTs may due to highest coating impedance that can retained coating from the NaCl solution to pass through the coated mild steel. Thus, addition of carbon nanotubes (CNTs) in acrylate improve the corrosion protection of the coating due to formation of CNTs as conducting materials would trap the ions from directly penetrate the mild steel and provides increment of barriers in the coating. Besides that, the barrier phenomenan can be related to the avoiding or blocking of the the free metal surface and decreasing the corrosion rate. In theory, abatement values of potential during measurement has shown scattered of electrotyle and corrosion activity occurs take place on defects or pin holes of the mild steel which is existed in the surface of the coated metal. According to A. Mostafaei et al. [27] state that the potential was shifted occurs due to the formation of the passive layer film on carbon steel substrate with assistance of conducting polymer.

Figure 4 has shown the Bode plot that evolutions of impedance modulus for uncoated mild steel with high value of impedance at the high frequency and slightly decreases in term of value of impedance at the low frequency. It is may due to the penetration of NaCl solution into the mild steel substrate without coating protection of acrylate:carbon nanocomposite thin film. Besides that, figure 5 shows that the Bode plots of the samples for fives different contain of CNTs found that almost similar with uncaoted mild steel where coating thin film started at high frequency with highest impedance and after a few times it is slightly decreases with frequency and the coating impedance become lowest at the end of immersed in NaCl solution. But, there is shifted for the sample contain 0.2 w/v % of CNTs which is highest in impedance at the lowest frequency then also decreases after a few times with frequency become smaller. In literature, the presence of nanoparticles in polymer coating comes with incorporating between filler and matrix can improve the barrier between coated metals and exposure solution that can protect from corrosion activity [28].

As shown in figure 6, optical photographs shows sample surface after was immersed in 3.5 w/v % of NaCl solution for investigation of corrosion performance of the coated mild steel. Circular surface around 2.85 cm^2 area exposure to the solution has been formed corrodes due to the electrolyte slightly penetrated the coating material and surface of mild steel started defect with no longer protected by the A:CNTs nanocomposite thin film. For the surface morphology of (A:CNTs) nanocomposite thin film were observed by using field emission scanning electron microscope (FESEM) with 5.00 kX magnification before this sample was immersed in 3.5 w/v % of NaCl solution. As shown in figure 7 has found that the formation of CNTs as filler is well distributed into acrylate but not even homogeneous and CNTs formed with black colour with tubes-likes and acrylate in colourless form.

IOP Conf. Series: Materials Science and Engineering **83** (2015) 012004 doi:10.1088/1757-899X/83/1/012004

Table 1. The impedance value (Z_{real} and $Z_{imaginary}$) that refer to the figure 2 and figure 3 of acrylate:carbon nanotubes nanocomposite thin film coated on mild steel

Contain of CNTs, (w/v %)	Z_{real}, (Ohm, Ω)	$Z_{imaginary}$, (Ohm, Ω)
Uncoated mild steel	371.8	89.59
0.2	1.051 k	182.1
0.4	4.454 k	1.439 k
0.6	1.689 k	369.2
0.8	1.804 k	450
1.0	1.374 k	226.1

Figure 6. The sample of acrylate:carbon nanotubes nanocomposite thin film deposited on mild steel after immersed in 3.5 w/v % of NaCl solution.

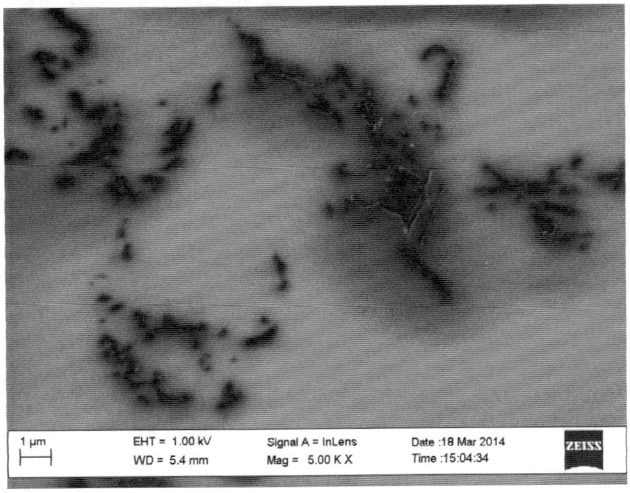

Figure 7. FESEM image of acrylate:carbon nanotubes nanocomposite thin film coated on mild steel.

4. Conclusion

As the conclusion, the acrylate:carbon nanotubes (A:CNTs) nanocomposite thin film were successful prepared by sol-gel technique and the results has shown addition of carbon nanotubes into acrylate improve the coating barrier and increase coating impedance that coated on mild steel. Electrochemical impedance spectrometer found that acrylate:carbon nanotubes nanocomposite thin film with 0.4 w/v % of CNTs contain possess the highest coating impedance provide good corrosion protection on coated mild steel that prevent from NaCl solution to pass through the substrate. Furthermore, the surface morphology were observed by FESEM shows that CNTs was formed in acrylate with good distribution but not even homogeneous. There is good improvement in corrosion protection properties in term of varies doped weight percentage of carbon nanotubes blended with acrylate and lengthen the corrosion protection on coated materials.

5. Acknowledgment

The authors (Muhammad Rashid Mahmud) are grateful to the Ministry of Education Malaysia (MOE) sponsored my studying with Mybrain15 scholarship. This work was supported by Research Grant from the Ministry of Science, Technology and Innovation of Malaysia and also supported by the Long-Term Research Grant Scheme for Nanostructure, Nanomaterials and Devices for Fuel Cells and Hydrogen Production (600-RMI/LRGS 5/3 (3/2013)) and Research Management Institute (RMI), Universiti Teknologi MARA (UiTM), Malaysia for their support. The authors also really appreciated the technical support by Faculty of Applied Sciences, NANO-SciTech Centre, NANO-ElecTronic Centre, Universiti Teknologi MARA (UiTM), Shah Alam, Electrochemical Impedance Laboratory at UiTM, Jalan Othman and Research Chair of Targeting and Treatment of Cancer Using Nanoparticles, Deanship of Scientific Research, King Saud University, Riyadh, Saudi Arabia for the Research Collaboration and Support.

References

[1] M. Shahid, "Corrosion protection with eco-friendly inhibitors," *Advances in Natural Sciences: Nanoscience and Nanotechnology,* vol. 2, p. 043001, 2011.

[2] M. K. Punith Kumar, T. V. Venkatesha, M. K. Pavithra, and A. Nithyananda Shetty, "The fabrication, characterization and electrochemical corrosion behavior of Zn-TiO2composite coatings," *Physica Scripta,* vol. 84, p. 035601, 2011.

[3] M. Rostami, S. Rasouli, B. Ramezanzadeh, and A. Askari, "Electrochemical investigation of the properties of Co doped ZnO nanoparticle as a corrosion inhibitive pigment for modifying corrosion resistance of the epoxy coating," *Corrosion Science,* vol. 88, pp. 387-399, 2014.

[4] F. Khelifa, M.-E. Druart, Y. Habibi, F. Bénard, P. Leclère, M. Olivier, *et al.*, "Sol–gel incorporation of silica nanofillers for tuning the anti-corrosion protection of acrylate-based coatings," *Progress in Organic Coatings,* vol. 76, pp. 900-911, 2013.

[5] T. Ç. Çanak, K. Kaya, and I. E. Serhatlı, "Boron containing UV-curable epoxy acrylate coatings," *Progress in Organic Coatings,* vol. 77, pp. 1911-1918, 2014.

[6] A. E. Eken, E. J. Tozzi, D. J. Klingenberg, and W. Bauhofer, "Combined effects of nanotube aspect ratio and shear rate on the carbon nanotube/polymer composites," *Polymer,* vol. 53, pp. 4493-4500, 2012.

[7] M. S. Shamsudin, M. R. Mahmud, F. S. Husairi, M. K. Harun, A. B. Suriani, S. Abdullah, *et al.*, "Micro-Raman, Optical and Impedance Characteristics of CNT-Substituted Acrylate/CNT Nanocomposite Thin Film," *Advanced Materials Research,* vol. 832, pp. 286-291, 2013.

[8] X. Chen, H. Wang, and J. He, "Synthesis of carbon nanotubes and nanospheres with controlled morphology using different catalyst precursors," *Nanotechnology,* vol. 19, p. 325607, Aug 13 2008.

[9] Z. Shi, Y. Lian, X. Zhou, Z. Gu, Y. Zhang, S. Iijima, *et al.*, "Mass-production of single-wall carbon nanotubes by arc discharge method," *Carbon,* vol. 37, pp. 1449-1453, // 1999.

[10] Y. P. Yu, "Preparation, Purification and Properties of Carbon Nanotubes," *Advanced Materials Research,* vol. 557-559, pp. 472-477, 2012.

[11] R. Saranya, G. Arthanareeswaran, and D. D. Dionysiou, "Treatment of paper mill effluent using Polyethersulfone/functionalised multiwalled carbon nanotubes based nanocomposite membranes," *Chemical Engineering Journal,* vol. 236, pp. 369-377, 2014.

[12] S. U. Khan, J. R. Pothnis, and J.-K. Kim, "Effects of carbon nanotube alignment on electrical and mechanical properties of epoxy nanocomposites," *Composites Part A: Applied Science and Manufacturing,* vol. 49, pp. 26-34, 2013.

[13] F. Soriano-Corral, L. F. Ramos-de Valle, F. J. Enríquez-Medrano, P. A. De León-Martínez, M. L. López-Quintanilla, and E. N. Cabrera-Álvarez, "Cast Nanostructured Films of Poly(methyl methacrylate-b-butyl acrylate)/Carbon Nanotubes: Influence of Poly(butyl acrylate) Content on Film Evaporation Rate, Morphology, and Electrical Resistance," *Journal of Nanomaterials,* vol. 2012, pp. 1-8, 2012.

[14] A. A. El-Hadad, V. Barranco, A. Samaniego, I. Llorente, F. R. García-Galván, A. Jiménez-Morales, *et al.*, "Influence of substrate composition on corrosion protection of sol–gel thin films on magnesium alloys in 0.6M NaCl aqueous solution," *Progress in Organic Coatings,* vol. 77, pp. 1642-1652, 2014.

[15] M. Szocinski and K. Darowicki, "Local impedance spectra of organic coatings," *Polymer Degradation and Stability,* vol. 98, pp. 261-265, 2013.

[16] Y. González-García, S. González, and R. M. Souto, "Electrochemical and structural properties of a polyurethane coating on steel substrates for corrosion protection," *Corrosion Science,* vol. 49, pp. 3514-3526, 2007.

[17] M. C. S. S. Macedo, I. C. P. Margarit-Mattos, F. L. Fragata, J. B. Jorcin, N. Pébère, and O. R. Mattos, "Contribution to a better understanding of different behaviour patterns observed with organic coatings evaluated by electrochemical impedance spectroscopy," *Corrosion Science,* vol. 51, pp. 1322-1327, 2009.

[18] C. H. Lin and J. G. Duh, "Electrochemical impedance spectroscopy (EIS) study on corrosion performance of CrAlSiN coated steels in 3.5wt.w/v % NaCl solution," *Surface and Coatings Technology,* vol. 204, pp. 784-787, 2009.

[19] M. Sabouri, T. Shahrabi, H. R. Faridi, and M. G. Hosseini, "Polypyrrole and polypyrrole–tungstate electropolymerization coatings on carbon steel and evaluating their corrosion protection performance via electrochemical impedance spectroscopy," *Progress in Organic Coatings,* vol. 64, pp. 429-434, 2009.

[20] V. S. Raja, A. Venugopal, V. S. Saji, K. Sreekumar, R. S. Nair, and M. C. Mittal, "Electrochemical impedance behavior of graphite-dispersed electrically conducting acrylic coating on AZ31 magnesium alloy in 3.5wt.w/v % NaCl solution," *Progress in Organic Coatings,* vol. 67, pp. 12-19, 2010.

[21] M. S. Shamsudin, M. F. Achoi, M. N. Asiah, L. N. Ismail, A. B. Suriani, S. Abdullah, *et al.*, "An Investigation on the Formation of Carbon Nanotubes by Two-Stage Chemical Vapor Deposition," *Journal of Nanomaterials,* vol. 2012, pp. 1-5, 2012.

[22] M. S. Shamsudin, N. A. Asli, S. Abdullah, S. Y. S. Yahya, and M. Rusop, "Effect of Synthesis Temperature on the Growth Iron-Filled Carbon Nanotubes as Evidenced by Structural, Micro-Raman, and Thermogravimetric Analyses," *Advances in Condensed Matter Physics,* vol. 2012, pp. 1-7, 2012.

[23] M. S. Shamsudin, S. Abdullah, and M. R. Mahmood, "Structural and Thermal Behaviors of Iron-Filled Align Carbon Nanotubes Formulated by Two-Stage Catalytic Chemical Vapor Deposition," *Advanced Materials Research,* vol. 364, pp. 191-195, 2011.

[24] M. Maryam, A. B. Suriani, M. S. Shamsudin, and M. Rusop Mahmood, "BET Analysis on Carbon Nanotubes: Comparison between Single and Double Stage Thermal CVD Method," *Advanced Materials Research,* vol. 626, pp. 289-293, 2012.

[25] E. Bakhshandeh, A. Jannesari, Z. Ranjbar, S. Sobhani, and M. R. Saeb, "Anti-corrosion hybrid coatings based on epoxy–silica nano-composites: Toward relationship between the morphology and EIS data," *Progress in Organic Coatings,* vol. 77, pp. 1169-1183, 2014.

[26] Q. Zhou and Y. Wang, "Comparisons of clear coating degradation in NaCl solution and pure water," *Progress in Organic Coatings,* vol. 76, pp. 1674-1682, 2013.

[27] A. Mostafaei and F. Nasirpouri, "Epoxy/polyaniline–ZnO nanorods hybrid nanocomposite coatings: Synthesis, characterization and corrosion protection performance of conducting paints," *Progress in Organic Coatings,* vol. 77, pp. 146-159, 2014.

[28] F. Brusciotti, A. Batan, I. De Graeve, M. Wenkin, M. Biessemans, R. Willem, *et al.*, "Characterization of thin water-based silane pre-treatments on aluminium with the incorporation of nano-dispersed CeO2 particles," *Surface and Coatings Technology,* vol. 205, pp. 603-613, 2010.

The Influence of Immersion Time to the Optical properties of ZnO Growth on PMMA-coated Substrate by Solution-Immersion Method

A. Aadila[1, 2, b], A. N. Afaah[1, 2], N. A. M. Asib[1, 2], M. R. Mahmud[1,2], Salman A.H. Alrokayan[4], Haseeb A. Khan[4], R. Mohamed[1, 3], M. Rusop[1, 3] and Z. Khusaimi[1, 2, a]

[1]Faculty of Applied Sciences, UniversitiTeknologi MARA, 40450 Shah Alam, Selangor, Malaysia

[2]NANO-SciTech Centre, Institute of Science, UniversitiTeknologi MARA, 40450 Shah Alam, Selangor, Malaysia

[3]NANO-Electronic Centre, Faculty of Electrical Engineering, UniversitiTeknologi MARA, 40450 Shah Alam, Selangor, Malaysia

[4]Research Chair of Targeting and Treatment of Cancer Using Nanoparticles, Department of Biochemistry, College of Science, King Saud University, Riyadh 11451, Saudi Arabia

Corresponding author: zurai142@salam.uitm.edu.my[a], aadilaazizali@gmail.com[b]

Abstract. We present the growth of ZnO on PMMA-coated substrate under influence of different immersion time by solution-immersion technique. PMMA polymer was coated on the substrate by spin-coating method. The PMMA-coated substrate was then prepared for non-immersed and immersed for four (4) and six (6) hours in zinc solution. The optical properties of the film obtained were characterized using Photoluminescence (PL), Raman Spectra and Ultraviolet-Vis (UV-Vis) spectroscopy. It was found that shorter time of immersion (4 hours) produced higher intensity of PL and Raman. However, the reverse was found for UV-Vis measurement. The low absorption that was shown from the film immersed for 4 hours may correspond to the random distribution and orientation of ZnOparticle growth on PMMA-coated substrate.

1. Introduction

In recent years, the synthesis of zinc oxide has received considerable attention owing to the many interesting and unique properties like a wide direct band gap of 3.37 eV and a large exciton binding energy of 60 meV[1-3]. So far, there have been many of studies the ZnO in the form of thin films. This is due to the ZnO thin films that is highly transparent in the visible region [1]. Transparent ZnO thin films are well-known for the application in electronic and optoelectronic devices like transparent thin film transistor (TTFTs), optical filters, transparent electrodes, back-reflector in solar cells thin film and space applications [2, 4]. There are various types of methods to fabricate ZnO such as

IOP Conf. Series: Materials Science and Engineering **83** (2015) 012005 doi:10.1088/1757-899X/83/1/012005

thermal oxidation, chemical vapor deposition, RF sputtering, sol-gel, spray pyrolysis and others [5, 6]. Among these methods, sol-gel has advantages due to its ability to tune the size of ZnO to μm and nm level, low-cost, lower crystallization temperature and large surface area coating capability [7].

Polymers have been used in various applications like power tool handles, cable, jackets, capacitor films, electronic packaging materials and others. Recently, the electrical and optical properties of polymers have been extensively investigated due to their application in optical devices [8].The polymer that be chosen in this work is poly (methyl methacrylate) (PMMA). PMMA is a transparent polymeric materials, which possesses many desirable properties such as light weight, high light transmittance, chemical resistance, high rigidity, good tensile strength and hardness, colorless, resistance to weathering corrosion and good insulating properties [8, 9]. Furthermore, PMMA has extensive application in the field of sensor, coating and others due to its good optical transparency and processability[10]. Commonly, the spin-coating methods has always been used to fabricate PMMA as the method is easy, fast, low-cost, controllable [11] and produced uniform thickness of PMMA.

Organic-inorganic combinations are of great interest, owing to their excellent structural and opticalproperties [12, 13]. Introducing ZnO on PMMA surface mainly focuses on improving the optical properties of the films that was acting as multifunction materials in UV shielding, antibacterial or bactericides [10]. Furthermore, there have been reports that the free surface energy of the films will be enhanced by introducing metal films into polymer [10].

Have been reported that 4 hours of immersion time successfully obtained to growth ZnOnanorods on the substrate. However, the ZnO was grown on the other seed layer like a gold or ZnO-seed [14]. Therefore, due to the optimization of 4 hours have been done before, we are increasing the immersion times until about 6 hours and on different seed layer. Thus, in this work we are investigating the growth of ZnO on PMMA-coated substrate where PMMA-coated substrate acting as a seed layer. Further, we are focused on the effect of different immersion times to the growth of ZnO on PMMA-coated substrate and the optical properties of the films were being studied.

2. Experimental work

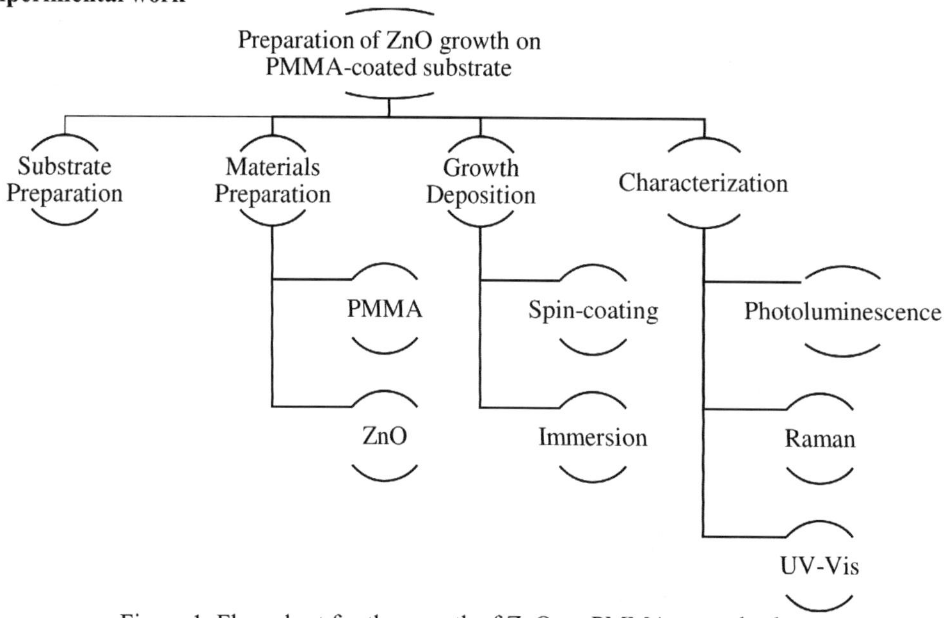

Figure 1. Flow chart for the growth of ZnO on PMMA-coated substrate

2.1. Substrate preparation
The glass substrate was prepared by cutting into 1.5 cm x 2 cm of dimension. After cutting process, the glass substrate was cleaned by using three different solutions; methanol, acetone and distilled water. The glass substrate was placed in the beaker and the solution was filled up until all the substrate fully immersed. Then, the beaker was placed in the sonicator bath for 10 minutes. The process was repeated for two times for each of the solutions. Finally, the clean substrate was dried by using argon gas and ready to be used.

2.2. Material preparation
The ZnO sol-gel was prepared as follows: zinc nitrate hexahydrate Zn $(NO_3).6H_2O$ and hexamethylenetetramine (HMTA) (CH_4N_2O) was dissolved in distilled water, and the both of the solution were mixed together at room temperature. At the same time, PMMA solution was prepared by dissolving PMMA powder with average molecular weight of 120,000 in a toluene solution. Toluene was acting as the solvent in order to prepare PMMA solution. All of the solutions prepared were stirred for 24 hours at 60 °C for aging process to make the solution homogeneously mingled.

2.3. Growth deposition
2.3.1. Spin coating. PMMA was deposited on the glass substrate using the spin - coating technique. The spin-coating time was 50 s. In the beginning 10 s, the spin-speed of the glass substrate was 3000 rpm, in the middle 30 s, the spin-speed was 6000 rpm and in the latter 10 s, the spin-speed was 2000 rpm. After coating, the sample was dried at 100 for 10 minutes.

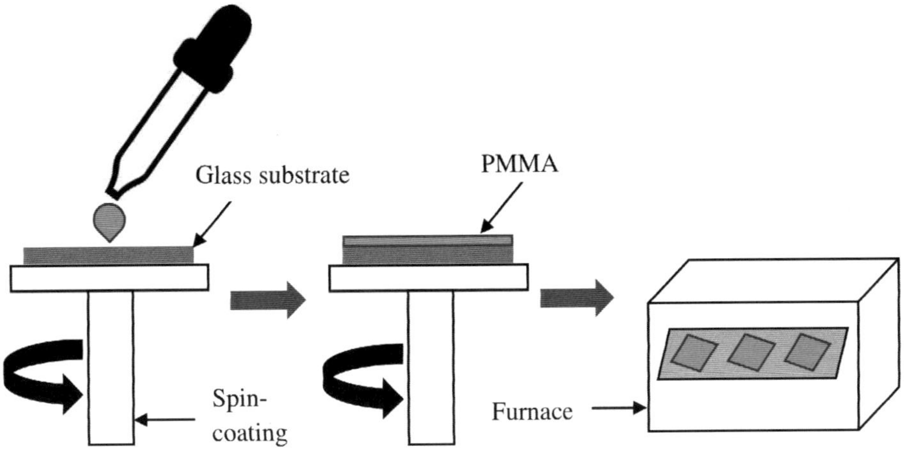

Figure 2. Schematic diagram of spin-coating process

2.3.2. Immersion. For non-immersion, the substrate with PMMA on it was not immersed in the zinc solution. The process continues for the PMMA-coated substrate that was immersed in a zinc solution by immersion method for 4 hours and 6 hours at temperature of 90 °C. After immersion, the films was annealed for 20 minutes at 100 °C. Subsequently, the samples were cooled at room temperature before taken to the characterization stage. The schematic diagram for growth deposition of ZnO on PMMA-coated substrate can be seen in Figure 2 and Figure 3 and the parameter was stated in Table 1.

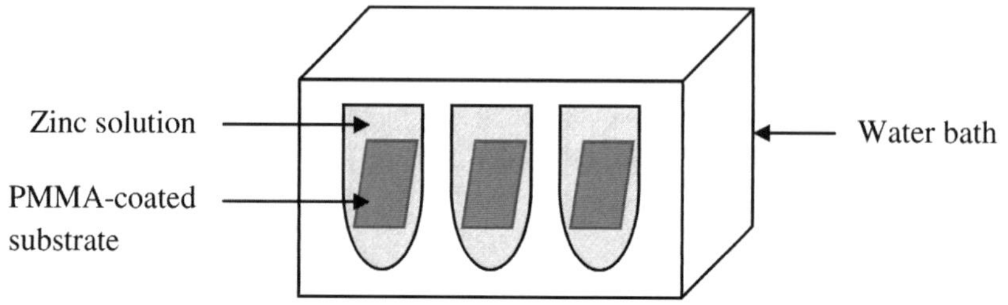

Figure 3. Schematic diagram of immersion process

Table 1. Parameter of ZnO growth on PMMA-coated substrate at different times of immersion

No. of sample	Time of immersion	Temperature of immersion (°C)	Annealing temperature (°C)	Annealing time (minutes)
(a)	Non-immersion	90	100	20
(b)	4 hours	90	100	20
(c)	6 hours	90	100	20

2.4. Characterizations

The optical properties were synthesized by using Photoluminescence Spectrometer, Raman Spectra and Ultraviolet-Visible (UV-Vis) Spectroscopy.

3. Results and discussions

The optical properties of ZnO growth on PMMA-coated substrate were analysis by using Photoluminescence Spectrometer (Horiba JobinYvon), Raman Spectra (Horiba JobinYvon) and UV-Vis (Carry 5000 UV-Vis-NIR Spectrophotometer).

3.1. Photoluminescence (PL) spectra

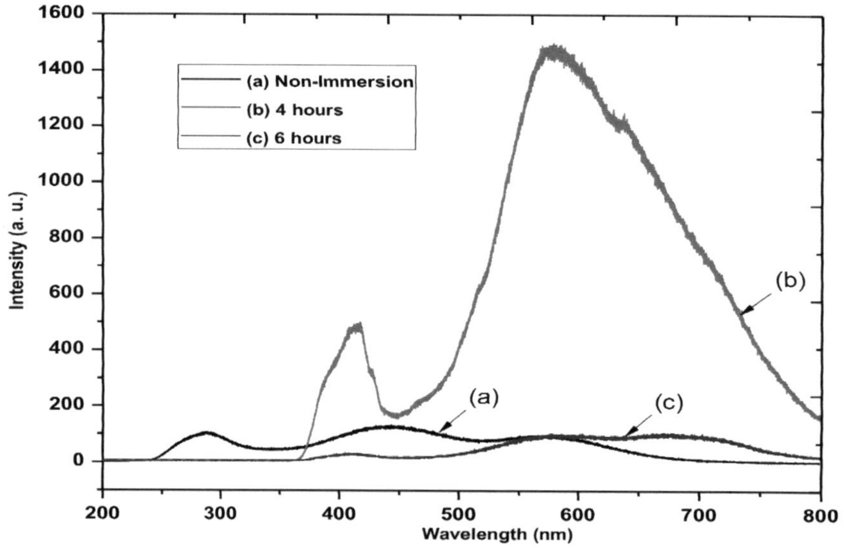

Figure 4. PL spectra of ZnO growth on PMMA-coated substrate at different time of immersion (a) non-immersion, (b) 4 hours and (c) 6 hours

Figure 4 demonstrated the PL spectra of ZnO growth on PMMA-coated substrate at different immersion times; 4 hours and 6 hours in the range wavelength of 300 - 800 nm. The spectra also show for the film that was non-immersed as shown in figure 4(a). In case of non-immersed, the film was prepared without ZnO growth on it. Therefore, the film present as pristine PMMA-coated on the substrate. In general the PL spectrum consists of two characteristic parts: a short-wavelength region located in the ultraviolet region and broad spectral bands in visible region [15]. The spectrum shows for the non-immersed film, two peaks of emission appeared centred at 286 nm and 441 nm of wavelength. After the films immersed for 4 hours and 6 hours, the peak of emission shifted to the longer wavelength, centred at 412 nm and 578 nm. The shifted wavelength may indicate that the films immersed in different immersion times were dominated by larger domains [16]. The peak in UV region called the UV emission or near band edge (NBE) which contributed to the free exciton recombination, while the visible emission originates from the defects[17].

The ZnO growth on PMMA-coated substrate at 4 hours immersion time depicts good intensity and sharp peak in the UV emission. As be reported that the appearance of sharp and strong the UV emission indicates the ZnO may have good crystalline structure in the samples[15]. However, after immersion time enhanced to 6 hours, the intensity reduced with broad and wide peak as shown in both UV and visible emission. The reason may due to the crystalline structure of the ZnO have possibility changes and defected. The longer periods of immersion time probably causing the ZnO underwent crystal restructuring during the immersion process in the aqueous solution [7]. As a result the films might be affected and have defect in its structure which give the lower intensity and broad band emission [18]. Therefore, 4 hours of immersion time present the films have good intensity and better crystalline structure of ZnO compared to the films that are non-immersed and immersed for longer immersion times.

3.2. Raman Spectra

Raman spectroscopy provides information on the vibrational properties of ZnO growth on PMMA-coated substrate [19]. The Raman spectra of ZnO growth on PMMA-coated substrate under influence of different immersion time are depicted in Figure 5.

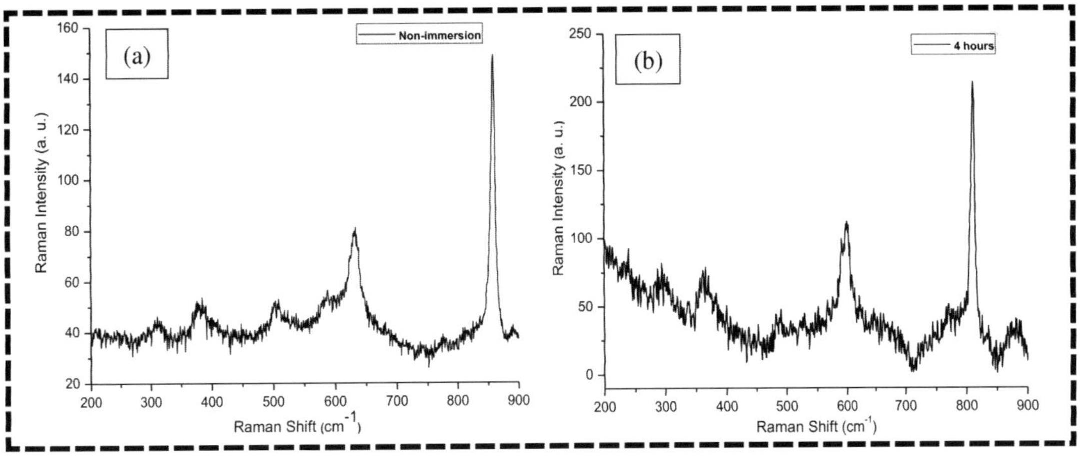

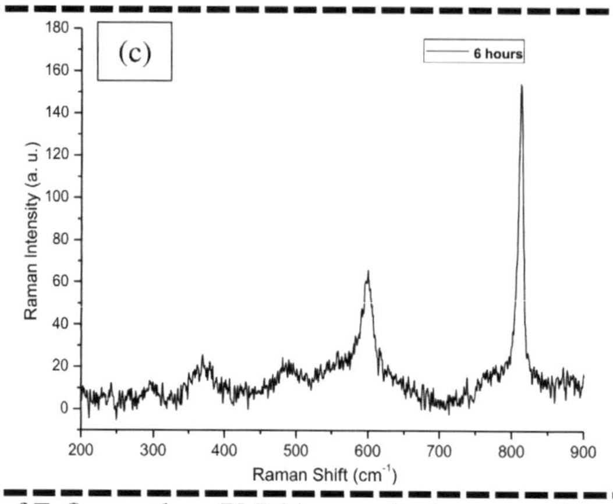

Figure 5. Raman spectra of ZnO growth on PMMA-coated substrate at different time of immersion (a) non-immersion, (b) 4 hours and (c) 6 hours

The spectra present the films figure 5(a) non-immersed and two different immersion times; figure 5(b) 4 hours and figure 5(c) 6 hours. The spectra covered in the range 200 – 900 cm^{-1}of Raman shift. Non-immersion represents the film of PMMA-coated substrate without ZnO as mentioned in the PL spectra analysis before. In the spectra, non-immersion film was observed as the lowest intensity with obvious peak appeared at 311 cm^{-1}, 376 cm^{-1}, 503 cm^{-1}, 632 cm^{-1}and 860 cm^{-1}of Raman shift. After the films were immersed, the spectra exhibit two peaks at 600 cm^{-1} and 811 cm^{-1} of Raman shift. It can be seen the film that was immersed for 4 hours give the higher Raman intensity compared to others. After the immersion time increase to 6 hours, the Raman intensity of the films decrease and no shifting peak is observed.

There are strong and weak peak present in the spectra shown in both graphs. A peak appeared at 600 cm^{-1}were contributed to the A_1 (LO) mode. This vibrational mode of A_1 (LO) was caused by defects such oxygen vacancies, zinc interstitials, the complexes or free carrier [16, 19, 20]. Furthermore, peak appeared at ~600 cm^{-1} probably indicates the films consists of c-axis of wurtzite ZnO are parallel to the sample surface [15, 19]. However, a weak peak was observed as the A_1 (LO) mode in the spectra. As be reported by M. F. Malek et al, the weakest intensity was contributing to the minimization of the effect of oxygen deficiency on the growth of ZnO[19]. Therefore, it means the structure of ZnO growth on PMMA-coated substrate for 4 hours of immersion time is better and crystalline compared to the film that was immersed for 6 hours as well as for non-immersion film. Thus, the spectra analysis is consistent with the PL result in figure 4 where the ZnO growth PMMA-coated substrate at 4 hours immersion time gives better intensity.

3.3. Ultraviolet-Vis (UV-Vis) spectroscopy

Figure 6 presents the absorption spectra of ZnO growth PMMA-coated substrate under influence of different immersion times in the wa
velength range of 280 - 900 nm. All the samples showed high absorption in the UV region and transparent in the visible region with the absorption edge around at ~330 nm.

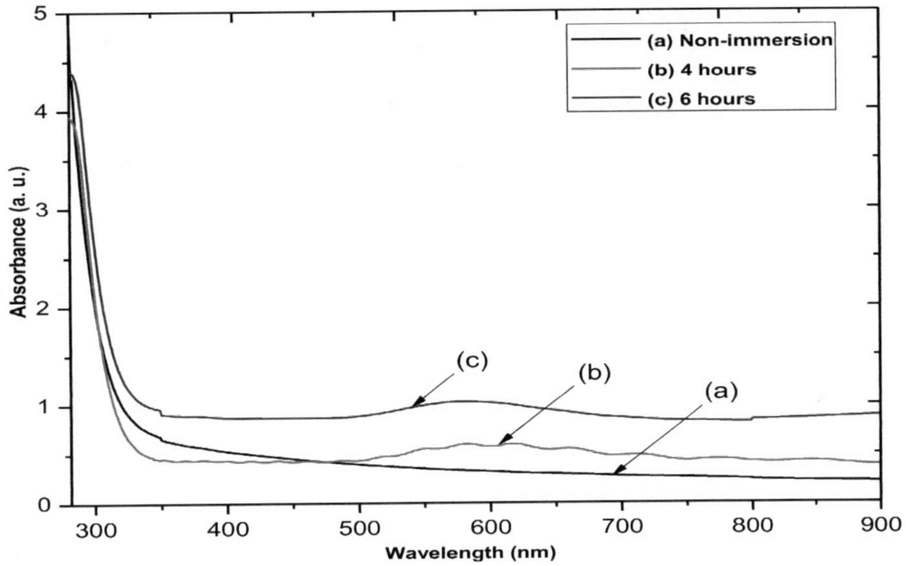

Figure 6. UV absorption of ZnO growth on PMMA-coated substrate at different immersion times (a) non immersion, (b) 4 hours and (c) 6 hours

From Figure 6, the ZnO growth on PMMA-coated substrate at 6 hours of immersion time gives good absorption. It can clearly see in the UV region, non-immersion film present higher absorbance than the film that was immersed at 4 hours. However, the absorbance intensity decreased and the film highly transparent in the visible region. The absorption of the film increased with the increasing of immersion times and the absorption edge shift to the lower wavelength. It has been reported that when the absorption edge shifted into the lower wavelength, the absorbance increase with decreasing size of particles [21, 22]. 6 hours of immersion time present higher absorption and the film probably contains the small particles size of ZnO in the sample. However, the analysis inconsistent with the PL and Raman result previously where the films that was immersed for 4 hours exhibit good PL and Raman intensity.

In the case of UV absorption, the reason may be due to the distribution and orientation of ZnO particles contains in the sample that was immersed for 4 hours were randomly scattered. The random distribution of ZnO particles caused the scattering of light not effectively scattered on the rough surface and low absorbance are produced. As a result, the UV absorption of the films that was immersed for 4 hours exhibit low absorbance compared to the films that was non-immersed and immersed for 6 hours in UV region. Even 6 hours give a better absorption, have been reported that the longer immersion time will be produced larger of ZnO nanorods[23]. The good absorption showed probably caused from the distribution of ZnO in this sample is better. However, 4 hours of immersion will be chosen in this UV measurement because the sample is more crystalline and may have small size of ZnO as shown in another analysis previously. From that, it can be seen that the optical properties was significantly affected by the surface morphology and the scattering light efficiently is very dependent on particle distribution of the films [24]. Therefore, further study need to be done to improve the morphology as well as the optical properties of the sample that was immersed for 4 hours.

4. Conclusion

In summary, the ZnO growth on PMMA-coated substrate under influence of immersion times was successfully prepared by the solution - immersion method. PL analysis and Raman spectra exhibit the film that was immersed for 4 hours have good intensity and might form better crystalline of ZnO particles on the PMMA-coated substrate. However, UV absorption of the film immersed for 4 hours give low absorbance due the sample may have a random distribution and orientation of ZnO particles. It was shown that the optical properties are significantly affected by the surface morphology and the scattering of light efficiency dependent on particle distribution. Thus, further studies need to be done to improve the distribution of particles as well as optical properties of ZnO growth on PMMA-coated substrate for the sample that was immersed 4 hours.

5. Acknowledgement

The researchers are grateful to Faculty of Applied Sciences of UniversitiTeknologi MARA (UiTM) Malaysia, Long-Term Research Grant Scheme for Nanostructures, Nanomaterials and Devices for Fuel Cells and Hydrogen Production (600-RMI/LRGS 5/3 (3/2013)), Research Management Institute (RMI) and Ministry of Education of Malaysia for financial support. This study was also supported by Research Chair of Targeting and Treatment of Cancer Using Nanoparticles, Deanship of Scientific Research, King Saud University, Saudi Arabia for the Research Collaboration and Support. Thanks to NANO-SciTech Centre and NANO-Electronic Centre colleagues for their kind support and encouragement.

References

[1] L. Xu, G. Zheng, J. Miao, and F. Xian, *Applied Surface Science,* vol. 258, pp. 7760-7765, 2012.
[2] S.K. Panda and C. Jacob, *Solid-State Electronics,* vol. 73, pp. 44-50, 2012.
[3] Shashidhara Bhat, Shrisha B. V., and K. Gopalakrishna Naik, *Archives of Physics Research,,* vol. 4, pp. 61-70, 2013.
[4] Libing Duan, XiaoruZhao, ZhujunZheng, YajunWang, WangchangGeng, and FuliZhang, *Journal of Physics and Chemistry of Solids,* vol. 76, 2014.
[5] K. Thongsuriwong, P. Amornpitoksuk, and S. Suwanboon, *Advanced Powder Technology,* vol. 24, pp. 275-280, 2013.
[6] R.Y. Hong, J.Z. Qian, and J.X. Cao, *Powder Technology,* vol. 163, pp. 160-168, 2006.
[7] Sharul Ashikin Kamaruddin, Mohd Zainizan Sahdan, Kah-Yoong Chan, Mohamad Rusop, and Hashim Saim, *Microelectronics International,* vol. 27, pp. 166-169, 2010.

[8] D. Dorranian, Z. Abedini, A. Hojabri, And M. Ghoranneviss *Journal of Non-Oxide Glasses,* vol. 1, pp. 217 - 229, 2009.
[9] F.-Z. Tighilt, N. Gabouze, S. Sam, S. Belhousse, and K. Beldjilali, *Surface Science,* vol. 601, pp. 4217-4221, 2007.
[10] Xi-Wen Du, Ying-Song Fu, Jing Sun, Xue Han, and Jim Liu, *Semicond. Sci. Technol.,* vol. 21, pp. 1202-1206, 2006.
[11] N.G. Semaltianos, *Microelectronics Journal* vol. 38, pp. 754–761, 2007.
[12] Akhmad Herman Yuwono, Junmin Xue, And John Wang, *Journal of Nonlinear Optical Physics & Materials,* vol. 14, pp. 282-297, 2005.
[13] P.K. Khanna and Narendra Singh, *Journal of Luminescence,* vol. 127, pp. 474-482, 2007.
[14] Mohamad Hafiz Mamat, Zuraida Khusaimi, Musa Mohamed Zahidi, and Mohamad Rusop Mahmood, InTech, 2012.
[15] Rui Zhang, Peng-Gang Yin, Ning Wang, and Lin Guo, *Solid State Sciences,* vol. 11, pp. 865-869, 2009.
[16] M.H. Mamat, M.F. Malek, N.N. Hafizah, Z. Khusaimi, M.Z. Musa, and M. Rusop, *Sensors and Actuators,* vol. 195, pp. 609-622, 2014.
[17] N.D. Md Sin, M.H. Mamat, Abdul Aziz. A, and M. Rusop, *International Journal Of Electrical And Electronic Systems Research,* vol. 6, pp. 11-17, 2013.
[18] J. Rouhi, M.Alimanesh, S.Mahmud, R.A.Dalvand, C.H.RaymondOoi, and M.Rusop, *Materials Letters,* vol. 125, pp. 147-150, 2014.
[19] M.F. Malek, M.H. Mamat, Z. Khusaimi, M.Z. Sahdan, M.Z. Musa, A.R. Zainun, A.B. Suriani, N.D. Md Sin, S.B. Abd Hamid, and M. Rusop, *Journal of Alloys and Compounds,* vol. 582, pp. 12-21, 2014.
[20] A. Ismail and M.J. Abdullah, *Journal of King Saud University – Science,* vol. 25, pp. 209-215, 2013.
[21] R. Sreeja, Jobina John, P.M. Aneesh, and M.K. Jayaraj, *Optics Communications* vol. 283, pp. 2908-2913, 2010.
[22] Soosen Samuel M, Lekshmi Bose, and George KC, *SB Academic Review,* vol. XVI, pp. 57-65, 2009.
[23] M. Awalludin, M.H. Mamat, M.Z. Sahdan, Z. Mohamad, and M. Rusop, *Advanced Materials Research,* vol. 667, pp. 375-379, 2013.
[24] Narges Kiomarsipour and R. S. Razavi, *Ceramics International40(2014)11261–11268,* vol. 40, pp. 11261-1268, 2014.

Effect of molarity of TiO_2 seeded-template to the growth of ZnO nanostructures

N.A.M. Asib[1,2,a], A. N. Afaah[1,2], A. Aadila[1,2], M. R. Mahmud[1,2], Lim Y. C.[1], Salman A.H. Alrokayan[4], Haseeb A. Khan[4], M. Rusop[1,3], Z. Khusaimi[1,2,b]
[1]Faculty of Applied Sciences;
[2]NANO-SciTech Centre, Institute of Sciences;
[3]NANO-ElecTronic Centre, Faculty of Electrical Engineering;
Universiti Teknologi MARA, 40450 Shah Alam, Selangor, Malaysia
[4]Research Chair of Targeting and Treatment of Cancer Using Nanoparticles, Department of Biochemistry, College of Science, King Saud University, Riyadh, Saudi Arabia

E-mail: [b]zurai142@salam.uitm.edu.my, [a]amierahasib@yahoo.com

Abstract. ZnO nanostructures were deposited by solution-immersion method on TiO_2 layers by sol-gel spin-coating technique. Seven layers of TiO_2 were coated on glass substrates at different molarities, followed by annealing treatment and solution-immersion process in aqueous solution containing 1:1 ratio of 0.05 M of zinc nitrate hexahydrate ($Zn(NO_3)_2.6H_2O$) and hexamethylenetetramine (HMTA) ($C_6H_{12}N_4$). FESEM images confirmed that ZnO nanostructures grown on TiO_2 seeded-template are in needle-like shape with smaller tip can be observed. Photoluminescence (PL) spectroscopy showed that as the molarity of TiO_2 seeded-template increase, the intensity of PL emission at UV region decrease. UV emission peak for 0.10 M and 0.25 M are higher compared to UV emission peak film of 0.50 M and 1.00 M, which are too low. Meanwhile, the UV absorption properties of the nanostructured thin film for 0.25 M of TiO_2 seeded-template was higher compared to other thin films.

1. Introduction

In the field of semiconductor nanostructures, one–dimensional (1D) Zinc oxide (ZnO) nanostructures (e.g. nanowires, nanorods, nanobelts) becoming the most promising candidates due to their important physical properties and application prospects. ZnO is a direct band gap semiconductor with hexagonal wurtzite crystal structure, having good optical and electrical properties including has a wide band gap of 3.37 eV, large exciton binding energy about 60 meV and has high refractive index of 2.01 [1]. These excellent properties of ZnO thin films, have increase the interest of many researches to study about the material especially in nanoscale as the nanoscale electronic devices have the potential to achieve faster response and higher sensitivity compared to bulk material [2]. ZnO can be widely applied in various fields such as in optical and optoelectronic devices [3], photocatalysts, solar cell, UV light emitting diodes and UV sensor.

In ZnO based UV sensor, many efforts have been made to improve the performance of the devices. It is known that the performance of the device depends on the surface condition, structural quality and the dimension of ZnO nanostructures. Surface modification or structural improvement is expected can increase the sensor performance. Where, the process of modifying the surface is related to the surface adsorption of gas molecules from the atmosphere and surface defects, which influence the characteristics of sensor [4]. Therefore, an approach of utilizing the catalytic seed layer to growth better ZnO was used as this seed layer can act as nucleation centres to prepare a suitable site to which ZnO nanostructures can adhere since it can provide an almost mistmach-free interfacial layer between the ZnO and the seed layer so that well aligned of ZnO nanostructures can grow. In addition, the seed layer can assits an epitaxial growth process on the seed layer coated on glass substrate.

Since titanium dioxide (TiO_2) is also a material that belongs to the group of metal oxide, it can be a good candidate of seed layer due to its excellent properties such as good semiconductor properties, high chemical stability, high photocatalytic activity, non-toxicity [5] and long thermal photostability [6]. Besides, TiO_2 has a wide direct band gap of 3.0 eV or 3.2 eV in the rutile or anatase crystalline phase which is similar to ZnO. However, ZnO is a direct band gap semiconductor but TiO_2 is an indirect band gap semiconductor with a relative lower cost [3, 7]. Moreover, TiO_2 can act as catalytic promoter during the involved reaction [8] and can greatly modified the physical and optical properties of thin films when we introduce it into ZnO matrices [9, 10].

Nanostructures of ZnO can be produced by several methods such as spray pyrolysis [11], chemical vapour synthesis (CVS) [12], pulsed laser deposition, and solution-based method. In particular, solution-based method has advantages of synthesizing nanostructured ZnO at low processing temperature and low cost. The simplicity of the experimental method of preparing a large-area thin film with excellent compositional control suitable for large-scale production of ZnO [13].

In this study, we prepared TiO_2 as a seeded-template for fabrication of ZnO nanostructures using sol-gel spin-coating and solution-immersion method, respectively. The effect of different molarities of TiO_2 as a seeded-template to the growth of ZnO nanostructures were investigated based on the changes of surface morphology and optical properties shown by the nanostructures thin films measured from Field Emission Scanning Microscope (FESEM), Photoluminescence (PL) spectroscopy and Ultraviolet-Visible (UV-Vis) spectrophotometer. These changes of properties are expected can improve the performances of cost-effective devices such as UV photoconductive sensor.

2. Experimental

2.1. Materials and chemicals
Glass substrate, Titanium (IV) butoxide, Ethanol, Triton-X-100, Glacial acetic acid (GAA), Zinc nitrate hexahydrate (Zn $(NO_3)_2.6H_2O$) and Hexamethylenetetramine ($C_6H_{12}N_4$) HMTA.

2.2. Preparation of TiO_2 seeded-template
Titanium (IV) butoxide was used as a starting material for TiO_2 seeded-template preparation. Prior to seed layer deposition, titanium (IV) butoxide dissolved in ethanol as solvent with triton-X-100 as stabilizer, under constantly stirring. Then, glacial acetic acid (GAA) within distilled water was added to the solution in order to accelerate the chemical reaction in the solution. The mixture was heated at 60 °C for 2 hours and left to aged for 24 hours with continuously stirring. To investigate the effect of different molarity of TiO_2 as seeded-template to ZnO nanostructures, molarity of TiO_2 solution were varied at 0.10, 0.25, 0.50 and 1.00 M. Coating layer of TiO_2 is performed by spin-coater at three steps of different speeds on a glass substrate, followed by drying process at 150°C for 10 minutes. For each layer, the same procedure was repeated until seven layers of coating. Finally, the TiO_2 seeded templates were annealed in annealing chamber for at 450 °C to crystallize the templates.

2.3. Preparation of needle-like ZnO
ZnO nanostructures were fabricated on TiO_2 seeded-template via solution-immersion method. Aqueous solution of zinc nitrate hexahydrate (Zn $(NO_3)_2.6H_2O$) was used as a precursor in 100 ml of deionized water. As a stabilizer, hexamethylenetetramine, HMTA ($C_6H_{12}N_4$) also was dissolved in 100 ml of deionized water and added to the Zn^{2+} solution with molar ratio 1:1. Then, another 50 ml of deionized water was added to increase the volume of the solution up to 250 ml. The solution was heated at 60 ^{0}C for 1 hour and left to aged for 24 hours at room temperature with continuous stirring.

The TiO_2 seeded-templates were suspended in centrifuged tubes which contained the prepared solution. The growth of ZnO nanostructures was take placed in water bath at 70-90 °C for 4 hours. After the deposition, the thin films were taken out and rinsed with deionized water several times, followed by annealing process at 500 ^{0}C for 1 hour

2.4. Characterization of TiO_2: ZnO nanostructures thin films

The change in surface morphology and optical properties between bare ZnO nanostructures and ZnO 7nanostuctures grown on different molarities of TiO_2 seeded-template were observed by using FESEM (JEOL JSM-7600F), PL spectroscopy (Horiba Jobin Yvon) and UV-Vis spectrophotometer (VARIAN 5000).

3. Results and discussion

3.1. Field emission scanning electron microscope (FESEM)

Figure 1 show FESEM images of the surface morphologies of needle-like ZnO nanostructures deposited on glass substrate without seed layer of TiO_2 (Figure 1 (a)) and with varied molarities of TiO_2 seed layer (Figure 1 (b-f)) that were fabricated via solution-immersion method. The images were captured at 5.0 kV of voltage at 10 kx magnification.

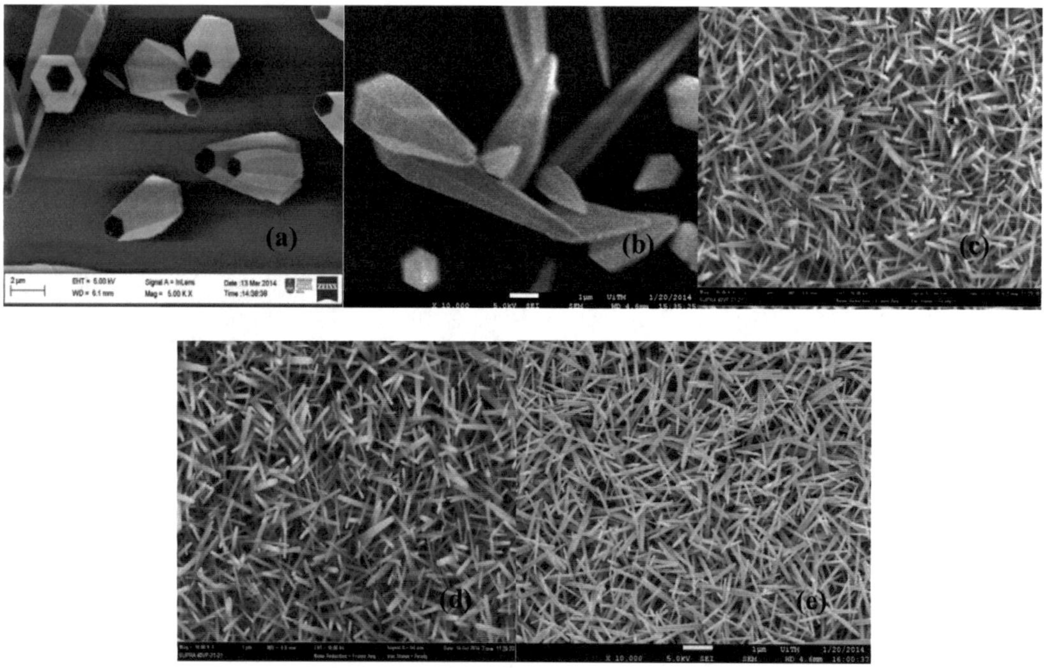

Figure 1. FESEM images of (a) bare ZnO and ZnO nanostructures deposited on (b) 0.10, (c) 0.25, (d) 0.50 and (e) 1.00 M of TiO_2 seeded-template.

Figure 1 (a) shows surface view images of ZnO nanostructures with hexagonal tip grown on glass substrate. As the TiO_2 seed layers were coated on the glass substrates, needle-like ZnO nanostructures with smaller tip can be seen on all the the TiO_2 templates, as shown in Fig. 1 (b) until Fig. 1 (e). It is revealed that the needle-like ZnO nanostructures are well grown with smaller diameter sizes and denser arrays on the TiO_2 coated-substrates compared to bare ZnO. Table 1 recorded the estimated average diameter of the needle-like ZnO nanostructures ranged between 61.2 nm to 1254.2 nm. A magnified FESEM images of ZnO as shown in Figure 1 (a) clearly confirms that needle-like ZnO with bigger sizes and less dense arrays were deposited on glass substrate. As the molarity of TiO_2 seeded templates are increase to 1.00 M, the average diameter sizes of needle-like ZnO nanostructures are significantly change to smaller sizes with denser distribution of ZnO. For 0.25, 0.50 and 1.00 M of TiO_2, the average diameter sizes of ZnO were about 61.2, 61.3 and 62.5 nm, respectively with good distribution of ZnO. The smallest diameter size of needle-like ZnO nanostructures with most dense arrays was deposited on 0.25 M of TiO_2. At low concentration, smaller sizes of particles is formed [14]. The decreasing in sizes of TiO_2 contribute to the sizes reduction of ZnO nanostructures grown on this seed layer [15]. The densification of the TiO_2 particles resulted from smaller size of grown particles causes formation of compact particles thus, resulted more nanostructures of ZnO with smallest sizes appeared in the thin film compared to other samples as shown and recorded in Figure 1 (c) and Table 1.

Table 1. Average diameter size of bare ZnO and needle-like ZnO nanostructures deposited on different molarities of TiO_2 seeded-template.

Sample	Average diameter size of ZnO (nm)
ZnO	1254.2
TiO_2(0.10 M): ZnO	533.3
TiO_2(0.25 M): ZnO	61.2
TiO_2(0.50 M): ZnO	61.3
TiO_2(1.00 M): ZnO	62.5

3.2. Photoluminescence (PL) study

PL spectroscopy was used to investigate the optical properties and quality of bare ZnO and ZnO nanostructures grown on different molarities of TiO_2 seed layer. When excited under laser source of 325 nm, there are difference were observed between the PL spectrums corresponding to different conditions of ZnO fabrications as shown in Figure 2.

The narrow UV band which appeared due to free-exciton recombination by the process of exciton-exciton collison [3, 16] at 394 nm and broad visible band within the range 600- 666 nm are observed for bare ZnO which grown on glass substrate only, without coated with TiO_2 seed layer. Meanwhile, when the molarity of TiO_2 seeded-template increases from 0.10 M to 1.00 M, the UV band are blue shifted to the left compared to the bare ZnO thin film and the broad visible band are extending from 550- 700 nm. The position of the UV band for each samples were recorded in Table 2. The blue shifted of UV peak emission to the lower wavelength might be due to the reduction of ZnO nanostructures sizes and typically, visible band arises due to structural and surface defects such as zinc interstitials (Z_{ni}), oxygen vacancies (V_O), oxygen interstitials (O_i), Zn vacancies (V_{Zn}) and the oxide antisite defect (O_{Zn}) [17, 18]. As the molarity of TiO_2 seeded-template increases, obvious decreases of UV peak emissions intensity are observed. The intensity of UV emission peak for 0.10 M and 0.25 M are clearly seen and higher compared to 0.50 M and 1.00 M, which are too low. Thus, 0.10 M and 0.25 M has higher crystallinity of ZnO nanostructures due to free exciton emission [19].

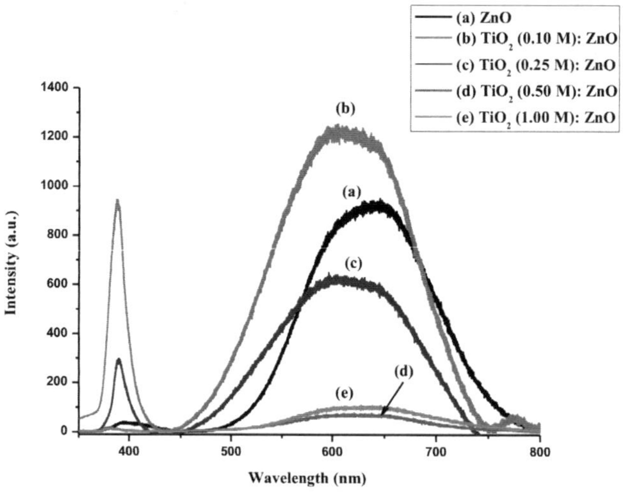

Figure 2. PL spectra of bare ZnO and ZnO nanostructures deposited on different molarities of TiO_2 seeded-template.

The crystal quality and optical properties are known to be enhanced when the ratio of UV emission peak intensity and the visible emission peak intensity, I_{UV}/I_{VIS} is increased [17, 20, 21]. The high peak intensity ratios indicates the decrease of defect concentrations and impurities in grown materials with the molarity decrease from 1.00 M to 0.10 M. The calculated ratio, I_{UV}/I_{VIS} were recorded to be 0.04, 0.76, 0.47, 0.26 and 0.14 for bare ZnO, 0.10, 0.25, 0.50 and 1.0 M of TiO_2 seeded-template, respectively as also recorded in Table 2. Therefore, the ZnO nanostructures grown on the TiO_2 seeded-template has better crystal and optical quality of the nanostructures thin films compared to bare ZnO.

Table 2. Position of UV emission peak and the ratio of the UV emission peak intensity and the visible emission peak intensity, I_{uv}/I_{vis} based on PL spectra for bare ZnO and ZnO nanostructures deposited on different molarities of TiO_2 seeded-template.

Sample	Wavelength of UV emission peak (nm)	I_{UV}/I_{Vis}
ZnO	395	0.04
TiO_2(0.10 M): ZnO	389	0.76
TiO_2(0.25 M): ZnO	390	0.47
TiO_2(0.50 M): ZnO	383	0.26
TiO_2(1.00 M): ZnO	383	0.14

3.3. Ultraviolet-visible (UV-Vis) study

Figure 3 displays the UV-Vis spectrum for bare ZnO and TiO_2: ZnO nanostructures thin films, at four different TiO_2 percursor concentrations (0.10, 0.25, 0.50 and 1.00 M). The fabricated TiO_2: ZnO nanostructures thin films show good transparency in the visible region (400-800 nm) and high absorption properties at UV region (below 400 nm).

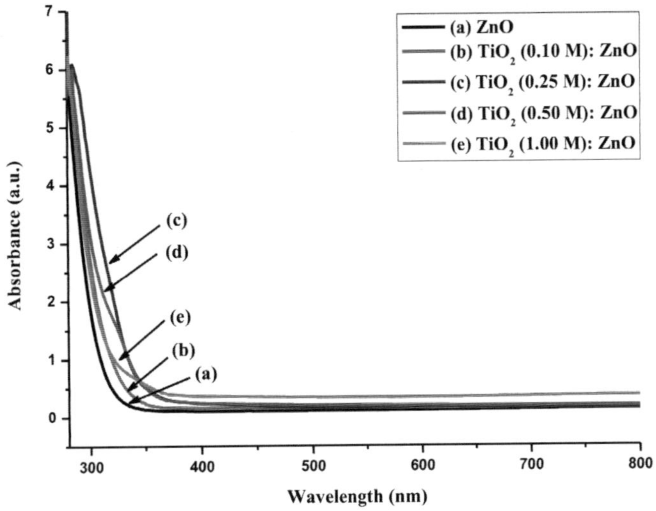

Figure 3. UV-vis absorbance spectra of bare ZnO and ZnO nanostructures deposited on different molarities of TiO_2 seeded-template.

In a comparison to bare ZnO, the ZnO nanostructures grown on TiO_2 seeded-template exhibit stronger absorption properties in UV and visible region. This can be attributed to the improvement of light-harvesting efficiency, which enhance the use of UV light compared with bare ZnO [22, 23]. It is suggested that the improvement are likely due to the existence of TiO_2 that modifies the ability of the samples in efficiently adsorbed the incident light reaching the thin films. This improvement also indicates the reduction in transmittance of the thin films and might be resulted from the trapped incident lights caused by scattering in the needle-like structure [18]. As seen in the FESEM images, needle-like ZnO nanostructures have denser distribution as the molarities of TiO_2 increase. The increasing of grown ZnO which were distributed vertically and horizontally towards the TiO_2 nanoparticles surface caused increasing of trapped incident light in the thin films. Hence, the optical path length inside the thin films might also increase and improved the light absorption. The red shift was observed on the absorption edges of the TiO_2: ZnO nanostructures compared to the ZnO thin film. The light scattering and the grain size of one-dimensional nanostructure might have influenced this red shift in longer-wavelength region [22].

For bare ZnO, although the thin film has weak UV absorption properties, it means that the sample has good transmittance which higher at both regions, UV and visible regions (300- 800 nm). This reason might be resulting from the smoother surface morphology and less grain boundaries of the thin film has, as confirmed by FESEM result above. From Figure 1 (a), bare ZnO thin film has less dispersion which contributes to the lower distribution of fabricated needle-like ZnO on the glass substrate and thus leads to the lower grain boundary. When the incident light scattered on the smoother surface with less denser distribution of needle-like ZnO, more light can escaped and passed through the thin film and caused higher optical transmittance values [24, 25].

4. Conclusion

ZnO nanostructures were successfully synthesized via solution-immersion method on TiO_2 seeded-template prepared by sol-gel spin-coating technique, on the glass substrates. FESEM images revealed that needle-like ZnO nanostructures with smaller tip can be seen on the templates as the TiO_2 seed layers were coated on the glass substrates. From PL spectra, when the molarity of TiO_2 seed layer increases from 0.10 M to 1.00 M, the UV band are blue shifted to the left compared to the bare ZnO nanostructures thin film and the broad visible band are extending from 550- 700 nm. The blue shifted of UV peak emission to the lower wavelength might be due to the reduction of ZnO nanostructures sizes. Besides, ZnO nanostructures grown on the TiO_2 seeded-template has better crystal and optical quality of the nanostructures thin films compared to bare ZnO. Furtherrmore, in a comparison to bare ZnO, the ZnO nanostructures grown on TiO_2 seeded-template exhibit stronger absorption properties in UV and visible region which was likely improved the light-harvesting efficiency of the thin films. As a result, ZnO nanostructures grown on 0.25 M of TiO_2 seeded-template, having smallest average size of ZnO nanostructures, better crystallinity, improved optical quality and high UV absorption properties might be promising for use in UV photoconductive sensor. Future work in observing the influence of number of TiO_2 layers to the grown ZnO nanostructures should be done for further investigation.

5. Acknowledgement

We would like to express our gratitude to Ministry of Education Malaysia and Research Management Institute, Universiti Teknologi MARA (UiTM), Shah Alam, Selangor, Malaysia and Long-Term Research Grant Scheme for Nanostructures, Nanomaterials and Devices for Fuel Cells and Hydrogen Production (600-RMI/LRGS 5/3 (3/2013)) for the financial support. This study was also supported by the Research Chair of Targeting and Treatment of Cancer Using Nanoparticles, Deanship of Scientific Research, King Saud University, Riyadh, Saudi Arabia.

6. References

[1] C.-C. 2012 *Nanorods* (Croatia: InTech) pp 33-50.
[2] P. K. G. Soumen Dhara 2012 *Nanorods* (Croatia: InTech) pp 1-32.
[3] T. A. Yinyan Gong, Gertrude F. Neumark, Stephen O'Brien, and Igor L. Kuskovsky 2007 *Nanoscale Res Lett.* **2** 297
[4] Z. K. Mohamad Hafiz Mamat, Musa Mohamed Zahidi and Mohamad Rusop Mahmood, O. Yalçın 2012 *InTech*. 250.
[5] S. T. Jirapong Arin and Titipun Thongtem 2013 *Materials Letters* **96** 78
[6] Baoshun Liu, Qi Hu Liping Wen, and Xiujian Zhao 2009 *Thin Solid Films* **517** 6569
[7] Hua-Jie Wang, Yuan-Yuan Sun, Ying Cao, Xue-Hong Yu, Xiao-Min Ji and Lin Yang 2011*Chemical Engineering Journal* **178** 8
[8] E. C. Davide Barreca, Angelo P. Ferrucci, Alberto Gasparotto, Chiara Maccato, Cinzia Maragno, Giorgio Sberveglieri, and Eugenio Tondello 2007 *Chem. Mater* **19** 5642
[9] Simin Janitabar-Darzi and Ali Reza Mahjoub 2009 *Journal of Alloys and Compounds* **486** 805
[10] M. A. Kanjwal, N. A. M. Barakat, F. A. Sheikh, S. J. Park, and H. Y. Kim 2010 *Macromolecular Research* **18** 233
[11] Bambang Veriansyah, Jae-Duck Kim, Byoung Koun Min, Young Ho Shin, Youn-Woo Lee, and Jaehoon Kim 2010 *J. of Supercritical Fluids* **52** 76
[12] F. A. A. Guvenc Akgul, Klaus Attenkofer, Markus Winterer 2013 *Journal of Alloys and Compounds* **554** 177
[13] S. A. Z. Khusaimi, H. A. Rafaie, M. H. Mamat, N. Abdullah, S. Abdullah, M. Rusop 2009 *Malaysian Journal of Science 28 (Special Edition)* **28** 197
[14] M. N. b. A. Mohd Faizal b Achoi, Mohamad Rusop, and Saifollah Abdullah 2011 *Transaction of the Materials Research Society of Japan* **36** 273.
[15] A. A. N.A.M. Asib, A.N. Afaah, M. Rusop, and Z. Khusaimi 2014 *Advanced Materials Research* **832** 596
[16] D. Basak, G. Amin, B. Mallik, G. K. Paul, and S. K. Sen 2003 *Journal of Crystal Growth* **256** 73

[17] Q. L. Zi Qin, Yunhua Huang, Lidan Tang, Xiaohui Zhang, and Yue Zhang 2010 *Materials Chemistry and Physics* **123** 811

[18] Z. K. M.H. Mamat, M.Z. Musa, M.F. Malek, and M. Rusop 2011 *Sensors and Actuators A: Physical* **171** 241

[19] A. A. F.S. Husairi, M. Rusop, and S. Abdullah 2013 *Microelectronic Engineering* **108** 145

[20] J. Lee, JooyoungChung, and SangwooLim 2010 *Physica E* **42** 2143

[21] M. H. Mamat, Z. Khusaimi, M. M. Zahidi, and M. R. Mahmood 2011 *Japanese Journal of Applied Physics* **50**

[22] J. T. Hong Zhu, Tao Wang and Jie Deng 2011 *Applied Surface Science* **257** 10494

[23] A. A. Chun Cheng, Chao Zhu, Zuli Xu, Haisheng Song and Ning Wang 2014 *Scientific Reports*

[24] C. M. Firdaus, M. S. B. S. Rizam, M. Rusop, and S. R. Hidayah 2012 *Procedia Engineering* **41** 1367.

[25] Prabitha B. Nair, V.B.Justinvictor, Georgi P.Daniel, K.Joy, V.Ramakrishnan, and P.V.Thomas 2011 *Applied Surface Science* **257** 10869

The effect of different molarities of precursor $Zn(NO_3)_2.6H_2O$ to the growth of ZnO by solution-immersion deposited on ZnO seeded template

A N Afaah[1,2,a], N A M Asib[1, 2], A Aadila[1, 2], K A Eswar[1, 2], M R Mahmud[1,2], Salman A H Alrokayan[4], Haseeb A Khan[4], R Mohamed[3], M Rusop[2, 3], Z Khusaimi[1,2, b]

[1]Faculty of Applied Sciences, Universiti Teknologi Mara (UiTM), 40450 Shah Alam, Selangor, Malaysia

[2]NANO-SciTech Centre (NST), Institute of Science, University Teknologi MARA (UiTM), 40450 Shah Alam, Selangor, Malaysia

[3]NANO-ElecTronic Centre (NET), Faculty of Electrical Engineering, Universiti Teknologi MARA (UiTM), 40450 Shah Alam, Selangor, Malaysia

[4]Department of Biochemistry, College of Science, Bldg. 5, King Saud University (KSU) P. O. 2454 Riyadh 11451, Kingdom of Saudi Arabia

[a]afaahabdullah@yahoo.com, [b]zurai142@salam.uitm.edu.my

Abstract. ZnO films were grown on glass substrates by mist-atomization, with the growth temperature of 300 ^{0}C to form ZnO seeded template. The ZnO seeded template then undergo second deposition method, which is solution-immersion method to grow ZnO on the ZnO seeded template thin film. The thin films were immersed in different molarities of starting materials; which are zinc nitrate hexahydrate and HMTA. The molarities were varied from 0.05, 0.10, 0.20 and 0.40 M. As a control to the study, the ZnO seeded template which undergoes mist-atomization only, denoted as MA only is prepared for comparison studies to the other samples. The properties of the products were examined by photoluminescence (PL) spectrophotometer, Raman spectrophotometer, and ultraviolet-visible (UV-Vis) spectrophotometer. From the PL measurement, it shows that 0.40 M sample have highest UV emission and from Raman measurements, it was found that the sample immersed in 0.40 M precursor have high crystal quality. From UV-Vis measurements, sharp UV emission at ~378 nm is observed and 0.40 M ZnO thin film shows a better absorption properties.

1. Introduction

Zinc oxide (ZnO) is a semiconducting material with a wide and direct band gap of 3.37 eV at room temperature and high exciton binding energy of 60 meV. ZnO consists of two interconnecting sublattices of Zn^{2+} and O^{2-}. Every of the Zn ions are surrounded by tetrahedral O ions, and vice versa. This semiconducting material received considerable interest due to its wide range of application such as gas sensor [1], solar cells [2], UV photoconductive sensor [3], and biomaterial [4]. Diverse synthetic approaches have been introduced for ZnO nanostructures such as RF magnetron sputtering [5], electron beam evaporation [6], and pulse laser deposition [7]. Among these, mist-atomization and solution-immersion method are appropriate for the synthesis of ZnO nanostructures due to low cost, suitable for large production fabrication and environmentally friendly. To the best of our knowledge, our study is the first to grow ZnO nanostructures by the following two-step deposition techniques; (i) mist-atomization and (ii) solution-immersion method. First, ZnO nanoseed were deposited on Au-seeded glass substrate by ambient mist-atomization method. Next, ZnO nanostructure was grown by facile solution-immersion method on the ZnO seed layer. The influence of precursor solution molarity to crystal quality of ZnO thin films obtained is studied.

2. Experimentals

2.1. Preparation of ZnO seed layer by mist-atomization

The starting materials used in this study were zinc nitrate hexahydrate ($Zn(NO_3)_2.6H_2O$) and hexamethylenetetramine, HMTA ($C_6H_{12}N_4$). Zinc nitrate hexahydrate of Riedel-de Haën brand was used as the source of zinc for the growth of zinc oxide nanostructures. As the stabiliser in the starting solution, HMTA of same brand with 99-100.5 % pure analytical grade was used. Cleaned glass substrates with size of 2 x 2 cm^2 were sputter-coated with gold to form Au seed layer as in Figure 1. Then, ZnO nanoseed were grown on the Au seed layer by mist-atomization method as in Figure 2. The ZnO seed layer were prepared by mist-atomization technique as reported in our previous work [8].

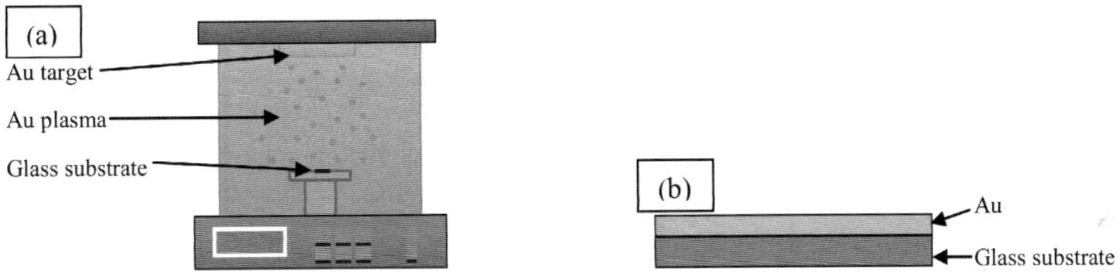

Figure 1. (a) Deposition of Au seed layer and (b) Au-seeded glass

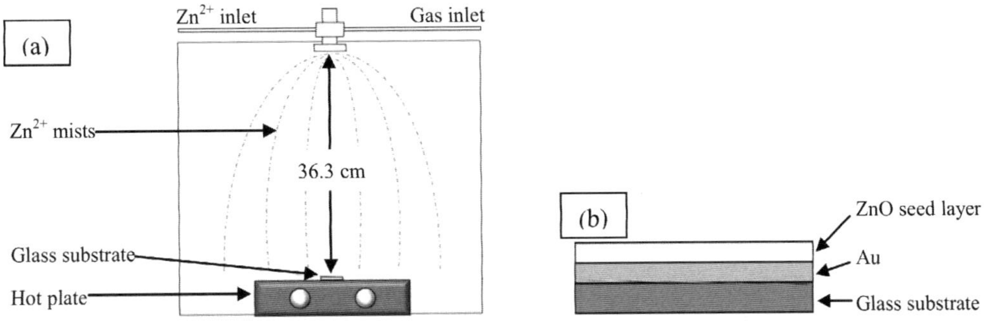

Figure 2. (a) Deposition of ZnO seed by mist-atomization method (b) ZnO/Au seeded glass

2.2. Synthesis of ZnO by solution-immersion method

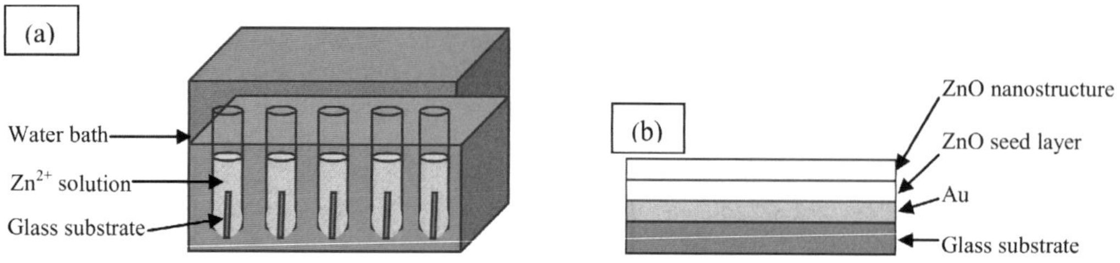

Figure 3. (a) Deposition of ZnO by solution-immersion method and (b) ZnO/ZnO/Au template

Next, the ZnO seeded templates undergo deposition method to grow well-aligned ZnO nanostructure by solution-immersion as in Figure 3. The details for the method are as explained elsewhere [9]. The ZnO seeded templates prepared by mist-atomization were immersed in different molarity of precursor of 0.05, 0.10, 0.20 and 0.40 M.

2.3. Characterizations

The photoluminescence (PL) and Raman properties of the synthesized films were deliberated using Horiba Yvon-79 DU420A-OE spectrophotometer with a helium-cadmium (He-Cd) excitation laser source operating at 325 nm for PL and 514 nm for Raman. The UV-Vis spectroscopy was completed using a Varian Cary 5000 UV-Vis-NIR spectrophotometer with the wavelength range between 300 and 8000 nm with a data interval of 1 nm.

3. Results & discussion

3.1. Photoluminescence studies

Figure 4 and 5 present the PL spectra of the ZnO thin films prepared at various molarities. The luminescence of the ZnO thin films is closely related to the crystalinity of the film. As the crystalinity of the film is improved, the density of the defects decreases. The films were measured using room temperature (RT) PL emission spectra in the wavelength range of 330-900 nm at an excitation wavelength of 325 nm. In general, ZnO films exhibit two emission peaks, which correlate to the UV emission and the visible emission. The ~370 nm UV emission has been well understood to the association of direct recombination at near band edge of photogenerated charge carriers (exciton emission), meanwhile the green wavelength emission is due to the structural defects and intrinsic defects such as oxygen vacancies [10], zinc interstitials [11] and zinc vacancies [12].

As shown in Figure 4, the PL spectra in all samples show UV emission at the near band edge centred at 380 nm. The peaks appear at ~380 nm is due to excitation transition of ZnO. It is associated with the free-exciton recombination [13]. The higher the concentration of the precursor solution, the higher the UV intensity except for the sample deposited by mist-atomization (MA) only. The sample immersed in 0.05 M of Zn^{2+} solution shows the lowest intensity amongst all. The noticeable UV emission of ZnO thin film with 0.40 M shows that the thin film has a low level of defects compared to the other samples. The intense UV emission could be related to the much higher crystal quality [14] of the sample with molarity of 0.40 M in comparison to the other molarities. Figure 4 depicts that the intensity of UV emission of ZnO thins films increases with increasing molarity. As the molarity increased, the thickness of the ZnO films observed is thicker. M. F. Malek et. al [14] proposed that the thicker film contains more zinc particle which caused the increase in carrier concentration and hence improving the UV emission paragraphs are indented.

The red shift (shifted to the higher number of wavelength) and broader width of the UV emission as observed in Figure 4 could be attributed to the inhomogeneous strain distribution due to large lattice mismatch between the ZnO thin film and the ZnO seed layer [15]. The samples with lower molarities showed red-shift properties and the range of UV emission is broader. From Figure 4, it can be concluded that ZnO thin film of 0.40 M molarity have lower lattice mismatch compared to ZnO thin films with other molarity.

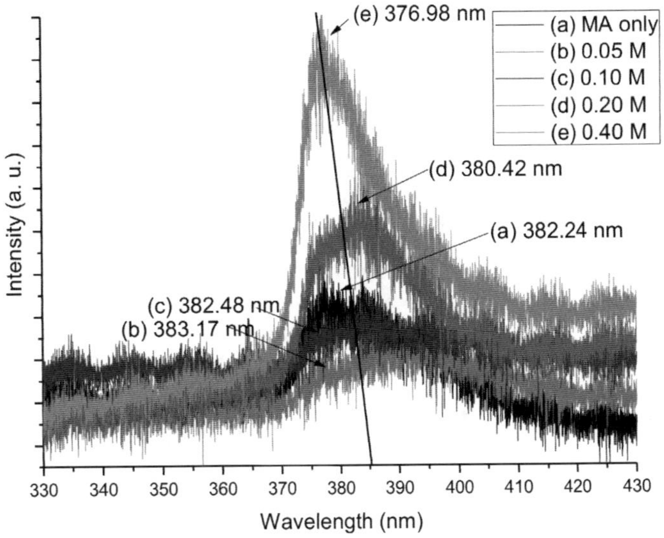

Figure 4. UV emission PL spectra of ZnO thin films with the variation molarities of 0.4 M seeded ZnO (MA only), 0.05 M, 0.10 M, 0.20 M and 0.40 M.

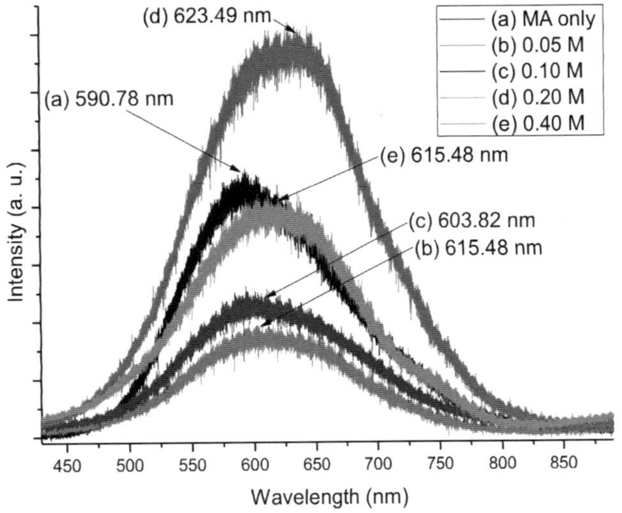

Figure 5. Visible emission PL spectra of ZnO thin films with the variation molarities of 0.4 M seeded ZnO (MA only), 0.05 M, 0.10 M, 0.20 M and 0.40 M

Figure 5 shows the broad visible emission in the wavelength range of 500 to 800 nm was observed. The peaks for the deep-level emission were observed centred at about 610 nm. As stated before, the green emission was attributed to structural or intrinsic defects of ZnO thin film. Some researchers reported below temperature of 1300 ^{0}C, Frenkel reaction occurs which result in zinc interstitials meanwhile at temperature above 1300 ^{0}C, there is Schottky reaction which led to oxygen vacancies [16, 17]. As the ZnO thin films in this work were annealed at 500 ^{0}C, it is show that Frenkel defects of zinc interstitials predominate the defects in the visible PL spectra.

Table 1 shows the UV emission to visible emission ratio, I_{uv}/I_{vis} of ZnO thin films with varied molarity. ZnO thin film synthesized from 0.40 M molarity shows the highest UV to visible emission ratio. This indicates that the 0.40 M ZnO thin film exhibit good crystalinity with lower defects.

Table 1. The UV emission to visible emission ratio, I_{uv}/I_{vis} of ZnO thin films with varied molarity

Molarity, M	Ratio, R = I_{uv}/I_{vis}
0.4 (MA only)	0.07334
0.05	0.11786
0.10	0.12726
0.20	0.07529
0.40	0.24500

3.2. Raman studies

Raman scattering is an effective technique to investigate the crystallization, structure and defects in the nanostructure materials. From the group theory, hexagonal wurzite ZnO belongs to C_{6v}^4 space group and the point of the Brilloun zone are $A_1+2B_1+E_1+2E_2$, all will have two modes at the center of Brillouin zone [18, 19]. Generally, the polar A_1 (LO) mode is associated with oxygen deficiency of the film. Meanwhile, the non-polar phonon modes of E_2 symmetry have two frequencies, E_2 (high) and E_2 (low). E_2 (high) is related with the oxygen atom and E_2 (low) related with the vibration of the heavy Zn sublattice [20]. Some researchers reported that E_2 (high) shifted due to intrinsic defect in ZnO films and agreed that the shift of E_2 (high) can give information on stress for the thin films [21].The film is under compressive stress when the E_2 (high) is up shifted, whilst a tensile stress will lead to its downshift [18, 22]. In the other hand, E_2 (low) determines the crytalinity of the films.

Figure 6 depicts the typical room temperature Raman spectra of ZnO growth by MA only, immersed in 0.05, 0.10, 0.20, and 0.40 M precursor. From the Figure 6, the sharpest and highest peak of E_2 (low) is owned by thin film prepared with 0.20 M. There is a small difference in intensity for the sample with 0.40 M. The thin film which undergoes deposition by mist-atomisation (MA) only shows the lowest Raman intensity, followed by the thin films immersed in 0.05 and 0.10 M of precursor solution. The presence of intense E_2 (low) peaks of thin films with 0.20 and 0.40 M precursor concentration portrayed the thin film exhibit best crystal quality with a hexagonal wurzite phase compared to the other samples with different molarities. According to the literature, the E_2 (high) wavenumber value for bulk ZnO is 437 cm^{-1}. From the Figure 6, the average Raman value for all of the samples is about ~432 cm^{-1}. All the samples showed downshift properties which revealed there is tensile stress exist in the films.

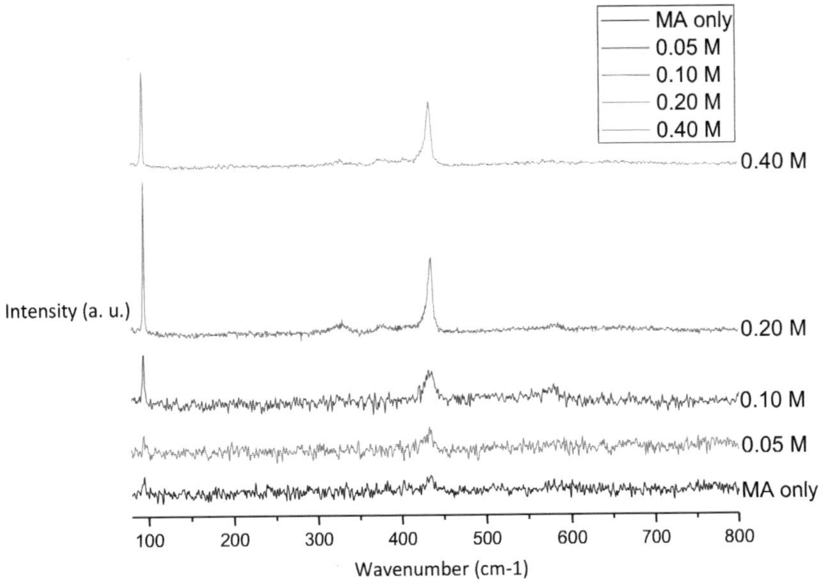

Figure 6. Raman spectra of ZnO thin films with the variation molarities of 0.4 M seeded ZnO (MA only), 0.05 M, 0.10 M, 0.20 M and 0.40 M

As stated before, A_1 (LO) is related with oxygen deficiency of the ZnO thin films. The A_1 mode is attributed to the vibration along the c-axis [23]. According to theoretical calculation, the A_1 (LO) wavenumber for bulk ZnO is 574 cm^{-1}. In this work, A_1 (LO) for ZnO thin films with 0.40 M has highest Raman shift (shifted to the right). A higher wavenumber proposed there is reduction in the defects of the ZnO films and better crystalinity of the films. The small and suppressed intensity of A_1 (LO) mode shows the reduction of oxygen deficiency effect on the ZnO films and the ZnO is grow along c-axis orientation. All the films generally and the thin film with 0.40 M precisely showed improvement in crystalinity compared to bulk ZnO and other set of samples.

Table 2. The resulting dataset for Raman energy compromises three main energies of ZnO thin films with varied molarity

Molarity, M	E_2 (low)	E_2 (high)	A_1 (LO)
0.4 (MA only)	93.80	432.44	NIL
0.05	93.81	432.44	NIL
0.10	94.52	432.44	578.95
0.20	93.58	433.86	579.18
0.40	94.52	433.62	579.66

3.3. UV-Vis studies

Figure 7 represents the absorbance spectra of ZnO thin films with the variation molarities of 0.4 M seeded ZnO (MA only), 0.05 M, 0.10 M, 0.20 M and 0.40 M in the wavelength range of ~290 nm to 800 nm. The range of inter-transition is 370-390 nm. The absorption edges observed for all ZnO thin films were about ~378 nm, corresponding to the intrinsic band gap of ZnO which is 3.26 eV. At this wavelength, there are transitions of electrons from the valence band to the conduction band. The different molarity shows that there is great influence to the absorbance properties of the ZnO thin films by varying the molarity.

Commonly, there are three main factors that are able to manipulate the absorbance of the ZnO films; which are surface roughness, defect centres and oxygen vacancies [24]. The higher absorbance of the samples with higher concentration could be attributed to the multiple scattering of the incident light within the thin film. As the concentration increases, more ZnO nanostructures produced and hence there is no free space for the light to transmit. From Figure 7, it can be seen that ZnO thin film with 0.40 M shows the highest UV intensity amongst all. The high intensity at wavelength ~378 nm of higher molarity indicates that there are more activities of direct transition of electrons; between the edge of valence band and the conduction band. The intensity increases with increasing molarity because the thicker film contains more zinc particles, leading to an increase in the carrier concentration and enhancing the UV emission.

As compared to the ZnO thin film with 0.40 M, the ZnO thin film with 0.4 M (prepared by MA only), 0.05 M, 0.10 M and 0.2 M experienced lower UV light absorption which considered due to higher defects in the structural properties of the ZnO thin films. I.Y.Y. Bu reported that well ordered nanostructure effectively traps the light and reduces transmittance; which means the film has enhanced probability of absorption [25]. The ZnO thin film of 0.40 M has better orientation compared to the other films, which supported by the photoluminescence and Raman studies.

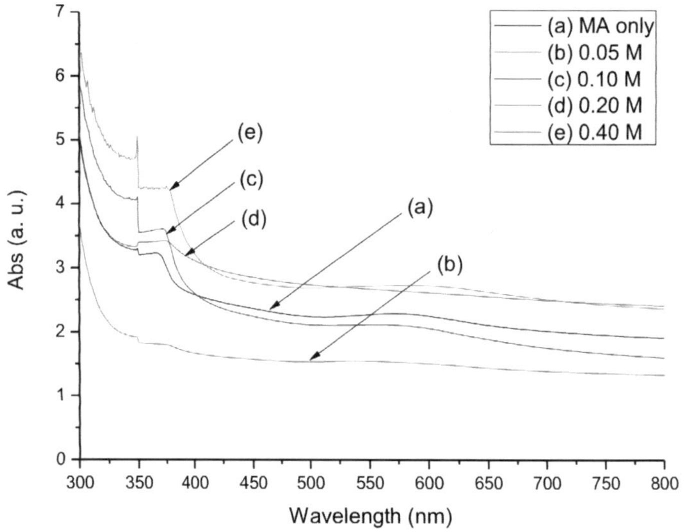

Figure 7. UV-Vis absorbance spectra of ZnO thin films with the variation molarities of 0.4 M seeded ZnO (MA only), 0.05 M, 0.10 M, 0.20 M and 0.40 M

4. Conclusion
In the present paper, we have investigated good crystal quality with high absorption properties ZnO formed on ZnO seed layer. The ZnO seed layer was synthesized by mist-atomization method, whilst the ZnO nanostructure was deposited by solution-immersion method. From PL UV emission, it can be concluded that ZnO with 0.40 M have lower lattice mismatch compared to the other samples, which proved that the sample have lower defects. From PL visible spectra, it is believed that the defect-related green emission was governed by Frenkel defects of zinc interstitials. From Raman studies, all the samples showed that tensile stress exist in the films. Based on three main energies of Raman energy, it is proved that ZnO with 0.40 M has highest crystal quality amongst all. UV-Vis studies shows that ZnO with 0.40 M has highest probability of absorption which indicates the film has better orientation compared to the other films. These results suggest that a suitable precursor molarity should be used to obtain higher crystal quality ZnO thin films for applications in optoelectronic devices such as sensors and solar cells.

Acknowledgments
The authors would like to thank Faculty of Applied Sciences of Universiti Teknologi MARA (UiTM) Malaysia, Research Management Institute (RMI) of UiTM and Ministry of Higher Education Malaysia (MOHE) for their support. This work was also supported by the Long-Term Research Grant Scheme for Nanostructures, Nanomaterials and Devices for Fuel Cells and Hydrogen Production (600-RMI/LRGS 5/3 (3/2013)) grant. This study was supported by the Research Chair of Targeting and Treatment of Cancer Using Nanoparticles, Deanship of Scientific Research, King Saud University, Riyadh, Saudi Arabia. The authors would also like to acknowledge College of Science, King Saud University, Kingdom of Saudi Arabia for the research collaboration and support.

References
[1] C. Wongchoosuk, S. Choopun, A. Tuantranont and T. Kerdcharoen 2009 *Materials Research Innovations* Au-doped zinc oxide nanostructure sensors for detection and discrimination of volatile organic compounds **13 (3)** 185-188
[2] M. Thambidurai, N.Muthukumarasamy, D. Velauthapillai and C. Lee 2013 *Materials Letters* Synthesis of garland like ZnO nanorods and their application in dye sensitized solar cells **92** 104–107
[3] M.H. Mamat, Z. Khusaimi, M.M. Zahidi and M.R. Mahmood 2011 *Jpn. J. Appl. Phys.* Performance of an Ultraviolet Photoconductive Sensor Using Well-Aligned Aluminium-Doped Zinc-Oxide Nanorod Arrays Annealed in an Air and Oxygen Environment **50** 06GF05
[4] B. Sadeghi 2014 *Molecular and biomolecular spectroscopy* Preparation of ZnO/Ag nanocomposite and coating on polymers for anti-infection biomaterial application, Part A **118** 787-792
[5] J.C. Li, Y.F. Li, T. Yang, B. Yao, Z.H. Ding, Y. Xu, Z.Z. Zhang, L.G. Zhang, H.F. Zhao and D.Z. Shen 2013 *Journal of Alloys and Compounds* Effects of S on solid solubility of Ag and electrical properties of Ag-doped ZnO films grown by radio frequency magnetron sputtering **550** 479-482
[6] D.R. Sahu, S.-Y. Lin and J.-L. Huang 2007 *Solar Energy Materials and Solar Cells* Deposition of Ag-based Al-doped ZnO multilayer coatings for the transparent conductive electrodes by electron beam evaporation **91** 851-855
[7] D. Yu, J. Li, L. Hu, H. Hu, H. Zhang, K. Sun and J. Zhu 2008 *Chemical Physics Letters* Synthesis and photoluminescence investigation of ZnO:P nanorods on an InP substrate by pulsed laser deposition **464** 69-72
[8] A.N. Afaah, N.A.M. Asib, A. Aadila, M. Rusop, R. Mohamed and Z. Khusaimi 2014 *Advanced Materials Research* Low Temperature Growth and Optical Properties of Zinc Oxide Nanostructure Prepared by Two-step Solution-based Deposition Method **832** 749-753

[9] A.N. Afaah, A. Aadila, N.A.M. Asib, M. Rusop, R. Mohamed and Z. Khusaimi 2014 *Advanced Materials Research* Influence of Annealing Temperature to Surface Morphology of Flowerlike ZnO Deposited on Seeded Glass **832** 754-758

[10] X.L. Wu, G.G. Siu, C.L. Fu and H.C. Ong 2011 *Applied Physics Letters* Photoluminescence and cathodoluminescence studies of stoichiometric and oxygen-deficient ZnO films **78** 2285-2287

[11] R. Bhaskar, A.R. Lakshmanan, M. Sundarrajan, T. Ravishankar, M.T. Jose and N. Lakshminarayan *Indian Journal of Pure & Applied Physics* Mechanism of green luminescence in ZnO **47** 772-774

[12] A. Janotti and C.G.V.d. Walle 2007 *Physical Review B* Native point defects in ZnO **76** 165202

[13] J. Rouhi, M. Alimanesh, S. Mahmud, R.A. Dalvand, C.H. Raymond Ooi and M. Rusop 2014 *Materials Letters* A novel method for synthesis of well-aligned hexagonal cone-shaped ZnO nanostructures in field emission applications **125** 147-150

[14] M.F. Malek, M.H. Mamat, Z. Khusaimi, M.Z. Sahdan, M.Z. Musa, A.R. Zainun, A.B. Suriani, N.D. Md Sin, S.B. Abd Hamid, and M. Rusop 2014 *Journal of Alloys and Compounds* Sonicated sol–gel preparation of nanoparticulate ZnO thin films with various deposition speeds: The highly preferred c-axis (002) orientation enhances the final properties **582** 12-21

[15] Fang Yi, Yunhua Huang, Xiaoqin Yan, Zhiming Bai, Zengze Wang, Xiang Chen, Qi Zhang and Y. Zhang 2014 *Physica E: Low-dimessional systems and nanostructures* Simple fabrication of a ZnO nanorod array UV detector with a high performance

[16] J. Han, P.Q. Mantas and A.M.R. Senos 2002 *Journal of the European Ceramic Society* Defect chemistry and electrical characteristics of undoped and Mn-doped ZnO **22** 49-59

[17] L. Schmidt-Mende and J.L. MacManus-Driscoll 2007 *Materials Today* ZnO – nanostructures, defects, and devices **10** 40-48

[18] W.J.K. Li, C. Y. Ruan, H. B. Qin, G. P. Huang, G. J. Yang, T. Y. Liang, W. W. Zhao, Y. H. Meng, X. D. Yu, P. Cui and Y. T. Fang, L. 2012 *Solid State Communications* Electrical properties and Raman scattering investigation of Ag doped ZnO thin films **152** 147-150

[19] Z. Peng, G. Dai, W. Zhou, P. Chen, Q. Wan, Q. Zhang and B. Zou 2010 *Applied Surface Science* Photoluminescence and Raman analysis of novel ZnO tetrapod and multipod nanostructures **256** 6814–6818

[20] I. Calizo, K.A. Alim, V.A. Fonoberov, S. Krishnakumar, M. Shamsa, A.A. Balandin and R. Kurtz 2007 *Proc. of SPIE* Micro-Raman spectroscopic characterization of ZnO quantum dots, nanocrystals and nanowires **6481** 64810N

[21] S.K.S.a.G.J. Exarhos 1997 *Solid State Phenomenon* Raman Spectroscopic Investigation of ZnO and Doped ZnO Films, Nanoparticles and Bulk Material at Ambient and High Pressure **55** 32-37

[22] L.Z.H. L.N. Wang, H.Q.Zhang, Y.Qiu, Y.Lang, G.Q.Liu, J.Y.Ji, J.X.Ma and Z.W.Zhao 2011 *Materials Science in Semiconductor Processing* Studying the Raman spectra of Ag doped ZnO films grown by PLD **14** 274–277

[23] H.-M. Cheng, H.-C. Hsu, Y.-K. Tseng, L.-J. Lin and W.-F. Hsieh 2005 *J. Phys. Chem.* Raman Scattering and Efficient UV Photoluminescence from Well-Aligned ZnO Nanowires Epitaxially Grown on GaN Buffer Layer B **109** 8749-8754

[24] I.Y.Y. Bu 2014 *Ceramics International* Optoelectronic properties of ZnO nanowires deposited under different zinc nitrate/hexamine ratio concentrations **40** 6345-6350

[25] I.Y.Y. Bu 2013 *Ceramics International* Rapid synthesis of ZnO nanostructures through microwave heating process **39** 1189-1194

Dynamism in a Semiconductor Industrial Machine Allocation Problem using a Hybrid of the Bio-inspired and Musical-Harmony Approach

Umi Kalsom Yusof [1] **and Mohd Nor Akmal Khalid** [1]

[1] School of Computer Sciences, Universiti Sains Malaysia, 11800 USM, Pulau Pinang, Malaysia

E-mail: umiyusof@cs.usm.my;mnak104041@student.usm.my

Abstract. Semiconductor industries need to constantly adjust to the rapid pace of change in the market. Most manufactured products usually have a very short life cycle. These scenarios imply the need to improve the efficiency of capacity planning, an important aspect of the machine allocation plan known for its complexity. Various studies have been performed to balance productivity and flexibility in the flexible manufacturing system (FMS). Many approaches have been developed by the researchers to determine the suitable balance between exploration (global improvement) and exploitation (local improvement). However, not much work has been focused on the domain of machine allocation problem that considers the effects of machine breakdowns. This paper develops a model to minimize the effect of machine breakdowns, thus increasing the productivity. The objectives are to minimize system unbalance and makespan as well as increase throughput while satisfying the technological constraints such as machine time availability. To examine the effectiveness of the proposed model, results for throughput, system unbalance and makespan on real industrial datasets were performed with applications of intelligence techniques, that is, a hybrid of genetic algorithm and harmony search. The result aims to obtain a feasible solution to the domain problem.

1. Introduction

The semiconductor industry, because of its technologies and complex manufacturing procedure, has been widely regarded as having one of the most complex manufacturing processes [1]. Its nature of business, which is highly dependent on fast-growing and highly competitive technologies, requires the industry to invest on advanced state-of-the-art machines and facilities. At the same time, it also faces a continuous demand to reduce prices for its customers. Consequently, the industry encounters a great challenge in balancing the capital investment and the risk of losing customers. To maintain or regain competitive advantages, many semiconductor manufacturers have put more emphasis on improving capacity planning as well as scheduling and production control aside from having a good method of forecasting product demands.

Capacity planning is a process of adjusting one's capability in response to the changing demands by maximizing machine utilization and allocating available resources to the right product mixtures while mitigating the technological constraints in meeting the customer demand. The ability to utilize the available resources to their utmost potential can be translated to a better managed and profitable company. Generally, the proposed approaches to solving the FMS machine allocation problem can be classified as (a) mathematical approaches, (b) heuristic approaches, and (c) artificial intelligence-based (AI) or metaheuristic approaches [2]. Although mathematical approaches are robust in their applicability, they tend to become impractical when the problem size increases. Heuristic and AI-based approaches are more promising in term of optimality and efficiency, even in more complicated problems, although sometimes the results are mainly dependent on the rules and constraints of individual problems.

These limitations motivate researchers to enhance the method and to look for innovative searching techniques to improve heuristic-based approaches further. One of the popular approaches is genetic algorithm (GA), which is known for its capability in intelligent probabilistic searching. [3] suggest the use of a GA-based scheduling tool to handle multiple resource constraints and a multiple-level product structure. [4] proposed a constraint-based GA in handling complex constraints in the FMS loading problem.

Another potential algorithm that can handle the machine allocation problem effectively is harmony search (HS). This heuristic algorithm is derived from an artificial phenomenon found in musical performance that always seeks for better harmony. HS has been applied to several computational optimization problems such as music composition [5] and timetabling problem [6]. According to [7], the main features of HS that make it different from other methods are as follows: HS makes a new vector after considering all existing vectors instead of considering only two (parents) as in GA, and HS does not require the setting of initial values of decision variables. These features increase the flexibility to find better solutions. In addition, [8] performed a study which concludes that HS poses a strong explorative power, which is a very important characteristic of evolutionary algorithms (EA).

According to [9], most studies associated with machine allocation problem are based on the assumption of a deterministic environment. However, in the actual manufacturing world, FMS operates in a dynamic environment where interruptions such as machine breakdown or introduction of new jobs are common and occur almost every day. Machine breakdown is one of the major undesirable inputs in the FMS environment. It causes not only additional maintenance cost but also increases the manufacturing lead time. According to an estimation by [10], maintenance cost alone can form 15% to 40% of the total production cost. In any manufacturing plant, the operating cost largely comes from the tools expenditure, which may take up to 30% of the total revenue [11]. Hence, it is important to minimize the negative effects of machine breakdowns because they may erode the profitability and productivity of FMS.

Consequently, it becomes critical to analyze and to minimize the adverse effect of failures on the objective measures of the machine allocation problem so that production goals can be achieved. Only a few studies have been conducted on the machine allocation problem that deals with machine interruption and control policies, mimicking the real environment of FMS. In this study, the real shop floor scenario is emulated, that is, the handling the machine allocation problem with the occurrence of a machine breakdown that aims to minimize the effect of the machine breakdowns on the overall performance of FMS. The model that encompasses an online machine monitoring scheme is proposed to handle the expected breakdowns due to wear-outs and manual machine monitoring, dealing with random failures that may involve the reloading of part types. In the online monitoring scheme, machines are continuously monitored to measure the failure potential and determine the action beforehand to avoid a potential breakdown. Conversely, manual monitoring ensures that the action is taken as soon as possible to minimize the effect due to a sudden breakdown.

In studying the well-known approaches to dealing with the conventional machine allocation problem, some of the researchers preferred using of a probabilistic search technique to gain optimum result. Meta-heuristic algorithms such as GA and HS are among the effective random search techniques, that can produce a near optimum result from a large solution space of the machine allocation problem in an acceptable computational time. Thus, these computational techniques have been employed to obtain quality solutions under machine breakdown conditions.

With the added advantage of GA and HS combined together in a hybrid; good results can be obtained.

Our work is an extension of the results obtained from the previous works on constraint-chromosome genetic algorithm (CCGA) [12], HS [13], and a hybrid of CCGA and HS (called H-CCGaHs) [14] performed on the benchmark problems available in [15]. For a detailed explanation on how the H-CCGaHs and the related operators work, refer to the previous related works. Based on the result, the proposed algorithm H-CCGaHs performs better than most of the heuristics available in the literature in the minimization of system unbalance as well as combined objective functions (COF).

The proposed H-CCGaHs can achieve the best solution for 7 out of 10 datasets tested as opposed to only 6 datasets for CCGA and HS. The average COF (1.8160) in the last row of Table IV in [14] shows that H-CCGaHs is superior in performance to the other heuristics considered in this evaluation. This paper focuses on the application of the proposed H-CCGaHs that incorporates machine breakdowns with the semiconductor industrial machine allocation problem using actual industrial datasets. As the nature of the datasets used in [14] is almost similar to that of the actual industrial datasets (except that the latter datasets are much bigger), this algorithm should be able to produce the same excellent result if experimented on actual industrial datasets.

The remainder of the paper is organized as follows. Section 2 describes the dynamic machine allocation, and Section 3 formulates the model formulation. Section 4 discusses the proposed solution to solve the problem, and Section 5 presents the results of the proposed solution. Section 6 concludes the paper.

2. Industrial Dynamic Machine Allocation Problem

One of the concerns of the dynamic machine allocation problem is the determination of the optimal sequence of part types considered for production. The solution should consider the unavailability of the machines that were previously allocated. Each part type is characterized by a set of operations with a pre-specified processing time. The objective is to allocate the operations to a set of machines such that some quantifiable measurements can be set up and optimized [4]. These measurements include the minimization of makespan and system unbalance and the maximization of throughput, considering the various technological constraints related to machining time.

Part type selection and machine allocation arrangement constitute two major components of the strategic planning of FMS. Part type selection deals with selection of a set of part types to be manufactured during the upcoming planning horizon, and the loading problem concerns the allocation of operations and the required machines and tools for the selected part types. Most of the earlier researchers managed these problems separately because of their complexities. Therefore, the solution of the part type selection problem may lead to an infeasible result for the loading problem.

There are two types of operations according to each part type. Essential operations of a part type indicate that these operations can be performed only on a particular machine using a certain number of tool slots, whereas optional operations imply that they can be carried out on a number of machines with the same or varying processing time and tool slots. As essential operations require performance at specific machine/s, flexibility lies in the selection of machines that are set for optional operations in which the machine allocation can be improved to give better results.

In this research, the dynamic machine allocation problem was solved in the context of random-type FMS. With the various part types being loaded and having a wide range of processing requirement variations, the decision on the machine allocation should be dynamic as well. The assembly line typically has machines that normally consist of multiple-functional machines, and fixed processing time is associated with it depending on the part types to be assembled. Depending on the product build specifications, certain operations of the part types can be essential, optional, or a combination of both. The essential operations are performed on a specified machine, whereas optional operations can be performed on any of the alternative machines. To denote the different handling of part types during machine breakdown, the part types in this study are categorized as follows [9]:

- Regain part type: In this category, the processing of interrupted operation (essential or optional) is resumed using the same machine immediately after it has been repaired. The processed portion of the job before the breakdown is considered valid.
- Repeat part type: In this case, as it involves the transfer of the job to a different machine, the job before the breakdown is discarded, and the whole operation (essential or optional) has to start from the beginning.

Machine breakdown is categorized into two types, namely, predetermined machine breakdown (PMB) and stochastic machine breakdown (SMB).

3. Model Formulation

3.1. Subscripts and parameters notations

The subscripts and parameters notations used to demonstrate the objective functions are as follows:

Subscripts	
N	Total number of parts
M	Total number of machines
H	Planning horizon
i	Part type, $i = 1, \ldots, N$
m	Machine, $m = 1, \ldots, M$
j	Operation type, $j = 1, \ldots, J$

Parameters	
UT_m	Under-utilized time on machine m
OT_m	Over-utilized time on machine m
β_i	Batch size of part type i
t^a_{im}	Time available on machine type m after the allocation of operation j of part type i
t_{imj}	Time required by machine type m for operation j of part type i
T^r_{im}	Number of machines for each machine type remaining on machine m
T^a_{im}	Number of machines for each machine type available on machine m after the allocation of operation j of part type i
T_{imj}	Number of machines for each machine type required by machine m for operation j of part type i
Y_ij	Set of machine types in which operation j of part type i can be performed after the allocation of operation j of part type i
R^p_m	Repair duration for predetermined breakdown on machine m
R^s_m	Repair duration for stochastic breakdown on machine m
Ts_{imj}	Start time of operation j of part type i on machine m
Te_{imj}	End time of operation j of part type i on machine m
Tsb_{imj}	Time at which stochastic breakdown occur processing of operation j of part type i on machine m
Ts	Slack time (Tsb_{imj} - Ts_{imj})

SU_{seq}	System unbalance for a particular sequence
SU_{max}	Maximum system unbalance
TH_{seq}	Throughput of a particular sequence
TH_{max}	Maximum throughput

3.2. Objective functions

The following equation calculates the cumulative processing time of a machine under breakdown circumstances:

$$CP_m = \sum_{i=1}^{N}\sum_{k=1}^{J} \alpha_i^c[(\alpha_i * t_{imj}) + (Bp_{imj} * R_m^p) + Bs_{imj}(\theta_i + R_m^s + \delta_i * (Ts + Bs_{imj} - t_{imj}))] \tag{1}$$

,

where CP_m (cumulative processing time) is the working time for machine m. Equation 1 clearly shows the negative effect of a breakdown, as it increases the working time. In the event of PMB or regain SMB, the makespan is increased because of the amount of repair time. In the event of repeat SMB, the makespan is increased because of repair time and slack time Ts (as the affected part types are transferred to the nearest available machine for reprocessing).

(i) System unbalance (SU) is the summation of the remaining time or idle time in all available machines. By minimizing SU, the utilization level of the machines will increased. This objective function is represented by notation F_1.

$$\arg\min_{F_1} F_1 = (\frac{SU_{seq}}{SU_{max}}) \tag{2}$$

which means

$$\arg\max_{F_1} F_1 = (\frac{SU_{max} - SU_{seq}}{SU_{max}}) \tag{3}$$

where

$$SU_{max} = \sum_{m=1}^{M} mtm_m \tag{4}$$

$$SU_{seq} = \sum_{m=1}^{M} rtm_m \tag{5}$$

$$rtm_m = mtm_m - CP_m \tag{6}$$

and mtm_m designates the total available time of machine m, and rtm_m indicates the remaining time on machine m. Equation 6 proves that, in the event of PMB and regain SMB, machine available time is reduced by both processing time and repair duration. In the event of repeat SMB, aside from the processing time and repair duration, available time in the machine also decreases because of slack time, as the disrupted part type is transferred to the nearest available machine within the same group for re-machining.

(ii) Throughput is defined as the summation of the batch sizes of the selected part types. More importantly, the maximization of throughput increases the productivity of manufacturing systems.

$$\arg\max_{F_2} F_2 = (\frac{TH_{seq}}{TH_{max}}) \tag{7}$$

where

$$TH_{seq} = \sum_{i=1}^{N} \beta_i * \alpha_i \tag{8}$$

$$TH_{max} = \sum_{i=1}^{N} \beta_i \tag{9}$$

The notation TH_{max} and TH_{seq} represent the throughput of a particular sequence of part types and the maximum throughput (summation of batch size of all part types), respectively.

(iii) The makespan or the cumulative processing time of machines, F_3, is defined as follows:

$$\arg\min_{F_3} F_3 = (\frac{Max[CP_m]}{\sum_{i=1}^{N}\sum_{j=1}^{J} t_{ij}}), \tag{10}$$

which means

$$\arg\max_{F_3} F_3 = (\frac{\sum_{i=1}^{N}\sum_{j=1}^{J} t_{ij} - Max[CP_m]}{\sum_{i=1}^{N}\sum_{j=1}^{J} t_{ij}}) \tag{11}$$

,

where t_{ij} denotes the processing time of operation j of part type i (either selected or rejected).

(iv) The fourth objective function is a combination of all the above objective functions:

$$\arg\max_{F} F = (\frac{W_1(F_1) + W_2(F_2) + W_3(F_3)}{W_1 + W_2 + W_3}) \tag{12}$$

,

where
W_1 Weight assigned to objective function F_1
W_2 Weight assigned to objective function F_2
W_3 Weight assigned to objective function F_3

For simplicity, 1 unit of weight is assigned to each of the objective although in real world, an individual objective may carry a different weight.

4. Proposed Solution

The part type sequence determines the processing arrangement of the long string of part types. The first part type in the sequence is allocated first, followed sequentially by the rest of the part types according to the sequence string. As the industrial dataset has a long sequence, maintaining the uniqueness of the part types takes a long processing time. Therefore, a method needs to be identified in which the processing time should not take too long or should be within the permissible processing time for it to be effective. One of the widely used methods is randomized algorithm in which the input (part type sequence array) is randomized through permutation array [16]. This random re-arrangement of the part types will not disturb or change their uniqueness, as each part type in the sequence is unique.

A sample of the part sequence for dataset CP is shown in Figure 1(a). Based on the sequence part type, the chromosome called the part-chromosome will be created. To ensure the feasibility of the solution, each gene in the part sequence will retrieve the operations assigned to it. As the operation sequence for each part type has to follow the sequence, indicating that the first

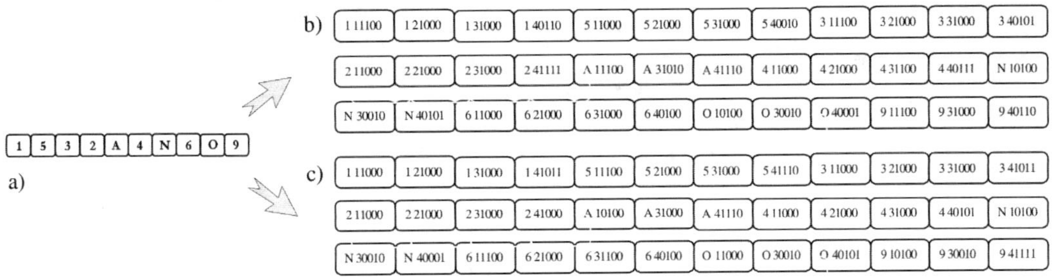

Figure 1. (a) A sample of a part type sequence; (b) and (c) Possible values of a part-operation chromosome for the part type sequence

operation needs to be allocated first, followed by the second operation and so forth, maintaining the sequence is required.

Part types in a dataset can carry a large quantity that should be allocated to either one or more machine types. This requirement enables the design of the gene representing the machines to be allocated as a 4-bit string (maximum of 4 machine types) with "1" denoting the corresponding machine type being allocated, and "0" indicating otherwise (e.g. 1 0 0 1 means machines 1 and 4 being allocated). The number of genes that can be created is 2^n - 1. For example, the operation with 2 machine types will have a maximum of three genes, and the operation with 4 machine types will have 15 possible genes.

Figures 1(b) and (c) two examples of the possible values of the part-operation chromosome based on the part sequence [1 5 3 2 A 4 N 6 O 9]. The first two digits of the values denote the part type and operation number, respectively, followed by the 4-bit string of machine types. As operation 1 of part type 4 is considered an essential operation and can only be performed on a particular machine, there is no change in gene that generates the part-operation-chromosome. Conversely, operation 4 of part type 2 is an optional operation that can be performed on a number of machines with the same or varying process times. In this example, operation 4 can be performed on all four machine types, in which the total processing times for these machine types could vary and lead to a different result.

5. Experimental Results and Discussion

The various combinations of the parameter values are used to conduct a sensitivity analysis of the sample dataset to identify the optimal control parameters for the proposed algorithms. [9] used a population size of 20 to 30 to handle 6 to 14 jobs. Using this as a baseline, for an average of 90 part types, the best population size is 300. In the same work, the crossover range between 0.7 and 0.8 produced better results for a large-sized problem. Therefore, 0.7 and 0.8 for the probability of crossover were tested. For the mutation, the range of 0.1 to 0.3 was taken. The number of generations was set to 75, as earlier experiments show that the increase in number of generations from 50 to 75 does improve the result. Table 1 shows the details of the control parameters for the algorithm.

Five datasets consist of 10 packages, denoted as CP, WP1, WP2, WP3, and WP4 containing 10, 105, 72, 105, and 140 number of parts, were used for the experiments. The CP dataset represents the production plan that intends to be loaded in four operations, and WP1, WP2, WP3 and WP4 represent the work-in-process lots that are already in production, requiring resources to be allocated for the processing of the corresponding operations. Owing to the nature of the dataset, the CP quantity becomes a lump sum of the quantity per package, and the machine processing hours are expected to be shared and loaded for the week (123.25 hours).

Table 1. Control parameters for the algorithm

Control Parameters	Option 1	2	3	4	5	6	7	8	9	10	11	12
Crossover Operator	Ordered Chromosome											
Mutation Operator	Reciprocal Exchange											
Selection Operator	RW						TN					
Probability of Crossover	0.7	0.7	0.7	0.8	0.8	0.8	0.7	0.7	0.7	0.8	0.8	0.8
Probability of Mutation	0.1	0.2	0.3	0.1	0.2	0.3	0.1	0.2	0.3	0.1	0.2	0.3
HMCR	0.85	0.9	0.95	0.85	0.9	0.95	0.85	0.9	0.95	0.85	0.9	0.95
PAR	0.01	0.1	0.2	0.2	0.01	0.1	0.01	0.1	0.2	0.2	0.01	0.1

Number of generations/iterations: 50; Population/Harmony Memory size: 300; RW: Roulette wheel; TN: Tournament
HMCR: Harmony memory consideration rate; PAR: Pitch adjustment rate

Datasets WP1 to WP4 are loaded in shift (7.25 hours). Thus, the same quantity will have the possibility of being allocated with more than one type of machines, considering the fact that one machine type consists of many machines. These part types are scheduled to be processed using 141, 67, 13, 31, and 20 machines, respectively. To demonstrate the robustness of the proposed model of the dynamic machine allocation problem, the proposed H-CCGaHs was used on these five industrial datasets.

Table 2 shows the machine resource summary in which each operation has a number of machine types, and each machine type has a number of machines. Based on the number of machines and processing hours, weekly and shifting remaining time hours (RTM) are calculated, and shifting and weekly predetermined machine breakdowns are calculated based on the probability of PMB. Note that the machine unit processing per hour (UPH) is based on the the operation number, macine type, and product type. The same machine type may vary its UPH if run on a different product type. The processing time allocated to perform the planned quantity is based on the total quantity over the UPH of the machine type.

Table 2. Machine resource summary

Operation No.	No. of Machine Types	Machine No (type)	No. of Machines	Weekly RTM*	Shiftly RTM*	Weekly Predetermined Machine Breakdown	Shiftly Predetermined Machine Breakdown
1	2	1	38	4683.5	275.5	212.89	12.52
		2	39	4806.75	282.75	218.49	12.85
2	1	1	13	1602.25	94.25	72.83	4.28
3	4	1	8	986	58	44.82	2.64
		2	8	986	58	44.82	2.64
		3	8	986	58	44.82	2.64
		4	7	862.75	50.75	39.22	2.31
4	4	1	5	616.25	36.25	28.01	1.65
		2	7	862.75	50.75	39.22	2.31
		3	4	493	29	22.41	1.32
		4	4	493	29	22.41	1.32
Total			141	17378.25	1022.25		

*RTM: Remaining time on machine

This random type FMS under consideration receives several part types with various processing requirements that are to be loaded on different machines. Consider that predetermined machine breakdown occurs on every 1.5 shifts for each machine on average, which is approximately 660 minutes. Under the online monitoring scheme adopted, the maintenance department is assumed to take 30 minutes to repair the failed machines.

SMB is also assumed to occur on probability of 15% of all the machines available and that the maintenance personnel takes a maximum of an hour to diagnose the problem and repair the machine. In the case of the repair time being more than an hour, the machine will be shut down for further investigation, and the part types that are running on that machine will be transferred to a different machine. The probability of this scenario happening is estimated to be 6% of total machines. The transfer time and the set up on the new machine, including buy off is assumed to be 45 minutes on average.

The proposed approaches were implemented using C# compiler. Table 3 shows the results obtained by our proposed algorithm. TH, SU and MS for each dataset, together with number of part types allocated and the number of part types that are not allocated, are shown along with the fitness values. The nature of the datasets causes SU and MS to be higher in dataset CP, although the percentage of TH is relatively low. The reason is that the quantity to be processed for each part type in dataset CP is large and requires a huge block of machine available hours to accommodate such as a big number. As stated previously, the main difference between CP and WPs is that CP is a production plan schedule that consists of a bulk quantity to be processed in the range of the operations, and WPs are the work-in-process (WIP) datasets that are currently awaiting to be processed in the next operation. Based on these results, TH is reduced due to machine downtime as less machine available hours are allocated. Moreover, the numbers of part types assigned (i.e., number of PA) is also reduced.

Table 3. Results on the dynamic machine allocation problem

Dataset	Total Quantity	Tot. Avail. Time	Tot. Proc. Time	TH	SU	MS	COF	No. of part types	No. of PA	No. of PU	Conver-gence	Time to run
CP	10493472	17378.3	19518.4	7766266	5287.2	16686.9	**1.5809**	10	8	2	48	02:17.6
WP1	746031	558.3	881.2	588300	1.0	710.7	**1.9802**	105	81	24	66	13:59.7
WP2	364956	94.3	667.4	85181	0.0	667.4	**1.2332**	72	18	54	44	0:06:36
WP3	681127	224.8	275.2	549521	120.4	223.1	1.4605	100	69	31	39	0:25:13
WP4	994708	145.0	251.5	653640	0.2	212.2	1.8116	140	87	53	43	0:37:31
	13280294			**9642908**			**1.6133**		263	164	47.96	0:17:08

Tot. Avail. Time: Total available time; Tot. Proc. Time: Total processing Time.
TH: Throughput; SU: System Unbalance; MS: Makespan; COF: Combined objective function value.
PA: Part assigned; PU: Part unassigned.

As reported from the previous study [14], H-CCGaHs is a promising approach in to solving the machine allocation optimization problem because it produces the best result with reasonable time taken to complete the task and converge within an acceptable number of generations. The strength of H-CCGaHs relies on the population swapping generated by both GA and HS and on the running against the other algorithm's operators after swapping, thus creating more chances of discovering the unvisited solutions and avoid the creation of a repeated or similar solution. This finding indicates that the time taken to find solutions is not only dependent on the size of the population but also on the tightness of the constraints and the approach used in designing the chromosome representation as well as on the effectiveness of the fitness function. The reason is that a substantial processing time is required to filter non-feasible solutions.

6. Conclusion

Applying the same algorithm to the manufacturing environment can be very challenging. It has its own constraints. The size of the industrial datasets used is typically large. In fact, the sample industrial datasets used in this experiments are only 5% to 10% of the actual manufacturing dataset size, and these are equivalent to more than 10 times of the benchmark datasets in [15]. Therefore, the chromosome representation tends to be very long. The part types in this experiment were denoted as a string of sequence, and no part type duplication was allowed. The sequence determines the processing arrangement of the part types during the initialization of the populations. As the industrial dataset has a long sequence, maintaining the uniqueness of the part types will take a large proportion of the processing time. To overcome this problem, RIP, one of the randomized array introduced by [16], was adopted. As presented in Table **??**, the estimated running time is reduced by up to 510 times compared with using a conventional randomized method.

Inspired by the successful application of intelligent search techniques in the field of combinatorial complex problems, a hybrid of the H-CCGaHs was employed to determine the robustness of the proposed model of the dynamic machine allocation problem. A comparative and analytical study was performed. The excellent results obtained from the case study show that the proposed model can provide a good, effective, and practical solution for industrial application. Extending the research, the superiority of the algorithm was also tested on a larger-sized problem adopted in this research. Moreover, various analyses were performed that would enable the manufacturers to consider the approaches most suitable for their environment to meet their business goals. The results demonstrate the superiority of H-CCGaHs among the applied algorithms in terms of its ability to find the highest overall COFs within a reasonable running time. This work is expected to aid the management in better managing the machine utilization as well as enabling more part types to be scheduled using the current resources.

Acknowledgments

The authors wish to thank Universiti Sains Malaysia for the support it has extended in the completion of the present research through Short Term University Grant No: 304/PKOMP/6313026.

References

[1] Huang D, Yan J and Qiao F 2004 *8th International Conference on Control, Automation, Robotics and Vision* **3** 2217–2222
[2] Nanvala H B and Awari G K 2011 *International Journal of Engineering and Technology* **3** 64–73
[3] Pongcharoen P, Hicks C and Braiden P M 2004 *European journal of Operational Research* **152** 215–225
[4] Kumar A, Prakash, Tiwari M K, Shankar R and Baveja A 2006 *European Journal of Operational Research* **175** 1043–1069
[5] Geem Z 2006 *Improved Harmony Search from Ensemble of Music Players* (*Lecture Notes in Computer Science* vol 4251) (Springer Berlin / Heidelberg)
[6] Al-Betar M and Khader A 2012 *Annals of Operations Research* **194** 3–31
[7] Geem Z W, Kim J H and Loganathan G 2001 *SIMULATION* **76** 60–68
[8] Mukhopadhyay A, Roy A, Das S and Abraham A 2008 *Third International Conference on Digital Information Management, 2008. ICDIM 2008.* pp 775 –781
[9] Mandal S K, Pandey M K and Tiwari M K 2010 *International Journal of Production Research* **48** 3535–3559
[10] Sheu C and Krajewski L 1994 *International Journal of Production Research* **32** 1365–1382
[11] Cakanyildirim M R R 2002 *IIE Transactions* **34** 449–465
[12] Yusof U K, Budiarto R and Deris S 2012 *International Journal of Innovative Computing, Information and Control* **8** 1591–1609
[13] Yusof U K, Budiarto R and Deris S 2011 *Data Mining and Optimization (DMO), 2011 3rd Conference on* pp 26–31 ISSN 2155-6938
[14] Yusof U K, Budiarto R and Deris S 2011 *Proceedings of Sixth International Conference on Bio-Inspired Computing: Theories and Applications, Penang, Malaysia* pp 89–96 ISBN 978-0-7695-4514-1
[15] Mukhopadhyay S K, Midha S and Krishna V M 1992 *International Journal of Production Research* **30** 2213–2228
[16] Cormen T H, Eliserson C E, Rivest R L and Stein C 2008 *Introduction to Algorithms* (Prentice-Hall, Inc.)

IOP Conf. Series: Materials Science and Engineering **83** (2015) 012009 doi:10.1088/1757-899X/83/1/012009

A design method with differentiability for optimal control problems with non-differential Hamilton-Jacobi equations

Joe Imae, Norito Kanda and Tomoaki Kobayashi

Osaka Prefecture University, Osaka, Japan

E-mail: `jimae@osakafu-u.ac.jp`

Abstract. As far as optimal feedback controllers for nonlinear systems are concerned, Hamilton-Jacobi(HJ) equations play a key role in the design process. However, we sometimes meet the HJ equations in non-differential situations, which are extremely difficult to solve analytically/numerically. This paper proposes a simple design method to attack such optimal control problems using differentiable solutions of the associated HJ equations. By simulation, we demonstrate the proposed method useful in a non-differentiable situation.

1. Introduction

As far as optimal feedback controllers for nonlinear systems are concerned, Hamilton-Jacobi(HJ) equations play a key role in the design process. However, we sometimes meet the HJ equations in non-differential situations, such as unicycle or car-like vehicle problems, underwater control problems, bilinear dynamical system problems, and so on. Unfortunately, those HJ equations are extremely difficult to solve analytically/numerically. A lot of researchers have been tackling such non-differentiable control problems. However, no powerful technique has been found so far. Note that, especially in the non-differentiable HJ equations, viscosity solutions [1] are generally used instead of differentiable solutions. As far as viscosity solutions are concerned, no effective computational tool exists so far in solving non-differentiable HJ equations.

This paper proposes a simple design method to attack the optimal control problems with non-differentiable HJ equations, based on two steps. The first step is to find non-differentiable points in the state space (say, Algorithm I). This is not an easy task. We propose a unique idea to find such points in an easy fashion. The second step is to divide a whole state space into subspaces, based on the result given by Algorithm I, where non-differentiable points are not contained. Now that we are in a differentiable situation, we apply the existing HJ solvers for each subspace.

The paper is organized as follows. The optimal control problem in question is formulated in Section 2. In Section 3, the simple but effective design method is proposed with differentiability assumptions. Section 4 demonstrates the usefulness of the proposed method. Section 5 gives the conclusion.

2. Problem formulation

Consider the following dynamical system (1), with the performance index (2) to be minimized.

$$\dot{x} = f(x) + g(x)u \tag{1}$$

$$J = \int_0^\infty (l(x) + u^{\mathrm{T}}Ru)dt \tag{2}$$

where, $x \in \mathbb{R}^n$ is a state vector, and $u \in \mathbb{R}^m$ is an input variable. f and g are smooth functions with $f(0) = 0$.

Even with a standard type of optimal control problems, we sometimes meet the non-differentiable HJ equations in deriving feedback controllers. Of course, there are some approaches to such non-differentiable situations. However, we propose a new technique which is completely different from the existing methods to attack such problems. What we are going to do is to select the non-differentiable points in the state space, and take away those points in order to apply the existing differentiable methods. How can we find these non-differentiable points? This is a major issue that we have to solve. For this reason, we give some preliminaries. In the sequel, we use "Distribution" and "Tangent space". Those are given as follows.

Tangent space: Given a manifold $\mathcal{M}$, construct a subspace at $x \in \mathcal{M}$, using all the tangent vectors of $x \in \mathcal{M}$. The subspace is called "Tangent space", denoted by "$T_x(\mathcal{M})$".

Distribution: Consider a linear variational system, $\delta\dot{x} = A\delta x + \sum_{i=1}^m b_i \delta u_i$, which is given around a point (x, u) with a nonlinear system $\dot{x} = f(x, u)$, and construct a subspace using the following vectors, b_1, Ab_1, $\cdots$, $A^{n-1}b_1$, $\cdots$, $b_m Ab_m$, $\cdots$, $A^{n-1}b_m$. The subspace is called "Distribution", denoted by "Span$(b_1, Ab_1, \cdots, A^{n-1}b_1, \cdots, b_m, Ab_m, \cdots, A^{n-1}b_m)$" or "Span$(\Delta(x))$". Here,

$$\Delta(x) = (b_1, Ab_1, \cdots, A^{n-1}b_1, \cdots, b_m, Ab_m, \cdots, A^{n-1}b_m). \tag{3}$$

Roughly speaking, Distribution is in a sense related to a controllability matrix. By using Distribution, we can check the reachability of the linearized variational systems.

3. Design method

First, we describe how to find the non-differentiable points in the state space. Next, based on these non-differentiable points, we propose our design method dividing the whole state space into the subspaces, where the subspaces do not contain the non-differentiable points. Because of that, we can apply existing methods which are developed for differentiable situations.

3.1. Non-differentiable points

We describe an algorithm to find non-differentiable points.

Algorithm I

Step 0: Set $D_1 = \varnothing$ and $D_2 = \varnothing$.

Step 1: Derive the following variational system around the point (x_l, u_l) with the nonlinear system (1).

$$\delta\dot{x} = A_l \delta x + B_l \delta u \tag{4}$$

Step 2: Construct "Distribution" at (x_l, u_l), using the variational system as mentioned in Section 2.

Step 3: Construct "Tangent space" at x_l, using the manifold with $l(x)$ as mentioned in Section 2.

Step 4: If Tangent space in Step 3 contains Distribution in Step 2, update D_1 by letting x_l become an element of D_1, and go to Step 5. Otherwise, go to Step 5.

Step 5: If every point of (x_l, u_l) is selected, go to Step 6. Otherwise go back to Step 1.

Step 6: If D_1 contains no element or the origin only, update D_1 by re-setting $D_1 = \varnothing$ and go to Step 7. Otherwise, go to Step 9.

Step 7: Construct Distribution at (x_l, u_l) as in Step 2. If Distribution is not equivalent to the whole state space, update D_2 by letting x_l become an element of D_2, and go to Step 8. Otherwise, go to Step 8.

Step 8: If every point of (x_l, u_l) is selected, stop. Otherwise, go back to Step 7.

Step 9: Update D_2 by re-setting $D_2 = \text{span}(\Delta(x))$ with $(x_l, u_l) = (0, 0)$.

3.2. Design Method

Now, we have found all of the non-differentiable points. Therefore, we are in a position to propose a design method for nonlinear systems with non-differentiable HJ equations, based on Algorithm I.

Design Method

Step 1: Apply Algorithm I to the nonlinear system (1) with the performance index (2).

Step 2: Choose a practical region Ω for the control design in the state space, in order to construct a feedback controller over the such a region.

Step 3: Based on D_1 and D_2, divide the practical region into subregions which do not contain non-differentiable points in each region.

Step 4: Apply the Galerkin approximation method [2] to each subregion, together with basis functions appropriately chosen.

Step 5: Construct a feedback controller over the whole practical region, based on differentiable solutions of subregions given in Step 4 of the design method.

4. Simulation

Four examples are given to show how Algorithm I works in a control design process. Two more examples are given to demonstrate how effective the proposed design method works. One of them is shown in this section.

Consider the following bilinear system (5) with the performance index (6)

$$\begin{bmatrix} \dot{x}_1 \\ \dot{x}_2 \end{bmatrix} = \begin{bmatrix} 1 & 0 \\ 0 & x_1 \end{bmatrix} \begin{bmatrix} u_1 \\ u_2 \end{bmatrix} \tag{5}$$

$$J = \int_0^\infty (5x_1^2 + 5x_2^2 + u_1^2 + u_2^2)dt \tag{6}$$

where a viscosity solution exists in the associated HJ equation. As mentioned before, it would be a difficult task to obtain viscosity solutions analytically/numerically. We tackle this problem, according to the design procedure given in section 3. Applying Algorithm I, we obtain D_1 and D_2, as follows.

$$D_1 = \{x_l | x_{l,1} = 0\}, \; D_2 = \{x_l | x_{l,2} = 0\} \tag{7}$$

Based on such D_1 and D_2, we attack an associated HJ equation.

$$-\frac{1}{4}\left(\frac{\partial V}{\partial x_1}\right)^2 - \frac{x_1^2}{4}\left(\frac{\partial V}{\partial x_2}\right)^2 + 5x_1^2 + 5x_2^2 = 0 \tag{8}$$

By the Galerkin approximate method with the practical region $\Omega = \{(x_l, u_l) \mid -1.5 \leq x_1 \leq 1.5, -1.5 \leq x_2 \leq 1.5\}$, we obtain the solution in Fig.1, where non-differentiable points can be shown on the line of $x_1 = 0$. A comparison with the optimal solution is given in Fig.2, where the optimal one is given numerically by the reference [3]. The results show that the proposed and optimal ones are sufficiently close from a practical point of view.

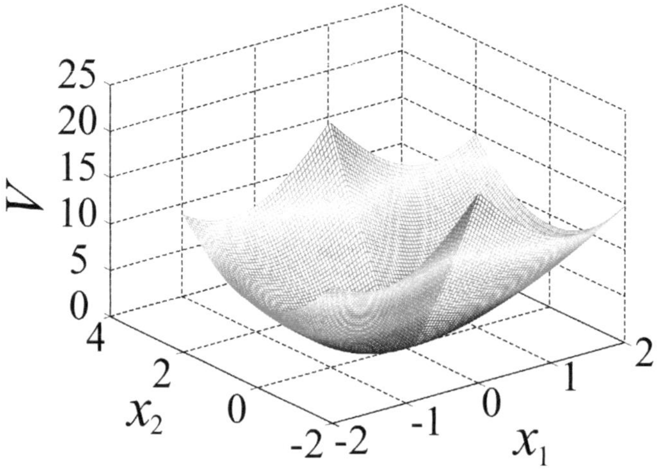

Figure 1. Value of Cost

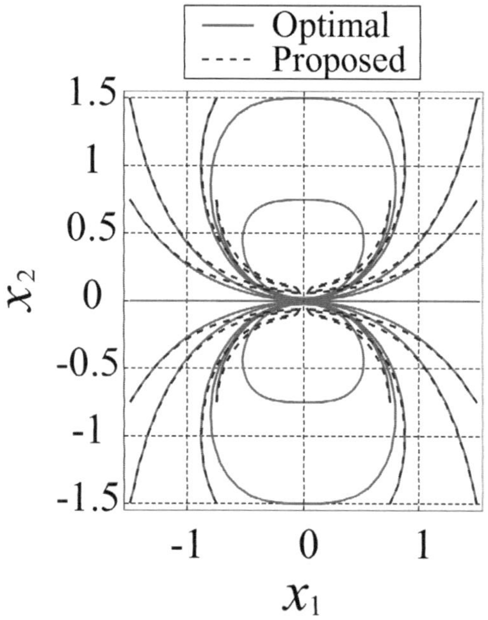

Figure 2. Phase Portrait

5. Conclusion

This paper has proposed an algorithm to find non-differentiable points (Algorithm I), and a design method based on the results given by Algorithm I. Numerical simulation is given in order to demonstrate the usefulness of the proposed method of control design.

This work is supported by JSPS KAKENHI Grant Number 24560551.

References

[1] M. G. Crandall and P. L. Lions: Viscosity Solutions of Hamilton-Jacobi Equations; *American Mathematical Society* (1983)

[2] R. Beard, G. Saridis and J. Wen: Galerkin Approximations of the Generalized Hamilton-Jacobi-Bellman Equation; *Automatica*, Vol. 33, No. 12, pp. 2159-2177 (1997)

[3] J. Imae, R. Torisu: A Riccati-Equation Based Algorithm for Nonlinear Optimal Control Problem; *Proceeding of the 37th IEEE Conference on Decision and Control*, pp. 4422-4427 (1998)

Practical Computation of Flat Outputs for Nonlinear Control Systems

Joe Imae, Yuki Akasawa and Tomoaki Kobayashi
Osaka Prefecture University, Osaka, Japan
E-mail: jimae@osakafu-u.ac.jp

Abstract. In this paper, we deal with a design method for nonlinear systems, focusing on flat outputs of the dynamical systems. One of the main reasons why flat outputs are focused on is as follows. Once flat outputs are found for nonlinear systems, all we have to do is to apply a linear design method to such nonlinear systems, because the nonlinear systems may be transformed identically into the associated linear systems by means of flat outputs. However, the existing computation methods are complicated and not suitable to a control design from a practical point of view. For example, we sometimes meet the integrability issue concerning the flat outputs. This paper proposes a new type of practical computation method about how to circumvent the integrability issue of the flat outputs. Simulation is given for illustrating the usefulness of the proposed design method.

1. Introduction

Most of the mechanical systems are nonlinear systems, such as automobiles, industrial robots, aircrafts, and so on. In this paper, we deal with a design method for nonlinear systems, focusing on flat outputs of the dynamical systems. One of the main reasons why flat outputs are focused on is as follows. Once flat outputs are found for nonlinear systems, all we have to do for a design is to apply a linear design method to nonlinear systems, because the nonlinear systems may be transformed identically into the associated linear systems by using such flat outputs.

However, it should be noted that it is extremely difficult to find flat outputs analytically/numerically. The existing computation methods for flat outputs are complicated and not suitable as a control design from a practical point of view. For example, we have the integrability problem in deriving flat outputs. This paper proposes a new type of computation method for flat outputs, where the computational process for flat outputs is dramatically simplified and easy to implement through Poincare lemma. And also a design example is given in order to demonstrate the practicability of the design method.

The outline of the paper is as follows. In Section 2, some preliminaries are given, such as what the implicit system is like, what the flatness is like, what the flat output is like, and so on. In Section 3, a new type of practical computation method for flat outputs is proposed. In Section4, a design example is shown to illustrate the usefulness of the method, and the conclusion is given in Section 5.

2. Preliminaries

Consider the following nonlinear system.

$$\dot{x} = f(x) + g(x)u, \quad g(x)u = \sum_{i=1}^{m} g_i(x)u_i \tag{1}$$

where a vector $x \in \mathbb{R}^n$ is a state, and a vector $u \in \mathbb{R}^m$ is an input. Here, f and g are smooth functions. And note that, for $i = 1, \cdots, n$, $\frac{d}{dt}x_i^{(j)} \overset{\text{def}}{=} \dot{x}_i^{(j)} = x_i^{(j+1)}$, and $x_i^{(0)} = x_i$.

Note that the equation (1) for nonlinear systems is common in the community of control engineers, because almost of the practical systems are represented in this fashion.

2.1. *Implicit System* [4]

Consider a control system described by implicit functions of $n - m$ equations.

$$F(x, \dot{x}) = 0 \tag{2}$$

where a vector $x \in \mathbb{R}^n$ is a state and $\text{rank}(\frac{\partial F}{\partial \dot{x}}) = n - m$. This system comes from the input elimination of the control system (1). By the implicit function theorem, this elimination is always possible if $\text{rank}\frac{\partial(f(x)+g(x)u)}{\partial u}$ has full column rank m.

2.2. *Flatness* [1], [7]

The system (1) is differentially flat, or shortly flat, if and only if there exists a vector $y = (y_1, \cdots y_m)$ such that

(i) y and its successive time derivatives $\dot{y}, \ddot{y}, \cdots$ are functionally independent,

(ii) y is a function of x , u and a finite number of time derivatives of the components of u,

(iii) x and u can be expressed as functions of the components of y and a finite number of their successive time derivatives

$$x = \varphi(y, \dot{y}, \cdots, y^{\alpha}), \quad u = \psi\left(y, \dot{y}, \cdots, y^{\alpha+1}\right)$$

for some multi-integer, $\alpha = (\alpha_1, \cdots, \alpha_m)$, using the notation

$$y^{(\alpha)} = \left(\frac{d^{\alpha_1}y_1}{dt^{\alpha_1}}, \cdots, \frac{d^{\alpha_m}y_m}{dt^{\alpha_m}}\right).$$

In other words, the implicit system (2) is differentially flat, or shortly flat, if and only if there exist integers s and r, and functions ψ and φ_0 such that

$$\begin{aligned}
&\text{(i)} \quad y = \psi(x, \dot{x}, \cdots, x^s) \\
&\text{(ii)} \quad x = \varphi_0(y, \dot{y}, \cdots, y^r) \\
&\text{(iii)} \quad F\left(\varphi_0, \frac{d\varphi_0}{dt}\right) = 0
\end{aligned}$$

A vector y having these properties is called a flat output. Flatness theory tells us that we can construct a nonlinear controller using the flat output.

2.3. Flat Output [3]
Consider the following matrix, polynomial with respect to the differential operator $\frac{d}{dt}$

$$P(F) = \frac{\partial F}{\partial x} + \frac{\partial F}{\partial \dot{x}}\frac{d}{dt} \tag{3}$$

with $P(F)$ of size $(n-m) \times n$. The matrix (3) constructs the (non-commutative) principal ideal ring of polynomials in $\frac{d}{dt}$, denoted by $\mathcal{M}_{r,s}\left[\frac{d}{dt}\right]$ for $r, s \in \mathbb{N}$. Note that matrices whose inverse belong to $\mathcal{M}_{r,r}\left[\frac{d}{dt}\right]$ are called unimodular matrices, denoted by $\mathcal{U}_r\left[\frac{d}{dt}\right]$. Also, note that $\frac{d}{dt} \cdot x = \dot{x} + x \cdot \frac{d}{dt}$ on the principal ideal ring.

Every matrix admits a Smith decomposition

$$VMU = (\Delta, 0_{r,s-r}) \text{ if } r \leq s, \text{and} \begin{pmatrix} \Delta \\ 0_{r-s,s} \end{pmatrix} \text{ if } r \geq s \tag{4}$$

with $V \in \mathcal{U}_r\left[\frac{d}{dt}\right]$, $U \in \mathcal{U}_s\left[\frac{d}{dt}\right]$, Δ is diagonal. U and V are indeed non-unique. We say that $U \in \mathrm{R-Smith}(M)$ and $V \in \mathrm{L-Smith}(M)$. Note that $M \in \mathcal{M}_{r,s}\left[\frac{d}{dt}\right]$is said hyper-regular if and only if its Smith decomposition leads to $\Delta = I$. From now on, we assume that $P(F)$ is hyper-regular. Also note that (2) is flat if, for $\Theta \in \mathcal{M}_{n,m}\left[\frac{d}{dt}\right]$,

$$P(F)\Theta = 0 \tag{5}$$

Here, $\Theta \in \mathcal{M}_{n,m}\left[\frac{d}{dt}\right]$is given as follows.

$$\Theta = T\begin{pmatrix} 0_{n-m,m} \\ I_m \end{pmatrix} \tag{6}$$

with $T \in \mathrm{R-Smith}(P(F))$. Eventually we obtain

$$P(F)T = (I_{n-m},\ 0_{n-m,m}) \tag{7}$$

which plays a key role in a design algorithm. Because Θ is hyper-regular, we also obtain by Smith decomposition

$$S\Theta = ST\begin{pmatrix} 0_{n-m,m} \\ I_m \end{pmatrix} = S\hat{T} = \begin{pmatrix} I_m \\ 0_{n-m,m} \end{pmatrix} \tag{8}$$

where $S \in \mathcal{U}_n\left[\frac{d}{dt}\right]$ and $\hat{T} = T\begin{pmatrix} 0_{n-m,m} \\ I_m \end{pmatrix}$. And we denote $\tilde{S} = (I_m, 0_{m,n-m})\, S$. These matrices play a key role in a design algorithm.

Remark: We denote by ω the m-dimensional vector 1-form, defined by

$$\omega = \begin{pmatrix} \omega_1 \\ \vdots \\ \omega_m \end{pmatrix} = \tilde{S}dx \tag{9}$$

where operator d of dx indicates "exterior derivative". In case of $d\omega = 0$, we can obtain the flat output y by integrating ω by Poincare lemma, because $dy = \omega$.

3. Design Method

As mentioned in Section 2, we have to pay attention to the principal ideal ring $\mathfrak{K}\left[\frac{d}{dt}\right]$, which is troublesome for control engineers from a practical point of view. In this section, we propose a simple and practical computation process for Smith decomposition. Moreover, we pay attention to Poincare lemma as follows. Define $\tilde{S}$ like this

$$\tilde{S} = \begin{pmatrix} S_{0(1,1)} & \cdots & S_{0(1,n)} \\ \vdots & \ddots & \vdots \\ S_{0(m,1)} & \cdots & S_{0(m,n)} \end{pmatrix} \tag{10}$$

and rewrite $d(\tilde{S}dx) = 0$ into

$$\left(\frac{\partial S_{0(k,q)}}{\partial x_p} - \frac{\partial S_{0(k,p)}}{\partial x_q}\right) = 0 \tag{11}$$

where, $p = 1, \cdots, n-1, q = 2, \cdots, n, q > p, k = 1, \cdots, m$. Then ω is integrable.

Consider the following nonlinear system.

$$\dot{x} = f(x) + g(x)u, \quad g(x)u = \sum_{i=1}^{m} g_i(x)u_i \tag{12}$$

A design algorithm is as follows.

Algorithm:

Step 1: Transform the nonlinear control system (12) into the implicit system $F(x, \dot{x}) = 0$.

Step 2: Set $P_0 = \frac{\partial F}{\partial x}$, $P_1 = \frac{\partial F}{\partial \dot{x}}$, denote$P(F)$ by $P(F) = P_0 + P_1\frac{d}{dt}$.

Step 3: $i = 1$.

Step 4: Set $n \times n$ matrices of $T_0, T_1, \cdots, T_i$ as follows.

$$T_0 = \begin{pmatrix} T_{0(1,1)} & \cdots & T_{0(1,n)} \\ \vdots & \ddots & \vdots \\ T_{0(n,1)} & \cdots & T_{0(n,n)} \end{pmatrix}$$

$$T_1 = \begin{pmatrix} T_{1(1,1)} & \cdots & T_{1(1,n)} \\ \vdots & \ddots & \vdots \\ T_{1(n,1)} & \cdots & T_{1(n,n)} \end{pmatrix}$$

$$\vdots$$

$$T_i = \begin{pmatrix} T_{i(1,1)} & \cdots & T_{i(1,n)} \\ \vdots & \ddots & \vdots \\ T_{i(n,1)} & \cdots & T_{i(n,n)} \end{pmatrix}$$

Using those matrices, denote the $n \times n$ matrix T by

$$T = T_0 + T_1\frac{d}{dt} + \cdots + T_i\frac{d^i}{dt^i}.$$

Step 5: Obtain $P(F) \times T$ like this

$$\begin{aligned} P(F) \times T &= \left(P_0 + P_1\frac{d}{dt}\right) \times \left(T_0 + T_1\frac{d}{dt} + \cdots + T_i\frac{d^i}{dt^i}\right) \\ &= P_0T_0 + P_1\dot{T}_0 + \left(P_0T_1 + P_1T_0 + P_1\dot{T}_1\right)\frac{d}{dt} + \left(P_0T_2 + P_1T_1 + P_1\dot{T}_2\right)\frac{d^2}{dt^2} \\ &+ \cdots + \left(P_0T_i + P_1T_{i-1} + P_1\dot{T}_i\right)\frac{d^i}{dt^i} + P_1T_i\frac{d^{i+1}}{dt^{i+1}} \end{aligned}$$

For simplicity of expression, we denote $P(F)T$ by

$$P(F)T = \lambda_0(T) + \lambda_1(T)\frac{d}{dt} + \lambda_2(T)\frac{d^2}{dt^2} + \cdots + \lambda_{i+1}(T)\frac{d^{i+1}}{dt^{i+1}}.$$

Step 6: Because system is flat, we obtain from (7) that

$$P(F)T \equiv (I_{n-m},\ 0_{n-m,m}).$$

Therefore, using

$$\lambda_0(T) = (I_{n-m},\ 0_{n-m,m}),\ \ \lambda_1(T) = 0_{n-m,n}, \cdots, \lambda_{i+1}(T) = 0_{n-m,n},$$

we obtain T_0, T_1, T_2, $\cdots$, T_i in Step 4, and go to Step 7. If not, go back to Step 4, resulting $i = i + 1$.

Step 7: Set $j = 1$ and compute $\hat{T}$ by $\hat{T} = T\begin{pmatrix} 0_{n-m,m} \\ I_m \end{pmatrix}$.

Step 8: Set $n \times n$ matrices of $S_0, S_1, \cdots, S_j$ as follows.

$$S_0 = \begin{pmatrix} S_{0(1,1)} & \cdots & S_{0(1,n)} \\ \vdots & \ddots & \vdots \\ S_{0(n,1)} & \cdots & S_{0(n,n)} \end{pmatrix}$$

$$S_1 = \begin{pmatrix} S_{1(1,1)} & \cdots & S_{1(1,n)} \\ \vdots & \ddots & \vdots \\ S_{1(n,1)} & \cdots & S_{1(n,n)} \end{pmatrix}$$

$$\vdots$$

$$S_j = \begin{pmatrix} S_{j(1,1)} & \cdots & S_{j(1,n)} \\ \vdots & \ddots & \vdots \\ S_{j(n,1)} & \cdots & S_{j(n,n)} \end{pmatrix}$$

Using those matrices, denote the $n \times n$ matrix S by

$$S = S_0 + S_1\frac{d}{dt} + \cdots + S_j\frac{d^j}{dt^j}.$$

Step 9: Obtain $S \times \hat{T}$. For simplicity of expression, we denote $S\hat{T}$ by

$$S\hat{T} = \bar{\lambda}_0(S) + \bar{\lambda}_1(S)\frac{d}{dt} + \bar{\lambda}_2(S)\frac{d^2}{dt^2} + \cdots + \bar{\lambda}_{i+j}(S)\frac{d^{(i+j)}}{dt^{(i+j)}}.$$

Step 10: Because Θ is hyper-regular from (8), we obtain that

$$S\hat{T} \equiv \begin{pmatrix} I_m \\ 0_{n-m,m} \end{pmatrix}.$$

Therefore, using

$$\bar{\lambda}_0(S) = \begin{pmatrix} I_m \\ 0_{n-m,m} \end{pmatrix}, \ \bar{\lambda}_1(S) = 0_{n,m}, \cdots, \bar{\lambda}_{i+j}(S) = 0_{n,m}, \left(\frac{\partial S_{0(k,q)}}{\partial x_p} - \frac{\partial S_{0(k,p)}}{\partial x_q} \right) = 0,$$

we obtain S_0, S_1, S_2, $\cdots$, S_j in Step 8, and go to Step 11. If not, go back to Step 8, resulting in $j = j + 1$.

Step 11: Compute $\tilde{S}$ by $\tilde{S} = (I_m, 0_{m,n-m})\, S$.

Step 12: Compute ω by $\omega = \tilde{S}dx$.

Step 13: Integrate ω to obtain the flat output.

Step 14: With the flat output, we design a controller according to the reference [6].

4. Simulation

In this section, we demonstrate how effective our design algorithm is.

Consider a satellite model [4]

$$\begin{aligned} \dot{x}_1 &= \kappa_1 x_2 x_3 + u_1 \\ \dot{x}_2 &= \kappa_2 x_1 x_3 + u_2 \\ \dot{x}_3 &= \kappa_3 x_1 x_2 \end{aligned} \tag{13}$$

where the coordinates x_i, $(i = 1, 2, 3)$ represent the angular velocity of i-axis, and the inputs u_i, $(i = 1, 2)$ represent the torque given by gas jets, together with parameters κ_1, κ_2, κ_3 being positive. Because $n = 3$, $m = 2$, $n - m = 1$, and rank$\frac{\partial(f(x)+g(x)u)}{\partial u} = 2$, the nonlinear system (13) can be transformed into an implicit system. Now, we apply to design algorithm.

Step1: Transform the nonlinear control system (13) into the implicit system

$$F(x, \dot{x}) = \dot{x}_3 - \kappa_3 x_1 x_2 = 0 \tag{14}$$

Step2: Set $P_0 = \frac{\partial F}{\partial x}$, $P_1 = \frac{\partial F}{\partial \dot{x}}$, and denote$P(F)$ by $P(F) = P_0 + P_1 \frac{d}{dt}$. with

$$P_0 = (-\kappa_3 x_2, \ -\kappa_3 x_1, \ 0), \quad P_0 = (0, \ 0, \ 1) \tag{15}$$

Step3: Set $i = 1$.

Step4: Set $n \times n$ matrices of T_0, T_1 as follows.

$$T_0 = \begin{pmatrix} T_{0(1,1)} & \cdots & T_{0(1,3)} \\ \vdots & \ddots & \vdots \\ T_{0(3,1)} & \cdots & T_{0(3,3)} \end{pmatrix}, \quad T_1 = \begin{pmatrix} T_{1(1,1)} & \cdots & T_{1(1,3)} \\ \vdots & \ddots & \vdots \\ T_{1(3,1)} & \cdots & T_{1(3,3)} \end{pmatrix} \tag{16}$$

Using those matrices, denote the $n \times n$ matrix T by

$$T = T_0 + T_1 \frac{d}{dt}.$$

Step5: Obtain $P(F) \times T$. For simplicity of expression, we denote $P(F)T$ by

$$P(F)T = \lambda_0(T) + \lambda_1(T)\frac{d}{dt} + \lambda_2(T)\frac{d^2}{dt^2}.$$

Step6: Because the system is flat, we obtain from (7) that

$$P(F)T \equiv (I_1,\ 0_{1,2}).$$

Therefore, using

$$\lambda_0(T) = (I_1,\ 0_{1,2}), \lambda_1(T) = 0_{1,3}, \lambda_2(T) = 0_{1,3},$$

we obtain T_0, T_1 in Step 4 as follows.

$$T_0 = \begin{pmatrix} -\frac{1}{\kappa_3 x_2} & -\frac{x_1}{x_2} & 0 \\ 0 & 1 & 0 \\ 0 & 0 & 1 \end{pmatrix}, \quad T_1 = \begin{pmatrix} 0 & 0 & \frac{1}{\kappa_3 x_2} \\ 0 & 0 & 0 \\ 0 & 0 & 0 \end{pmatrix} \tag{17}$$

Step7: Set $j = 1$ and $\hat{T}$ by $\hat{T} = T \begin{pmatrix} 0_{n-m,m} \\ I_m \end{pmatrix}$. We obtain

$$\hat{T} = \begin{pmatrix} -\frac{x_1}{x_2} & \frac{1}{\kappa_3 x_2}\frac{d}{dt} \\ 1 & 0 \\ 0 & 1 \end{pmatrix}. \tag{18}$$

Step8: Set 3×3 matrices of S_0, S_1 as follows.

$$S_0 = \begin{pmatrix} S_{0(1,1)} & \cdots & S_{0(1,3)} \\ \vdots & \ddots & \vdots \\ S_{0(3,1)} & \cdots & S_{0(3,3)} \end{pmatrix}, \quad S_1 = \begin{pmatrix} S_{1(1,1)} & \cdots & S_{1(1,3)} \\ \vdots & \ddots & \vdots \\ S_{1(3,1)} & \cdots & S_{1(3,3)} \end{pmatrix} \tag{19}$$

Using those matrices, denote the 3×3 matrix S by

$$S = S_0 + S_1\frac{d}{dt}.$$

Step9: Obtain $S \times \hat{T}$. For simplicity of expression, we denote $S\hat{T}$ by

$$S\hat{T} = \bar{\lambda}_0(S) + \bar{\lambda}_1(S)\frac{d}{dt} + \bar{\lambda}_2(S)\frac{d^2}{dt^2}.$$

Step10: We obtain from (8) that

$$S\hat{T} \equiv \begin{pmatrix} I_2 \\ 0_{1,2} \end{pmatrix}. \tag{20}$$

Therefore, using

$$\bar{\lambda}_0(S) = \begin{pmatrix} I_2 \\ 0_{1,2} \end{pmatrix}, \ \bar{\lambda}_1(S) = 0_{3,2}, \ \bar{\lambda}_2(S) = 0_{3,2}, \ \left(\frac{\partial S_{0(k,q)}}{\partial x_p} - \frac{\partial S_{0(k,p)}}{\partial x_q}\right) = 0,$$

we obtain S_0 and S_1 in Step 8 as follows.

$$S_0 = \begin{pmatrix} 0 & 1 & 0 \\ 0 & 0 & 1 \\ -\kappa_3 x_2 & -\kappa_3 x_1 & 0 \end{pmatrix}, \quad S_1 = \begin{pmatrix} 0 & 0 & 0 \\ 0 & 0 & 0 \\ 0 & 0 & 1 \end{pmatrix} \tag{21}$$

Step11: Compute $\tilde{S}$ by $\tilde{S} = (I_2, 0_{2,1})\, S$.

$$\tilde{S} = \begin{pmatrix} 0 & 1 & 0 \\ 0 & 0 & 1 \end{pmatrix} \tag{22}$$

Step12: Compute ω by $\omega = \tilde{S}dx$.

$$\begin{pmatrix} \omega_1 \\ \omega_2 \end{pmatrix} = \begin{pmatrix} dx_2 \\ dx_3 \end{pmatrix} \tag{23}$$

That is,

$$d\omega_1 = d(dx_2) = 0$$
$$d\omega_2 = d(dx_3) = 0$$

Step13: We obtain the flat output by integrating ω.

$$y_1 = x_2, \quad y_2 = x_3 \tag{24}$$

Step14: According to reference [6], we obtain the controller as follows.

$$\begin{pmatrix} u_1 \\ u_2 \end{pmatrix} = \beta'(v - \alpha) \tag{25}$$

while, $(\kappa_1, \kappa_2, \kappa_3) = (1, 1, 1)$, and

$$\begin{aligned} \alpha &= \begin{pmatrix} x_1x_3 \\ x_2^2x_3 + x_1^2x_3 \end{pmatrix}, \quad \beta' = \begin{pmatrix} -\frac{x_1}{x_2} & \frac{1}{x_2} \\ 1 & 0 \end{pmatrix}, \\ v &= \begin{pmatrix} \dot{y}_1^* - a_{10}(y_1 - y_1^*) \\ \ddot{y}_2^* - a_{21}(\dot{y}_2 - \dot{y}_2^*) - a_{20}(y_2 - y_2^*) \end{pmatrix}. \end{aligned} \tag{26}$$

Simulation is shown in Fig. 1 and Fig. 2, where initial state $x_0 = [2, 2, 2]$, and target state $x^* = [0, 1, 0]$. This shows trajectories converge to the target state. This means our design algorithm is powerful in a practical situation.

5. Conclusion

A simple and practical computation method for flat outputs of nonlinear systems is proposed. Compared with the existing methods, our method is dramatically easy to implement and so surprisingly suitable to control engineers from a practical point of view. Note that the integrability with flat outputs is tackled. Simulation shows the practicability of our method through the satellite control problem.

References

[1] F. Antritter, J. Lévine: Towards a Computer Algebraic Algorithm for Flat Output Determination, *In Proc. of ISSAC'08*, Hagenberg, Austria, (2008)

[2] J. Lévine: On necessary and sufficient conditions for differential flatness, *Applicable Algebra Eng. Commun. Comput.*, Vol. 22, No. 1, pp. 47-90, (2011)

[3] F. Antritter, J. Lévine: Flatness Characterization: Two approaches, *Lecture Notes in Control and Information Sciences*, Vol. 407, pp. 127-139, (2010)

[4] M. Franke, K. Röbenack: On the Computation of Flat Outputs for Nonlinear Control Systems, *2013 European Control Conference*, pp. 167-172, (2013)

[5] K. Schlacher, M. Schöoberl: Construction of flat outputs by reduction and elimination, *in Proc. IFAC Symposium in Nonlinear Control Systems*, NOLCOS, Pretoria, Vol. 7, (2007)

[6] H. Sira-Ramírez, S. K. Agrawal: Differentially Flat Systems;*CRC Press* (2004)

[7] J. Lévine: Analysis and Control of Nonlinear Systems: A Flatness-based Approach; *Springer*, (2009)

IOP Conf. Series: Materials Science and Engineering **83** (2015) 012010 doi:10.1088/1757-899X/83/1/012010

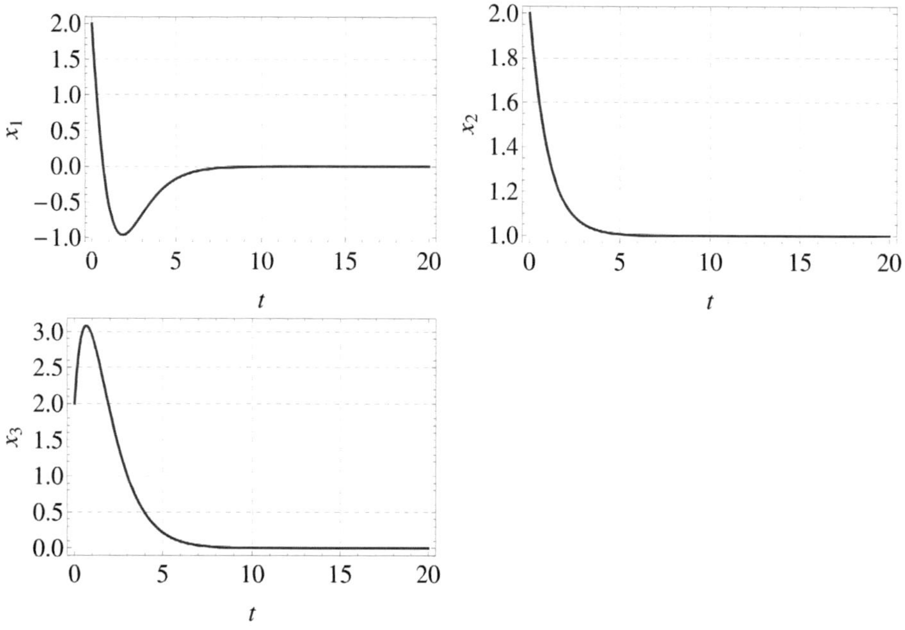

Figure 1. State trajectories.

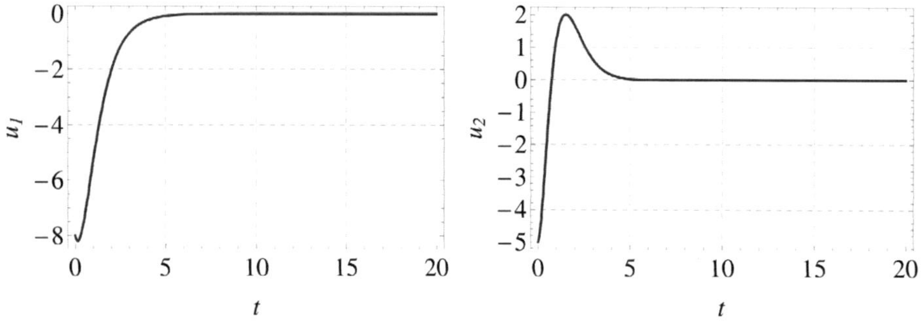

Figure 2. Inputs.

Identification and reduction of vibration and noise of a glass tempering system

M. S. Ashhab

Mechanical Engineering Department, Hashemite University
Zarqa 13115, Jordan

E-mail: sami@hu.edu.jo

Abstract. The vibration and noise of a glass tempering machine at a factory are studied. Experiments were conducted to identify the sources of vibration and noise. It was found that main sources for vibration and noise are two air barrels, the air pipes from the fans to the glass tempering machine and the fans location. Solutions were suggested to reduce vibration and noise from these three main sources. One of the solutions that were implemented is placing rubber dampers beneath the air barrels and pipes which almost cancelled the horizontal vibrations in the building structure and reduced the vertical vibrations to a low value most likely coming from noise. There are two types of noise, namely, radiation noise from the fans through the fans room walls and transmitted noise through the pipes caused by turbulence. A glass wool noise insulating layer was installed on the wall between the fans room and factory to reduce radiation noise through this wall. Part of the air pipe system in the factory is made of a light material which produced the highest levels of noise above 110 dBA. These air pipes were wrapped by glass wool rolls and the noise level near them was reduced to below 100 dBA which comes from other machine parts. In addition, noise levels were reduced between 2 and 15 dBA at different points in the factory.

1. Introduction

One of the major limitations at many industries is vibration and noise produced by machines. The problem of vibration and noise even extends to all kinds of buildings and automotive applications. Therefore, analysis of vibration and noise has been always an important issue for researchers. The solution or the reduction of these two factors is a mind bugling problem. The research presented in [1] proposes a novel tuned mass-damper-inerter passive vibration control configuration, to suppress the oscillatory motion of a support excited mechanical chain-like systems. The suggested technique improves performance. Identification of noises coming out from modern household refrigerators using Fourier analyses has been conducted in [2]. Solutions to reduce the noise level were suggested. Smart damping materials such as magnetorheological elastomers for noise and vibration in automotive applications were explored in [3]. In the research presented in [4], a new method is proposed in order to quickly estimate the noise emission in free space based on an approximated acoustic power of brake system. Recent research in the area of two-phase flow induced vibration in piping system can be found in [5].

The glass tempering system at a factory produced huge and undesired amount of noise and vibrations. This was really a dangerous situation because it could over the long run cause hearing damage to the factory staff and structural damage to the glass tempering system and the building containing the system. In addition, the performance of the glass tempering system could be deteriorated substantially and the glass could break or its properties would correspondingly differ from the desired and expected optimum ones. Therefore, it was extremely important to reduce the noise and vibrations to a minimum level through applying the appropriate and possible structural changes and noise control techniques.

The glass tempering system consists of three fans, pipe structure and rollers structure over which glass plates moves. The fans are located in a separate room and they pump air through the pipe structure to the glass. The pipe structure passes through two large barrels (air tanks) to regulate the air flow. Air reaches the glass through many small holes from both top and bottom.

Fans, pipe structure, air tanks, glass tempering system structure and air path were all possible noise and vibration sources. The huge amount of noise and vibrations could have a single source or a combination of sources. Therefore, the cause of the noise and vibration problem is being found and isolated appropriately. Possible noise and vibration sources were tested and analyzed. The solution to the problem requires knowledge and experience in engineering analysis and involves mechanical vibrations, control systems, noise control, instrumentation, fluid mechanics and dynamics.

The factory map including fans room, air tanks, air pipes, glass tempering machine and offices is shown in Figure 1. Note that the offices are located on the top floor and other parts are located on the bottom floor. Noise and vibration is generated in the bottom floor and propagates to the top floor.

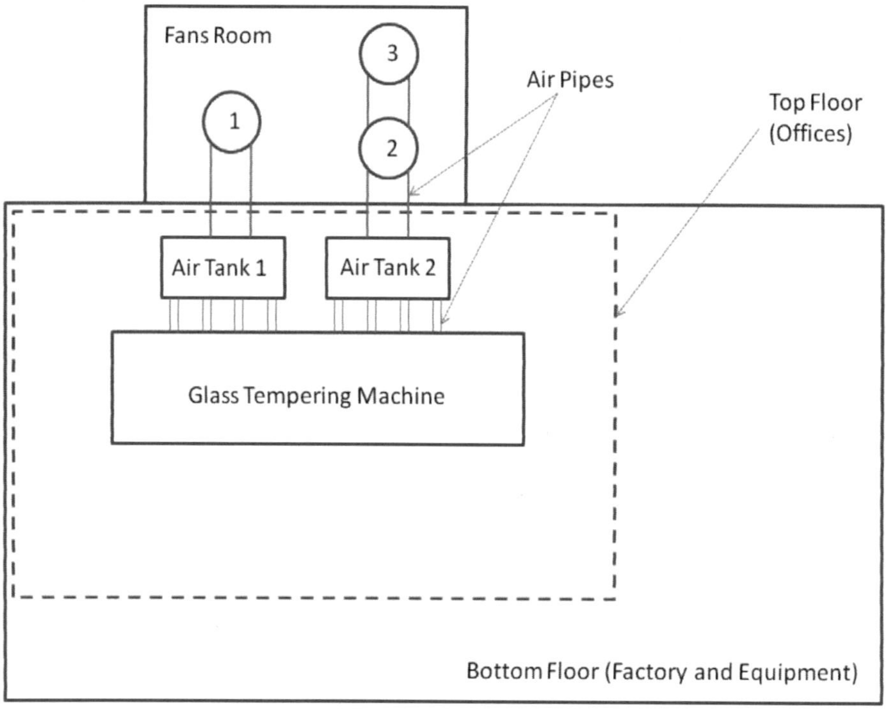

Figure 1. Factory map.

2. Identification of vibration and noise sources

It is extremely important to reduce the noise and vibration at the factory to a minimum level through applying the appropriate and possible structural changes and noise control techniques. The causes of the noise and vibration problem were found and are being isolated appropriately. This goal was achieved by applying several different techniques. Using noise and vibration measuring instruments to locate the problem sources through appropriate tests was vital. However, the instruments locations and how to operate the system during the tests were two important issues to solve. This depended on experience, understanding the system structure and engineering analysis which involves mechanical vibrations, control systems, noise control, instrumentation, fluid mechanics and dynamics. Another approach used was common sense which was supported by strong and logical reasoning. As testing and analysis of the noise and vibration sources proceeded new ideas and approaches were concluded towards solving the problem.

The following are the noise and vibration sources that were studied and analyzed

1) Different pressure at different locations

The air pressure out of the small holes and hitting the glass could vary with location. This would generate noise and vibrations due to the different forces at different locations of the glass tempering machine structure. This problem was tackled by measuring the air pressure at the different outlet air pipes. The pressure was equal at all outlets and therefore there is no discrepancy and this was not considered as a possible cause for noise and vibrations.

2) Different fans geometry including air path

Each fan was operated alone while separating its pipe structure from the rest of the fans. The problem persisted for each fan operating separately. Therefore, all the fans were causing noise and vibrations.

3) Pipe structure imbalance

The path through which air flows from the fan to the glass suffered from severe vibrations due to the air motion and inappropriate path shape and structure. It was found that the two barrels in the air path were a major cause of vibration. A measuring device (accelerometer) was used to determine the sources of mechanical vibrations in this case. Engineering analysis was used to suggest and implement appropriate solutions to reduce the vibrations to a minimum level.

4) Fan rotating imbalance

Rotating imbalance occurs due to imperfect load distribution of the fan propeller and it causes huge amount of vertical vibrations. It was found that all the fans had dampers at their bases or mounts (contact points with ground) which absorbed the large amount of vibrations. This was verified by the experiments discussed later.

5) Fan horizontal vibrations

Fan horizontal vibrations occur due to the conservation of momentum principle. When the fan pumps air out it is being pushed backwards. In addition, the pressure built up in the air pipes puts a force on the fans which could shake them severely. Two solutions were suggested to this vibration and noise cause. The first suggestion was to remove the connections between the factory building and fans that are affected by horizontal vibrations. This step was possible and it was implemented. The second suggestion was using vibration and noise reduction materials at the walls between the factory building and fans room which should reduce the noise levels. Note that the main source for noise is the fans since they contain the actuating parts, the motors.

3. Experimentation and measurements

Data about the glass tempering system relevant to the noise and vibration problem were collected. The data included the weights and dimensions of the system components in addition to the rotational speeds of the fans. The collected data along with engineering principles helped in finding solutions to the noise and vibration problem.

Experiments were conducted based on measuring devices, engineering analysis and logical reasoning to locate the vibration and noise sources precisely so that appropriate actions could be taken to reduce the levels of vibration and noise to the minimum possible situation. The measuring devices used were an accelerometer to measure the vibration level and a pressure gage to measure the air pressure. Other required data was available and did not need any measuring devices.

In the first experiment the air pressure was measured out of the small holes and hitting the glass. There are 19440 holes which is a very large number. We selected several different locations covering the whole area of the system to measure the air pressure and decided if the pressure did not change at the measured locations then it would not change at all locations. Indeed, the pressure was equal at all measured outlets and therefore there was no discrepancy and the pressure was not considered as a possible cause for vibration and noise.

In the second experiment the vibration level at many locations of the structure of the glass tempering system was measured using the accelerometer as shown in Figure 2. The locations included points on the ground and points at upper levels. The points on the ground are more important for determining mechanical vibrations transfer to the building. On the other hand the points on upper levels help in understanding how the vibrations propagate and in case the vibrations are large noise could be produced.

Figure 2. Picture showing vibration measurement using the accelerometer.

The vibration measurements were all very small (zero or very close to zero) except at the ground mounts (or bases) of the two large barrels through which air passes from the fans to the glass. The barrels task is to regulate the air flow to the glass through adjustable valves inside them. The ground mounts on which the barrels sit have rectangular shapes, made of steel and stand directly on the ground without any vibration attenuating components as can be seen in the picture of Figure 5. It was concluded that the barrels were the main source of mechanical vibrations in the building.

The accelerometer measures the mechanical vibration amplitude in m/s^2. Each fan has 12 mounting points which are isolated with dampers. There are three fans and therefore we have 36 mounts. The values recorded at the mounts are 0 at half of the mounts, 0.1 m/s^2 at 2 mounts, 0.2 m/s^2 at 2 mounts, 0.4 m/s^2 at 3 mounts, 0.5 m/s^2 at 3 mounts, 0.8 m/s^2 at 2 mounts, 0.9 m/s^2 at 2 mounts, 1.0 m/s^2 at 3 mounts, 1.1 m/s^2 at 1 mount, and 1.2 m/s^2 at 1 mount. In addition, the mechanical vibration level was measured at 7 different location at the barrels bases. All the readings were 17 m/s^2 on average which indicates clearly that the mechanical vibration problem is due to the barrel shaking.

4. Reduction of vibration and noise

In order to eliminate the vibrations from the barrels we needed to place appropriate dampers under the barrels ground mounts so that the vibrations are minimized and not transferred to the building. In order to obtain the appropriate dampers the weights of the two barrels and the sizes of the ground mounts were measured. A top view of the two barrels is shown in Figure 3. The first barrel weighs 1.1 tons and has two rectangular mounts each of width 8 cm and length 1.09 m. The second barrel weighs 1.2 tons and has two rectangular mounts each of width 48 cm and length 2.19 m.

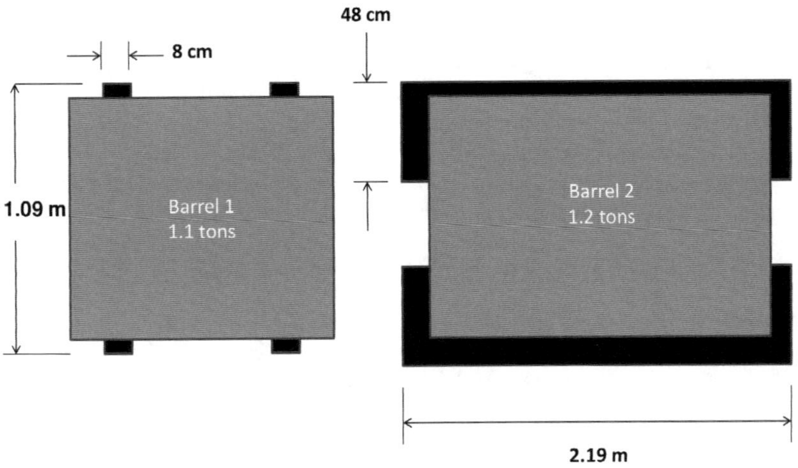

Figure 3. Top view of the two barrels of the glass tempering system.

Dampers suppliers were contacted to check for available dampers and their properties. Suitable rubber dampers were found and bought. The dampers come in two sizes. The first size is 40x40x5 cm and can carry up to 5 tons. The second size is 10x53x5 cm and can carry up to 3 tons. These rubber dampers could be cut to have the appropriate sizes for the glass tempering system barrels. For barrel 1 two dampers of the size 10x53x5 cm were bought and cut into four pieces of size 10x26.5x5 cm which form four legs for the barrel. For barrel 2 three dampers of the size 40x40x5 cm were bought and cut into six pieces of size 40x20x5 cm which form six legs for the barrel. A picture of the two different sizes of dampers after cutting is shown in Figure 4.

The dampers were placed under the two barrels mounts as shown in the pictures of Figures 5 - 7. It was a challenging problem to lift the two barrels that were connected to other glass tempering system components. The best way to lift them was to use the available overhead crane and strong cables which were hooked to the barrels. First we placed the 10x26.5x5 cm rubber dampers beneath tank or barrel 1 and then we repeated the same procedure for tank 2 with the 40x20x5 cm dampers. After lifting each barrel, placing the dampers and releasing the barrel on the dampers the barrel did not go back to the leveled position and we had to level the barrel using the overhead crane with cables. Eventually, we had the two barrels leveled on the dampers.

The vibration level was measured on the barrel bases after placing the rubber dampers and was reduced from an amplitude of 17 m/s^2 to 7 m/s^2, which is a substantial reduction. Note that the mechanical vibration to the ground will be much less that 7 m/s2 since the damper is located below the base at which we measured the mechanical vibrations. In case of the fans mounts the metal base where the mechanical vibrations were measured is below the damper and this is why we get smaller vibrations. After placing the dampers beneath the barrel bases the mechanical vibrations in the factory were not felt near the tempering machine but some shaking was still felt at offices of the factory which are located on the top floor.

The factory is made of steel structure composed of hollow circular steel beams which are covered by concrete at many locations. One of these locations is the wall between the fans and the barrels. The air path from the fans to the barrels is made of steel and used to sit on the concrete without damping components. This was obviously a major vibration source that transfer vibrations through the steel structure and had to be isolated. The mechanical vibrations were measured on the uncovered steel circular bars in the factory and offices using the accelerometer to see how the vibrations propagate. All the vertical vibrations on the circular bars were measured as 0 m/s2 which meant there were no vertical vibrations at all. We also measured the vertical vibrations on the offices floor and it was also 0. On the other hand, the circular steel bars were under horizontal vibrations. These vibrations were measured as 2.5 m/s^2 in the factory where the steel bars are away from the wall, and 12 and 16.5 m/s^2 at the offices whose wall is an extension of the wall between the fans and the glass tempering machines. Note that the steel structure is uncovered with concrete in the offices and this is why we could measure the horizontal and vertical vibrations there. It was concluded that the mechanical vibrations source was in the horizontal direction from the steel structure.

Figure 4. Picture of the two different sizes of rubber dampers after cutting.

In order to eliminate the horizontal vibrations from the steel structure two 40x40x5 cm rubber dampers were bought. Each damper was split into two 20x40x5 cm parts that were placed beneath the air path at the connection point with the wall as shown in Figure 8. Other surrounding areas have a small gap between the air path and wall. After placing these dampers the horizontal mechanical vibrations were measured as 0 m/s^2 in the factory (was 2.5 m/s^2 previously) and 0 and 2.5 m/s^2

(previously were 12 and 16.5 m/s^2, respectively). The situation became more comfortable in the offices and the mechanical vibrations were reduced a lot.

There is still something causing some shaking in the offices and the noise level was loud. The fans for the glass tempering machine produce huge amount of noise and are located in a closed room outside the factory building that shares one wall with the factory. This is the same wall through which the air pipe to the barrels pass and which extends to the offices. It is believed that since the sound level in the room is high, the room is closed and there are no noise insulators, the noise spreads through the factory and offices and could be also propagating through the hollow beams of the steel structure. The noise levels were measured by a sound level meter and were found as more than 110 dBA (or 130 dB based on linear combination of frequencies) in the factory and 73 dBA (or 85 dB based on linear combination of frequencies) in the offices, respectively, which are large. These amounts exceed the allowed limits of 85 dBA in industry and 55 dBA in offices. It became clear that the fans room should be noise isolated.

There are two types of noise, namely, radiation noise from the fans through the fans room walls and transmitted noise through the pipes caused by turbulence. A glass wool noise insulating layer of density 60 kg/m^3 was installed on the wall between the fans room and factory containing the glass tempering machine to reduce radiation noise through this wall. After implementing this step the largest amount of noise was trapped in part of the air pipe system made of a light material which produced noise levels of above 110 dBA. These air pipes were wrapped by glass wool rolls of density 32 kg/m^3 as shown in Figure 9 and the noise level near them was reduced to below 100 dBA which comes from other machine parts. In addition, noise levels were reduced between 2 and 15 dBA at different points in the factory. Transmitted and radiation noise are present partially from other locations which were identified. Reduction of noise levels from these locations requires covering the open areas of the glass tempering machine and thin air pipes with glass wool in addition to possibly glass wool insulation of the other fans room walls.

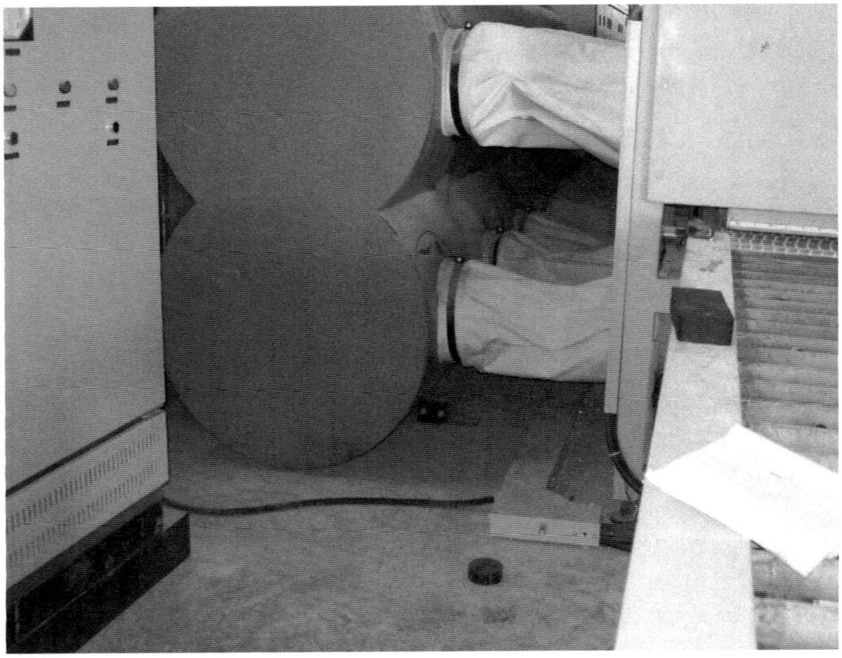

Figure 5. Picture of the barrels without dampers.

Figure 6. Picture of lifting the barrel.

Figure 7. Picture of the barrel base with the damper placed beneath it.

Figure 8. Picture of dampers beneath the air path at the connection point with the wall.

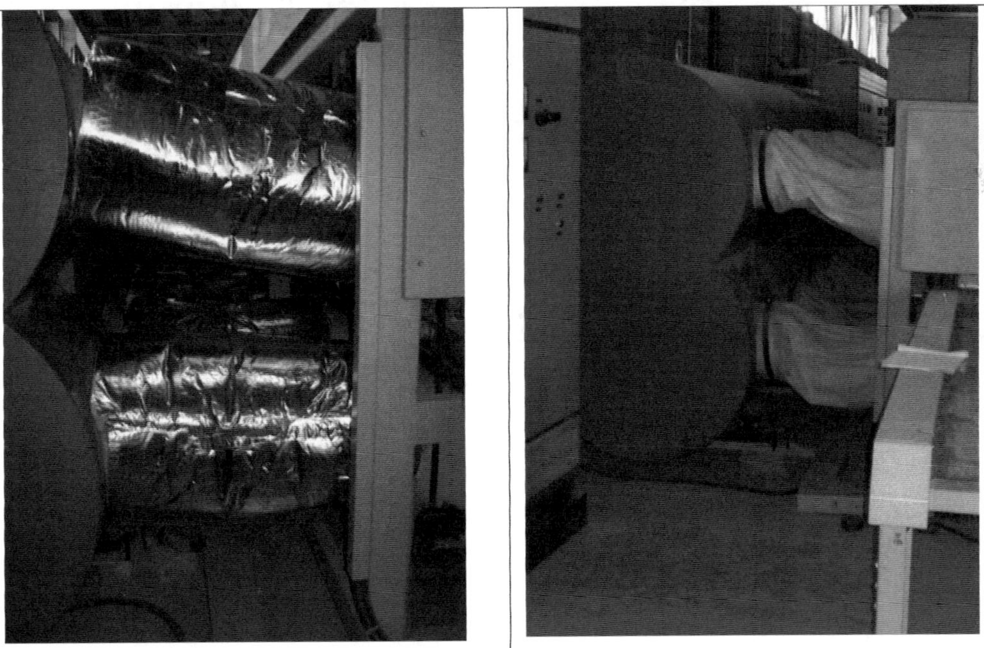

After glass wool insulation	Before glass wool insulation

Figure 9. Picture of glass wool insulation of pipes made of light material.

5. Conclusion and future work

Placing the rubber dampers beneath the two air tanks (barrels) and air pipes was a very important step in alleviating the mechanical vibrations transfer to a factory building and reducing the noise level in a glass tempering system. In addition noise insulation of the wall shared between the fans room and factory, and the light air pipes from the barrels to the glass tempering machine was a crucial step to reduce the noise levels in the range 2 – 15 dBA at different locations in the factory. In regards to the vibration of the factory building, the main causing source was the transmission of vibrations through the two large air tanks and the air pipes sitting on the wall. These sources were eliminated. Nevertheless, it is worthwhile to look for other minor sources of mechanical vibrations and if they exist, they will be eliminated by applying the appropriate engineering solutions.

It was concluded from experiments done so far and logical reasoning that remaining noise sources are the open areas of the glass tempering machine and the thin parts of the air pipes. The open areas will be sealed with glass wool and the thin pipe parts will be insulated with glass wool rolls. This step should reduce the noise levels substantially. Other walls of the fans room will be insulated with glass wool if noise levels are still not acceptable.

References

[1] Marian, L., Giaralis, A., Optimal design of a novel tuned mass-damper-inerter (TMDI) passive vibration control configuration for stochastically support-excited structural systems, Probabilistic Engineering Mechanics, 38, 2015, pp. 156–164.

[2] Koruk, H., Arisoy, A., Identification of crack noises in household refrigerators, Applied Acoustics, 89, 2015, pp. 234-243.

[3] Kumbhar, S. R., Maji, S., Kumar, B., Automotive vibration and noise control using smart materials: a state of art and challenges, World Journal of Engineering, 11, 2014, pp. 413-420.

[4] Soobbarayen, K., Besset, S., Sinou, J. J., A simplified approach for the calculation of acoustic emission in the case of friction-induced noise and vibration, Mechanical Systems and Signal Processing, 50-51, 2015, 732-756.

[5] Miwa, S., Mori, M., Hibiki, T., Two-phase flow induced vibration in piping systems, Progress in Nuclear Energy, 78, 2015, 270-284.

Application of Ester based Drilling Fluid for Shale Gas Drilling

Arina Sauki[1], Mohamad Safwan Zazarli Shah[2], Wan Zairani Wan Bakar[3]

Faculty of Chemical Engineering, Universiti Teknologi MARA, 40450 Shah Alam, Selangor, Malaysia

E-mail: [1]arina_sauki@salam.uitm.edu.my, [2]safwanzazarli@gmail.com, [3]zairani@salam.uitm.edu.my

Abstract. Water based mud is the most commonly used mud in drilling operation. However, it is ineffective when dealing with water-sensitive shale that can lead to shale hydration, consequently wellbore instability is compromised. The alternative way to deal with this kind of shale is using synthetic-based mud (SBM) or oil-based mud (OBM). OBM is the best option in terms of technical requirement. Nevertheless, it is toxic and will create environmental problems when it is discharged to onshore or offshore environment. SBM is safer than the OBM. The aim of this research is to formulate a drilling mud system that can carry out its essential functions for shale gas drilling to avoid borehole instability. Ester based SBM has been chosen for the mud formulation. The ester used is methyl-ester C12-C14 derived from palm oil. The best formulation of ester-based drilling fluid was selected by manipulating the oil-water ratio content in the mud which are 70/30, 80/20 and 90/10 respectively. The feasibility of using this mud for shale gas drilling was investigated by measuring the rheological properties, shale reactivity and toxicity of the mud and the results were compared with a few types of OBM and WBM. The best rheological performance can be seen at 80/20 oil-water ratio of ester based mud. The findings revealed that the rheological performance of ester based mud is comparable with the excellent performance of sarapar based OBM and about 80% better than the WBM in terms of fluid loss. Apart from that, it is less toxic than other types of OBM which can maintain 60% prawn's survival even after 96 hours exposure in 100,000 ppm of mud concentration in artificial seawater.

1. Introduction

It is estimated that there are large reserves of unconventional gas reservoirs throughout the world such as coalbed methane, shale gas and tight gas sand. Generally, reservoir that contains only free gas as the hydrocarbon source is called as gas reservoir. Gas in the reservoir can be classified as 'dry', 'wet' or 'condensate'. The condition is depends on its composition, pressure and temperature at which the accumulation occurred [1]. Shale gas is generally stored in the form of free compressed gas in open pores and cracks. It is stored as adsorbed gas on the surfaces of kerogen or clay minerals. Besides that, the gas has been stored as diffused gas within solid organic matter [2].

---

[1] arina_sauki@salam.uitm.edu.my

Nowadays, drilling in shale gas formation is expanding rapidly especially in the south central of United States. Some examples of shale gas fields in United States are Barnett shale, Appalachian basin and Rocky Mountain basin. While outside the United States, there is no commercial shale gas projects currently exist but the work for exploration of new shale gas reservoirs is continues to growth [3]. Drilling fluid is used in order to assist the drilling of borehole into the earth. A proper formulation of drilling fluid will enable operator to drill into a desired geological aim at low cost. Drilling fluid should be able to enhance penetration rates, reduce hole problems and minimize damages of formation.

The primary functions of drilling fluid are to control formation pressure, transport cuttings and maintain stable wellbore. Force exerted by a fluid column is called hydrostatic pressure and it depends on the mud density and true vertical depth. Drilling fluid controls subsurface pressure by this hydrostatic pressure. One of the essential functions of drilling fluid is cleaning the drilling hole. During drilling operation, cuttings must be removed from the wellbore. Drilling fluid will be circulated down through the drill bit and it will transport cuttings up through the annulus to the surface. Generally, drill solids have a specific gravity (SG) of 2.3 until 3.0 SG and assuming the average is 2.5 SG. When the mud used during drilling operations is lighter than the drill solids, the drill solids will be slipped downward through the mud [4]. Drilling fluid can be categorized into water-based, oil-based and synthetic-based mud. Each type of based mud has its own advantages and disadvantages. Generally the uses of drilling fluid depend on the well condition.

The formation which is water-sensitive may require oil-based and synthetic-based mud. A proper formulation of OBM can prevent water movement from the mud into the shale. Despite of its effectiveness, oil-based mud can give negative impact to environment when the pollutant is discharged and subsequently dispersed to the sea. The cuttings from oil-based mud do not disperse as much as water-based mud when it is discharged under water. It will form piles of cuttings that blanket parts of seabed. This condition may affect the bottom-dwelling organisms close to the rig [5,6]. The aim of this research is to formulate a synthetic based mud for shale gas drilling. Ester is used as the drilling fluid based.

2. Methodology

The methodology for the measurement of drilling fluid properties was constructed based on American Petroleum Institute (API) Recommended Practice 13B. The recommended practice provides standard procedures for determining the physical and rheological properties of oil-based mud. The measurable properties are drilling fluid density, viscosity, gel strength and filtration test. Other formulation and tests were conducted based on past literature.

2.1. Formulation of Ester based Mud

Three different formulations were designed to determine the best combination of Ester based mud at different oil-water ratio. The formulated ester based mud consists of ester as the base fluid, emulsifier, viscosifier, salt as salinity, lime as alkalinity, fluid loss control additive and weighting agent. The density of the samples was kept constant which is about 10.7 ppg and the formulations are shown in Table 1. Among the three formulations, the best formulation was selected based on the comparison of rheological performance of sarapar-based mud and water-based mud which was designed about the same density. The formulation of SBM and WBM were prepared as in Table 2.

Table 1. Formulation of Ester Based Mud

	Oil Water Ratios (OWR)		
Component	70/30	80/20	90/10
Ester	245.00 ml	280.00 ml	315.00 ml
Water	105.00 ml	70.00 ml	35.00 ml
Primary Emulsifier	10.00 g	10.00 g	10.00 g
Secondary Emulsifier	3.00 g	3.00 g	3.00 g
Viscosifier	5.00 g	5.00 g	5.00 g
Calcium chloride	26.25 g	17.50 g	8.75 g
Lime	10.00 g	10.00 g	10.00 g
Barite	195.00 g	195.00 g	195.00 g

Table 2. Formulation for other types of mud

	Sarapar-Based Mud
Sarapar-147	280.00 ml
Primary Emulsifier	10.00 g
Secondary Emulsifier	3.00 g
Viscosifier	5.00 g
Lime	10.00 g
Water	70.00 ml
Calcium chloride	17.50 g
Barite	195.00 g
	Water-Based Mud (WBM)
Water	330.00 g
Starch	4.20 g
Xantham Gum	2.00 g
Potassium Chloride	9.90 g
Sodium Hydroxide	4.00 g
Barite	185.00 g

2.2. *Rheological Test*

Prior to rheological test, mud was mixed using high shear blade mixer. The rheological test was then carried out using Fann viscometer to find the plastic viscosity (PV), yield point (YP) and gel strength of the mud samples in Table 1 and 2. Yield Point (YP) is the stress required to move drilling fluid. It is used to determine the ability of mud to lift cuttings out of the annulus. Plastic viscosity (PV) is a measurement of resistance of fluid to flow. For PV and YP analysis, the sample was stirred at 600 rpm. The reading was stabilized at 600 rpm before proceed to 200, 100, 6 and 3 rpm speed respectively. The results from viscometer reading were used to measure PV and YP value using Eq. 1 and Eq. 2.

$$\text{Plastic Viscosity, Centi Poise (CP)} = \text{Reading at 600rpm} - \text{Reading at 300rpm} \quad (1)$$

$$\text{Yield Point, } (\text{lb}/100\text{ft}^2) = \text{PV} - \text{Reading at 300rpm} \quad (2)$$

Gel strength was measured at time 10 seconds and 10 minutes. It is the shear stress of drilling mud that is measured at low shear rate after the mud had been static for certain time period. The properties are important to investigate the ability of drilling mud to suspend cuttings and weighting material when circulation is stopped. For gel strength analysis, the sample was stirred at 600 rpm for 30 seconds before taking 10-second gel strength at 3 rpm. The mud was stirred again at 600 rpm and let it undisturbed for 10 minutes. Then, the gel strength reading for 10 minutes was taken at 3 rpm.

2.3. API Filter Press Test

Filter press test was carried out in order to determine the fluid loss of mud at ambient temperature and 100 psi pressure. The equipment used is API filter press that consists of filtration cell, OFI special hardened filter paper, low pressure supply and measuring cylinder. The apparatus must be assembled correctly before the test is carried out. The filter paper must be ensured in a good condition. Mud sample must be put into the cell until it is ¾ full. Pressure of 100 psi was applied to the cell and the fluid loss reading was taken in measuring cylinder every three minutes period of time until it reached 30 minutes.

2.4. Shale Dispersion Test

The test was designed to simulate a shale dispersion scenario when drilling into shale gas zone. The test was started by drying the shale sample in the oven at 60°C. Then, the shale sample was screened through 6.73 mm and 2.38 mm sieve sizes respectively. The shale sample retained on 2.38mm sieve screen was weighed and put into a glass jar containing sample of mud for 48 hours. After 48 hours, the mud containing shale was poured onto the sieve screen. Then, the shale samples retained at 2.38 mm sieve screen were washed and blotted. The samples were then dried in the oven at 60°C. The percent of sample recovery was calculated using Eq. 3:

$$\%\,Recovery = \frac{Weight\ of\ sample\ retained}{Initial\ weight\ of\ sample} \times 100\% \quad (3)$$

2.5. Toxicity Test

The purpose of the test is to investigate the impact of ester based mud on marine life. Local marine life namely tiger prawn (*Penaeus Monodon*) was used. The Lethal Concentration (LC_{50}) was identified in which mud concentration killed 50% of exposed prawns after 96 hours of exposure. The LC_{50} value less than 100 ppm is considered as toxic material. The formulated mud was diluted with a ratio of 1:9 (mud : artificial seawater, v/v). It was mixed for about 30 minutes before settling for one hour. The mixture was used to make different concentrations of toxicant environment for the prawns such as 100, 1000, 10000 and 100000 ppm respectively. Ten prawns were exposed to each concentration for 96 hours and the observation was done throughout the period. The methodology used for this test was done according to past literature work [6].

3. Results and Discussion

3.1. Rheological Properties for Different OWR

Rheological properties are important to determine the effectiveness of drilling fluid in order to carry out its main functions. Table 3 shows the comparison of rheological properties of ester based mud at different oil-water ratio (OWR). From the results obtained, it was observed that the mud with oil-water ratio of 80/20 is the best candidate for the selection. It shows good rheological properties as the mud has a medium range of viscosity which is not too thick as compared with 70/30 OWR and not too thin as 90/10 OWR mud system. Besides, the plastic viscosity and yield point values of the 80/20 mud system are about the range of commonly used drilling mud with the plastic viscosity/yield point ratio of between 0.8 to 1.5 [7]. Even though the 70/30 OWR mud system has the lowest fluid loss (Figure 1), but the viscosity was too high and unstable to be used in drilling operation. The system cannot tolerate more contaminants and restrict extra loading of weighing material for higher mud density [8]. Mud with high gel strength is not favorable in drilling operation as it requires high pumping pressure in order to break circulation after the mud is static for long period. The required 10 min gel strength is typically less than 20 lb/100ft^2 in which the 80/20 and 90/10 systems passed the limit [7]. The knowledge of gel strength development with time is so important to detect any possible rheological problem in a mud system [9]. Progressive gels are undesirable condition occurred when there is wide different range between the 10-second and 10-minute gel readings. Besides, high-flat gels are also undesired circumstance in which both the 10-second and 10-minute readings are high with no significant difference. Progressive gel can be seen in 70/30 mud system and it was waived for selection. Excessive gel strength may increase the swabbing and surging effect during tripping process of drillpipe. Yield point is used to determine the ability of mud to lift cuttings out of the annulus. From Table 3, YP values decrease dramatically with decreasing of water ratio. The mud with high YP can carry cuttings better than mud with similar density but lower YP [9]. The finding shows YP value of 80/20 mud system was higher than 90/10 mud system. Thus, 80/20 OWR system is the best candidate and was used in the next experiments.

Table 3. Rheological properties for different OWR

OWR	70/30	80/20	90/10
600 rpm	95.0	51.0	35.0
300 rpm	69.0	36.0	27.0
PV	26.0	15.0	8.0
YP	43.0	21.0	19.0
Gel 10 s	18.0	9.0	5.0
Gel 10 min	24.0	16.0	6.0
Mud Weight (ppg)	10.7	10.7	10.7
Fluid Loss @ 30min (ml)	2.8	3.9	5.4

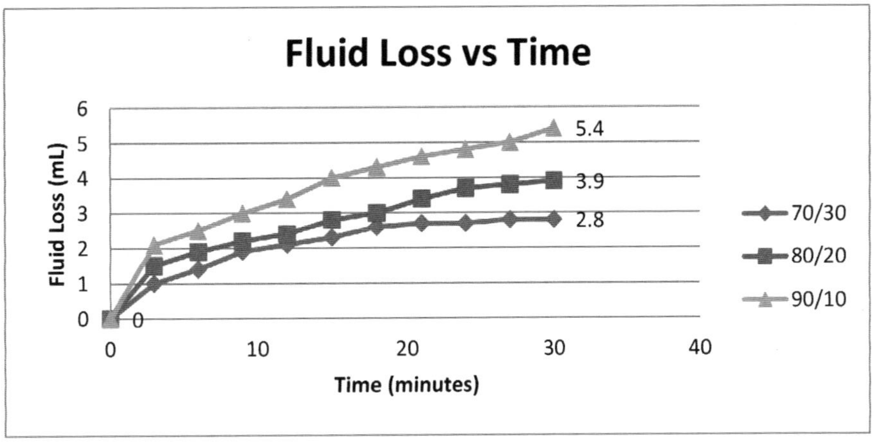

Figure 1. Fluid Loss for Different OWR

3.2. Comparison with Other Types of Mud

One of the objectives of this research is to design drilling fluid for shale gas drilling using ester as the base fluid. The research is focusing on the performance and compatibility of ester to act as the base fluid. The performance of ester-based fluid was compared with other base fluids i.e. water-based mud (WBM) and sarapar-based mud which are usually used for drilling wells and the results are shown in Table 4.

Table 4. Rheological properties for three different types of mud

Types of Mud	Ester-based	Sarapar-based	Water-based
600 rpm	51.0	63.0	55.0
300 rpm	36.0	45.0	39.0
PV	15.0	18.0	16.0
YP	21.0	27.0	23.0
Gel 10 s	9.0	14.0	15.0
Gel 10 min	16.0	21.0	17.0
Mud Weight (ppg)	10.7	10.9	10.6
Fluid Loss @ 30min (ml)	3.9	3.5	17.8

From Table 4, the performance of all three muds is nearly same. However, the PV of ester-based mud is slightly lower than sarapar-based and water-based. A low PV value shows the capability of the mud to assist drilling rapidly which will reduce the time consumption for drilling a well. Overall, the rheological properties of ester-based mud did not show significant different as compared to sarapar-based mud. Sarapar is commonly used drilling mud base in drilling industries. As such, it proved that ester can also be used as an alternative for drilling fluid base. Nevertheless, large amount of fluid loss (17.8 ml) was found in WBM as observed in Figure 2. It can be postulated that the WBM is not suitable for shale gas drilling as the filtrate will react with shale and may cause shale swelling. Fluid loss performance in ester based and sarapar based mud is about the same and competitive.

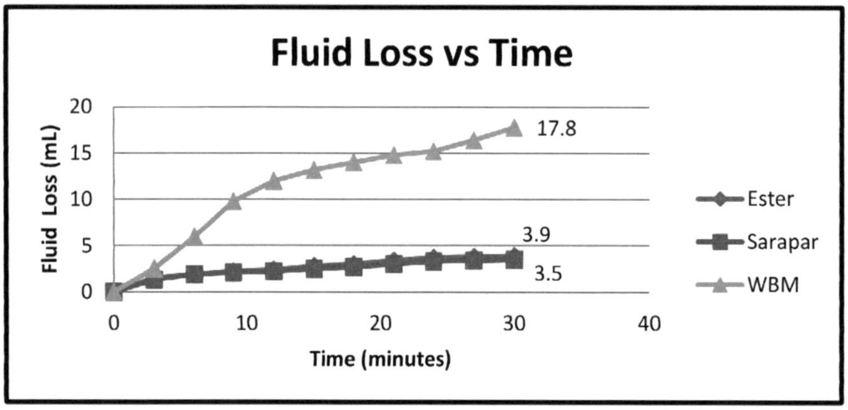

Figure 2. Fluid loss for different types of mud

3.3. *Shale Dispersion Test*

There have been numerous techniques were constructed to develop this test at different condition and feasibility [10]. In this paper, a new method was proposed to simulate shale-mud reaction in the borehole. Small size of shale cutting was used to promote larger surface area for the reaction and prevent the aggregation effect. Table 5 shows the percent recovery of shale cuttings in ester-based mud is slightly higher than the sarapar-based mud. It can be claimed that the higher recovery was associated with lesser reactivity of mud with shale. The findings revealed that ester based mud can reduce about 2.25% of shale reactivity as compared with sarapar based mud and is more effective when dealing with water-sensitive shale. It can also prove that the added Emulsified brine phase (calcium chloride) acts synergistically with ester based mud to preserve cuttings integrity and reduce water activity by creating osmotic forces in order to prevent adsorption of water by shale [5,8].

Table 5. Shale dispersion test

	Ester	Sarapar
Weight of sample used, gram (Size between 2.38 mm to 6.73 mm)	16.34	14.69
Weight of retained sample, gram (On 2.38 mm sieve screen)	15.89	13.95
Percent of Recovery, %	97.25	95.00

3.4. *Toxicity Test*

Toxicity test was carried out in order to observe the effect of ester-based mud towards the environment. Past research had been carried out to investigate the toxicity of diesel mud and palm oil mud. The methodology of the past research was used in this experiment [6]. The results from this experiment were compared with the past research result as shown in Table 6.

Table 6. Survival of prawns from toxicity test

	Survival (%)		
Mud Concentration	**Ester Mud**	**Palm Mud[6]**	**Diesel Mud[6]**
100,000 ppm	60	70	0
10,000 ppm	100	100	0
1,000 ppm	100	100	0
100 ppm	100	100	30

The result shows the comparison of prawn's survival when exposed to different types of mud and concentrations. Ester-based and palm oil based mud show slightly different result than diesel mud. The prawns started to die when exposed to 100,000 ppm concentration. After 96-hours period of time, four prawns had been observed dead and result in 60% of survival. When compared with past research, ester-based mud has nearly same percentage of survival with palm oil mud. The palm oil mud has 70% of prawn's survival after 96 hours of exposure in mud. The expected LC_{50} value could be more than 100,000 ppm which can be considered as non-toxic. In contrast, diesel mud shows the highest toxicity result with zero percent of survival starting from 1,000 ppm concentration. It could achieve LC_{50} value less than 100 ppm and can be considered as toxic material. Nowadays, some of the countries have banned the use of Oil based Mud for their E&P activities such as USA, United Kingdom, Holland, Norway, Nigeria, European countries, Saudi Arabia, and Qatar [8]. In some areas, the use of oil based mud is permitted if the used mud and cuttings are disposed in an approved disposal site. This research contributes to the preliminary knowledge of using ester based drilling fluid as a green drilling fluid for shale gas drilling.

4. Conclusion

As for conclusion, the best ester mud formulation has OWR of 80/20 which contains 80% ester and 20% water. The rheological properties of ester-based mud have no significance different with sarapar-based and water-based mud which is suitable to be used as a base fluid for drilling process. Minimal fluid loss can be observed in ester based formulation as compared to water based mud although nearly the same as sarapar-based mud. Ester based mud could be the most excellent mud to be used in shale gas drilling as it is less dispersed in shale as compared to sarapar based mud. Moreover, it was proved to be non-toxic with the expected LC_{50} of more than 100,000 ppm of mud concentration. It can be concluded that ester can be an alternative base fluid for drilling fluid especially for gas shale drilling.

5. Acknowledgment

The authors are grateful to the Faculty of Chemical Engineering, Universiti Teknologi MARA (UiTM) for the support and facilities. A million thank to our sponsor, RAGS Grant (RAGS/2012/UITM/TK05/11) for the financial support throughout the research project. Also special thanks to SCOMI GRTC and Emery Oleochemicals (M) Sdn. Bhd. for the facilitation and support on the mud additives and Ester.

6. References

[1] Chen, S., Zhu, Y., Wang, H., Liu, H., Wei, W., & Fang, J. (2011). Shale gas reservoir characterisation: A typical case in the southern Sichuan Basin of China. *Energy, 36*(11), 6609-6616.

[2] Hua Tian, Shuichang Zhang, Shaobo Liu, & Jianping Chen, R. I. o. P. E. a. D., PetroChina. (2013). *Overmature Shale Gas Storage Capacity Evaluation*. Paper presented at the 6th International Petroleum Technology Conference, Mar 26 - 28, 2013 2013, Beijing, China.

[3] Creties D. Jenkins, S., DeGolyer and MacNaughton, & Charles M. Boyer II, S., Schlumberger. (2008). Coalbed- and Shale-Gas Reservoirs. *Journal of Petroleum Technology, Volume 60, Number 2*, 92-99

[4] Nasser, J., Jesil, A., Mohiuddin, T., Ruqeshi, M. A., Devi, G., & Mohataram, S. (2013). Experimental Investigation of Drilling Fluid Performance as Nanoparticles. *World Journal of Nano Science and Engineering, 03*(03), 57-61Knaus, E., Killen, J., Biglarbigi, K., & Crawford, P. (2010). An Overview of Oil Shale Resources. *1032*, 3-20.

[5] Seang, A. t. k., Ismail, I., Ismail, A. R., & Rashid, K. A. (2001). The Comparison of Saraline, Sarapar and Diesel Performances as Base Oil at High Temperature and High Pressure.

[6] Dr. Abu Azam Md. Yassin et al.,(1991). Formulation Of An Environmentally Safe Oil Based Drilling Fluid

[7] Eckhout, D., Dolan, S., Gogan, R., Ledgister, H., Mowat, C., Tipton, P., Dye, W. (2000). *Development Process and Field Applications of a New Ester-based Mud System for ERD Wells on Australia's Northwest Shelf.* Paper presented at the IADC/SPE Asia Pacific Drilling Technology, Kuala Lumpur, Malaysia.

[8] R.A.M. Amin, D.K. Clapper, J.E. Norfleet, M.J. Otto, T. Xiang, C.A. Hood, J.E. Goodson, D.P. Gerrard. (2010). *Joint Development of an Environmentally Acceptable Ester-Based Drilling Fluid.* Paper presented at the Trinidad and Tobago Energy Resources Conference held in Port of Spain, Trinidad, 27-30 June 2010.

[9] M. Amani, M. Al-Jubouri, A. Shadravan. (2012). *Comparative Study of Using Oil-Based Mud Versus Water-Based Mud in HPHT Fields.* Advances in Petroleum Exploration and Development.Vol. 4, No. 2, 2012, pp. 18-27.

[10] Khodja, Mohamed and Canselier, Jean-Paul and Bergaya, Faiza and Fourar, Karim and Khodja, Malika and Cohaut, Nathalie and Benmounah, Abdelbaki. (2010). *Shale problems and waterbased drilling fluid optimisation in the Hassi Messaoud Algerian oil field.* Applied Clay Science, vol. 49. pp. 383-393. ISSN 0169-1317.

IOP Conf. Series: Materials Science and Engineering **83** (2015) 012013 doi:10.1088/1757-899X/83/1/012013

Hata's Path Loss Model Calibration for Prediction DTTV Propagation in Urban Area of Southern Thailand

P Keawbunsong[1] P Supanakoon[2] and S Promwong[3]

[1]D.Eng Student, Faculty of Engineering (Telecommunication), KMITL, Thailand
[2]Asst. Professor, Faculty of Engineering (Telecommunication), KMITL, Thailand
[3]Professor, Faculty of Engineering (Telecommunication), KMITL, Thailand

E-mail : pitak_k1@hotmail.com

Abstract. This article presents Hata's path loss model calibration in order to predict a design of the Digital Terrestrial Television (DTTV) Propagation in an urban area of the south of Thailand through measuring power signal of the network operators' broadcasting in 4 channels within Haadyai urban area, Songkla Province. The chosen location is a density area, a distance of 2.5-6.5 km. from the broadcasting station. The calibration was conducted through a statistical method of Root Mean Square Error (RMSE) from received power signal and compared with a path loss model from the prediction, followed by looking for relative errors to indicate the efficiency of the calibration model. The RMSE analytical result of CH 26 with the frequency of 514 MHz; CH 42 with the frequency of 642 MHz; CH 46 with the frequency of 674 MHz; CH 54 with the frequency of 738 MHz shows that Hata's path loss model calibration is closer to the measured data than the original and the other model whereas the relative errors are closer to zero than the predicted path loss model. This makes the Hata's path loss model calibration become more accurate in the prediction and subsequently more suitable for use in planning the network design.

1. Introduction

Thailand inaugurated the broadcasting of the standard Digital Terrestrial Television of DVB-T2 on April 1,2014 under the supervision of the Office of the National Broadcasting and Telecommunication Commission (NBTC) who then organized and planned the frequency network covering the whole nation. The main station was designed as a Multi Frequency Network (MFN) and the sub stations within each main station were a Single Frequency Network (SFN) [1]. It becomes necessary that the design of the network must consist of sub stations in the areas of low signal and the existence of the gap filler stations in the areas of dead spots causing from obstructive surroundings [2].

An urban area is normally encountered with a dead spot causing from various obstructions i.e. high-rise buildings built-in in one area. Therefore, in order to plan the efficient gap filler stations covering areas properly while using the cost reduction transmitted power, it is essential that an accurate prediction path loss model is used for a network design. This path loss model offers the suitable accuracy for many areas in various researches [3-4] etc. And a statistical method of Root Mean Square Error (RMSE) applied in many researches [5-6] etc.

This paper presents Hata's path loss model calibration for prediction DTTV propagation in an urban area of southern Thailand through a method of looking for RMSE statistical value as well as relative errors from the data of power signal obtained from the measurement. The second part of the paper explains the path loss model while the third part is about the details of a signal transmitted stations, a signal measurement method, and the parameter of the DTTV signal transmission. The fourth part describes the findings whereas the fifth is the research conclusion.

2. Path Loss Models

2.1. Free Space Model

The free space model is the signal loss in a form of the line of sight (LOS). The received power signal (P_r) is related to the transmitted power signal (P_t), as shown in the following equation of Friis's Transmission [7].

$$P_r = \frac{P_t G_t G_r \lambda^2}{(4\pi R)^2} \tag{1}$$

G_t, G_r are the transmitted antenna gain and the received antenna gain, λ is the wave length of used frequency, and the free space path loss model equation is as followed:

$$L(dB) = 10\log\frac{P_t}{P_r} = -10\log\frac{G_t G_r \lambda^2}{(4\pi R)^2} = 32.44 + 20\log f_c + 20\log R \tag{2}$$

f_c is the used frequency, with a unit of MHz and R is the distance between the transmitter and the receiver, with a unit of km.

2.2. Okumura Model

Okumura's path loss model [8] is a system of wave propagation for urban areas. This model was initiated using the information of Tokyo City in Japan and it can be used in an area with a frequency of (f_c) 150 – 1920 MHz at the transmitted antenna height of (h_t) 30 – 1000 metre, the received antenna height of (h_m) 1 – 10 metre, and the distance (R) between the transmitter and the receiver (TX – RX) of 1 –100 km. An equation of this model is:

$$L(dB) = L_F + A_{MU}(f,d) - G(h_{te}) - G(h_{me}) - G_{Area} \tag{3}$$

L_F is a free space path loss model, with a unit of dB; A_{MU} is a loss amount through the free space, with a unit of dB; $G(h_{te})$ is a factor of gain for the transmitted antenna height, with a unit of dB; $G(h_{me})$ is the factor of gain for the received antenna height, with a unit of dB; G_{Area} is the gain that is related to the environment.

$$G(h\ ) = \begin{cases} 20\log(h_{te}/200) & ,1000m > h_{te} > 30m \\ 10\log(h_{me}/3) & ,h_{me} \leq 3m \\ 20\log(h_{me}/3) & ,10m > h_{me} > 3m \end{cases} \quad (4)$$

2.3. Hata Model

Hata's path loss model [9] was developed by Y. Okumura and M. Hata based on the signal measurement in the urban and suburban areas of Japan in 1968. The model is set for work in an area where the frequency is (f_c) 150 – 1500 MHz at the transmitted antenna height (h_t) of30 – 200 metre, the received antenna height (h_m) of 1 – 10 metre and the distance between the transmitter and the receiver f 1 – 20 km. This model is divided based on urban, suburban and rural areas.

Urban areas

$$L(dB) = 69.55 + 26.16\log f_c - 13.82\log h_t + (44.9 - 6.55\log h_t)\log R - E \quad (5)$$

Large urban areas

$$E = 3.2[\log(11.7554h_m)]^2 - 4.97 \quad f_c \geq 400 \text{ MHz} \quad (6)$$

$$E = 8.29[\log(1.54h_m)]^2 - 1.1 \quad f_c \leq 200 \text{ MHz} \quad (7)$$

Medium and small urban areas

$$E = [1.1\log f_c - 0.7]h_m - [1.56\log f_c - 0.8] \quad (8)$$

Suburban areas

$$L(dB) = 69.55 + 26.16\log f_c - 13.82\log h_t + (44.9 - 6.55\log h_t)\log R - 2\left[\left(\log\frac{f_c}{28}\right)^2 + 5.4\right] \quad (9)$$

Rural areas

$$L(dB) = 69.55 + 26.16\log f_c - 13.82\log h_t + (44.9 - 6.55\log h_t)\log R - 4.78(\log f_c)^2 + 18.33\log f_c + 40.94 \quad (10)$$

2.4. Egli Model

Egli's path loss model [10] was presented by John Egli in 1957 as the model that is used in an area where the frequency(f_c)ranges between 40 – 900 MHz and the distance (R) between the transmitter and the receiver is less than 60 km. The predicted equation of wave propagation is as:

$h_m \leq 10$

$$L(dB) = 20\log f_c + 40\log R - 20\log h_t + 76.3 - 10\log h_m \quad (11)$$

$$h_m \geq 10$$
$$L(dB) = 20\log f_c + 40\log R - 20\log h_t + 85.9 - 10\log h_m \quad (12)$$

f_c is the used frequency, with a unit of MHz; h_t is the transmitted antenna height, with a unit of m; h_m is the received antenna height, with a unit of m; R is the distance between the transmitter and the receiver, and a unit of metre (m).

3. DTTV Propagation

3.1. Signal Transmitted Station

NBTC specifies that the propagation of DTTV in Thailand is organized from one station and shares the infrastructure of transmission lines and antennas. The transmitter of each network operator connects with a combiner system to commonly use a transmission line and an antenna, as shown in the figure 1. The DTTV transmission in Songkla Province, in this research, is located on KhaoKorhong at the latitude of 7 0' 57.95", the longitude of 100 31' 12.17", and 366m above sea level whereas the transmitted height is 66m.

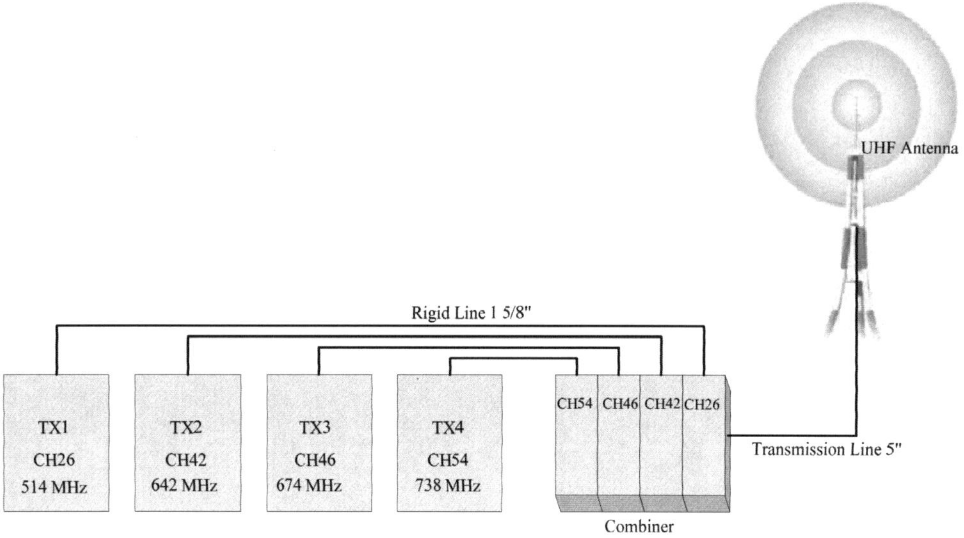

Figure 1. The transmission structure of a DTTV station

Digital transmitter broadcast in CH46 using a frequency of 674 MHz, NEC brand, DTL-10/1R0S model, and broadcasting power of 1KW; CH26 with the frequency of 514 MHz; CH42 with the frequency of 642 MHz and the latter two stations use the same transmitter of HRRIS, UAX-2000T2HEmodel, transmitted power of 1.3KW whereas CH54 with the frequency of 738 MHz, NEC brand, DTL-30/1R4SD model also uses the transmitted power of 1.3KW. The transmitted power of every channel is conducted through a 1 5/8" rigid line onto a combiner using a Spinner brand, CCS-6WAY model with the loss (Cl_1) of0.55 dB. The output of the combiner is sent through the transmission line, RFS brand, Flexwell HF 5"model with the total loss (Cl_2) of1.186 dB and onto the horizontal antenna, RFS brand, PHP48U model with the total gain (G_t) at 18.35 dBi. The relation between the received power signal (P_r) and the signal path loss (L) can be described as:

IOP Conf. Series: Materials Science and Engineering **83** (2015) 012013 doi:10.1088/1757-899X/83/1/012013

$$EIRP = P_t + G_t - Cl_1 - Cl_2 \tag{13}$$

$$L(dB) = (EIRP + G_r) - P_r \tag{14}$$

EIRP is the effective isotropic radiated power, with a unit of dBm; G_t G_r are the transmitted antenna gain and the received antenna gain, with a unit of dBi; P_t , P_r are the TV transmitted power and the received signal power, with a unit of dBm; Cl_1 , Cl_2 are the transmitted loss in the combiner and in the transmission line, with a unit of dB.

3.2. Signal Measurement

The measurement of the signal is performed by a signal analyzing tool, DVB-T2 PROMAX, HD RANGER+ model, which consists of GPS and USB Drive as a data collector. The received antenna is a commercial one from a Korean company, SPECTRUM.CO.LTD, Omni-Saturn model with the gain (G_r) of-3dBi being installed on top of the pickup at 2 meter height off the ground. Drive Test function is used to measure the received signal power of the pathways where a high number of build-ins are located on both sides of the street in the urban of Haadyai City, Songkla province and the speed limit of 20km/hr.

Figure 2. DTTV signal measurement equipment

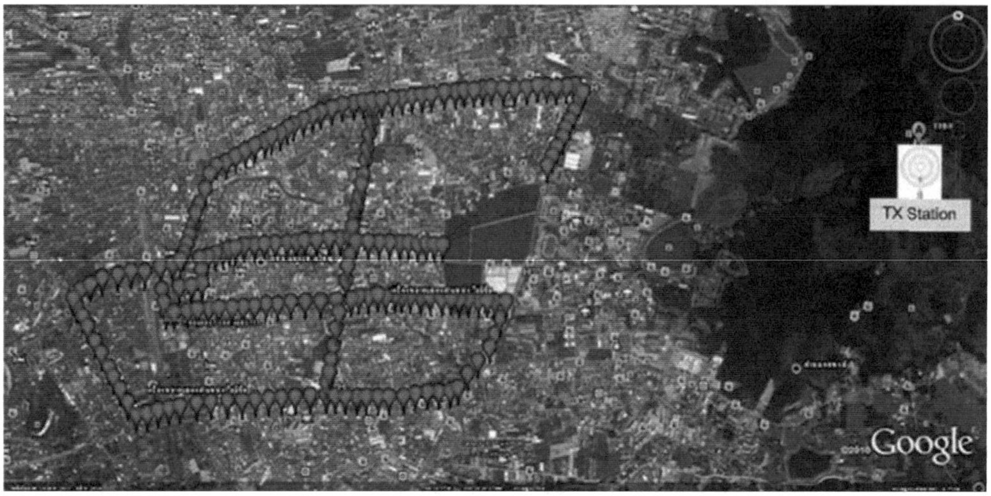

Figure 3.Haadyai Pathway, Songkla Province where the drive test function was used for data collection.

4. Analysis Findings

The path loss model calibration is done by using RMSE from the received power signal and compared with the predicted path loss model. The following equation shows how to get RMSE[6].

$$RMSE = \sqrt{\frac{\sum \left(PL_m - PL_{approx}\right)^2}{N-1}} \tag{15}$$

PL_m is the path loss obtained from the measurement; PL_{approx} is the predicted path loss; N is the data amount obtained from the measure which, in this research, is 2300 points from each channel. The comparison result of the measured path loss is close to Hata's path loss model that the calibrated Hata's path loss model can be described as:

$$Calibrated \quad Hata = Hata \quad Model + RMSE(dB) \tag{16}$$

The relative error, to indicate the efficiency of the calibrated model, is calculated from the following equation [10].

$$\delta = \frac{\left|PL_m - PL_{approx}\right|}{\left|PL_m\right|} \tag{17}$$

$$Accuracy = 1 - \delta \tag{18}$$

From the equation, the RMSE of each frequency is obtained. Table 1 shows that RMSE of a low frequency channel is higher than the one of a high frequency channel. RMSE value of Hata's path loss model is lesser than other models which shows the suitability of using the model to calibrate. Using the RMSE value in the calibration showing in figure 4, 5, 6, 7, it can be observed that the calculation result of Hata's path loss model is closer to the measured information than those of other models. The figures also show the compared result between Hata's path loss model calibration and the predicted path loss model of each channel. It is clear that the Hata's path loss model calibration is the closest to the measured data and its relative error is closer to zero than the original Hata's path loss model, as shown in Table 2.

Table 1. The result of RMSE for path loss model calibration

Frequency	RMSE (dB)			
	Hata	Okumura	Egli	Free space
CH26 (514 MHz)	15.40	20.90	34.80	51.20
CH42 (642 MHz)	11.07	16.57	30.29	46.64
CH46 (674 MHz)	10.16	15.60	29.03	45.48
CH54 (738 MHz)	9.57	15.40	28.60	45.40
Average	11.55	17.12	30.68	47.18

Table 2. The comparison of relative errors and accuracy value

Frequency	Relative Errors		Accuracy (%)	
	Calibrated Hata	Hata	Calibrated Hata	Hata
CH26 (514 MHz)	0.010	0.104	99.0	89.6
CH42 (642 MHz)	0.015	0.068	98.5	93.2
CH46 (674 MHz)	0.019	0.058	98.1	94.2
CH54 (738 MHz)	0.017	0.055	98.3	94.5

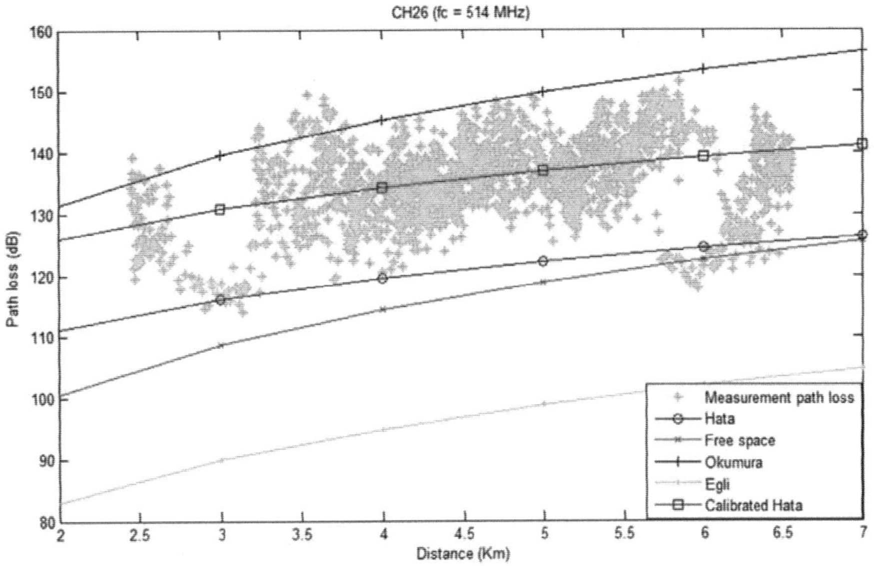

Figure 4. The comparison between Hata's path loss model calibration and the predicted path loss model of CH 26, the frequency of 514 MHz

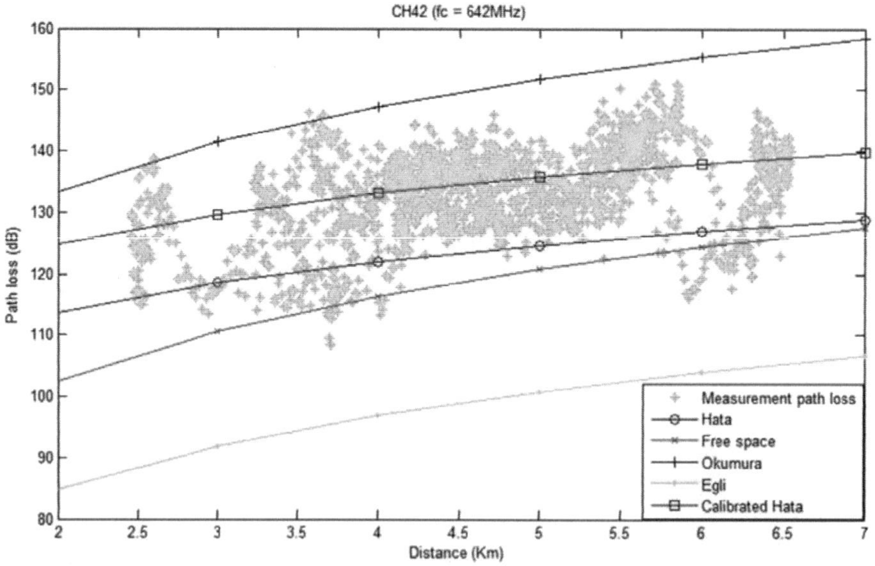

Figure 5. The comparison between Hata's path loss model calibration and the predicted path loss model for Channel 42, the frequency of 642 MHz

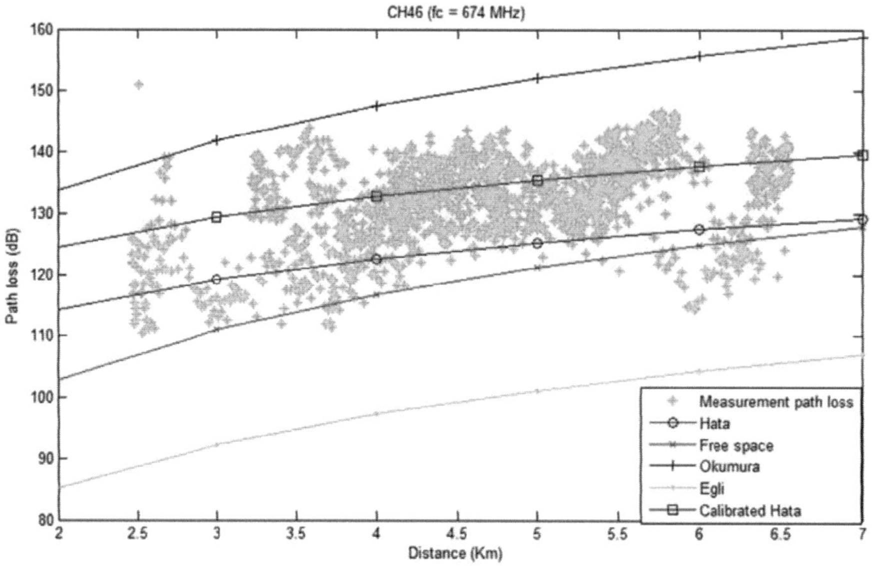

Figure 6. The comparison between Hata's path loss model calibration and the predicted path loss model for channel 46; the frequency of 674 MH

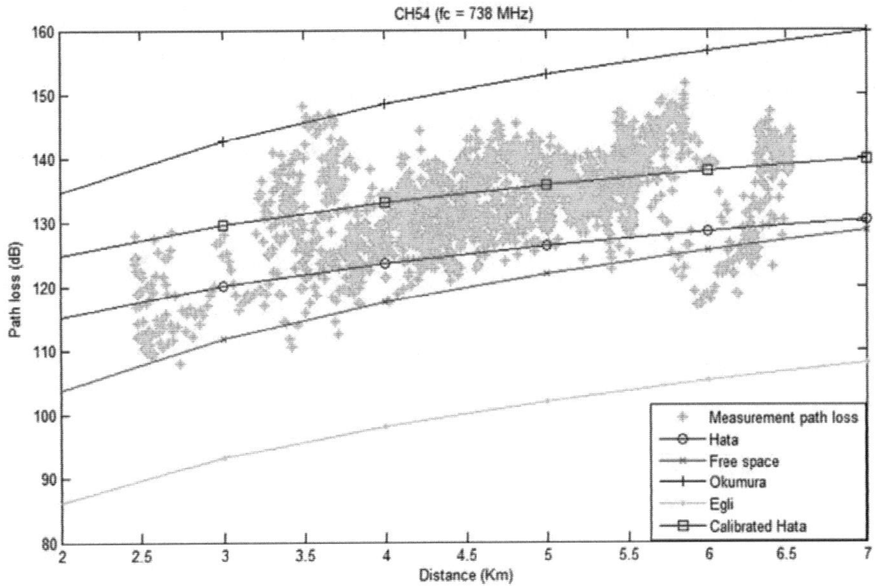

Figure 7. The comparison between Hata's path loss model calibration and the predicted path loss model of CH54; the frequency of 738 MHz

5. Conclusion

The obtained result of Hata's path loss model calibration is appropriate for designing the DTTV propagation in Haadyai City, Songkla Province where there is a dead spot. The result is also referred to use in other urban areas in the southern part of Thailand in order to properly cover different areas as well as to use the suitable power of the cost reduction transmitter. The calibration is done through finding the statistic of RMSE from the received power signal compared with the predicted path loss model in order to indicate the efficiency of the calibration. The analytical result of RMSE in CH26, the frequency of 514 MHz ; CH42, the frequency of 642 MHz; CH46, the frequency of 674 MHz; CH54, the frequency of 738 MHz shows the value of 15.40,11.0710.16 and 9.57dB, respectively. The result also shows that the Hata's path loss model calibration is closer to the measured data than the predicted model whereas the relative error is closer to zero than the original path loss model.

References

[1] Office of The National Broadcasting and Telecommunication Commission 2012 *Radio Frequency for Digital Terrestrial Television in Thailand* NBTC, December 18, 2012

[2] S.R.M. de Carvalho, Yuzo Iano and Rangel 2011 *ISDB-TB Field Trials and CoverageMeasurements with Gap-Filler in Suburban Environments* IEEE International Symposium on Broadband Multimedia Systems and Broadcasting, June 8-10, 2011

[3] Jalel Chebil, Ali K. Lwas, Md. Rafiqul Islam and Al-Hareth Zyoud2011*Adjustment of Lee Path Loss Model for Suburban Area in Kuala Lumpur-Malaysia* International Conference on Telecommunication Technology and Applications, Proc. Of CSIT, vol.5, pp.252-257

[4] Aied Khalaf Mohammed and Ahmed A. Jaafar2012*Performance Evaluation of Path Loss in Mobile Channel for Karada District in Baghdad City*Eng.&Tech. Journal, vol.30, November 17,2012

[5] Wiyada Bhupuok and Kobchai Dejhan 2012*A New Method for Prediction 3G Path Loss*

Propagation in Suburban of Thailand ECTI-CON 2012, Phetchaburi Thailand, May 16-18, 2012

[6] Liyth A. Missirat, Mahamod Ismail, Mahdi Nisirat and Mandeep Singh 2011*Lee's Path Loss Model Calibration and Prediction for Jiza Town, South of Aman City, Jordan at 900MHz*IEEE International RF and Microwave Conference, Seremban, Malaysia, December 12-14, 2011

[7] Rappaport, T.S. 2002. *Wireless Communications Principles and Practice* Prentice Hall

[8] R. Mardeni and L. Yih Pey 2010*The Optimization of Okumura's Model for Code Division Multiple Access (CDMA) System in Malysia* European Journal of Scientific Research, vol.45, pp.508-528

[9] M. Hata 1980 *Empirical Formula for Propagation Loss in Land Mobile Radio Service* IEEE Transaction on Vehicular Technology, vol.VT-29, No.3, pp.317-325

[10] R. Mardeni and K.F. Kwan 2010 *Optimization of Hata Propagation Model in Suburban Area in Malaysia* Progress in Electromagnetics Research, vol.13, pp.91-106

IOP Conf. Series: Materials Science and Engineering **83** (2015) 012014 doi:10.1088/1757-899X/83/1/012014

Using Grey Wolf Algorithm to Solve the Capacitated Vehicle Routing Problem

L Korayem, M Khorsid, S S Kassem[1]

Operations Research and Decision Support Department, Faculty of Computers and Information, Cairo University, Giza, Egypt

[1]E-mail: s.kassem@fci-cu.edu.eg

Abstract. The capacitated vehicle routing problem (CVRP) is a class of the vehicle routing problems (VRPs). In CVRP a set of identical vehicles having fixed capacities are required to fulfill customers' demands for a single commodity. The main objective is to minimize the total cost or distance traveled by the vehicles while satisfying a number of constraints, such as: the capacity constraint of each vehicle, logical flow constraints, etc. One of the methods employed in solving the CVRP is the cluster-first route-second method. It is a technique based on grouping of customers into a number of clusters, where each cluster is served by one vehicle. Once clusters are formed, a route determining the best sequence to visit customers is established within each cluster. The recently bio-inspired grey wolf optimizer (GWO), introduced in 2014, has proven to be efficient in solving unconstrained, as well as, constrained optimization problems. In the current research, our main contributions are: combining GWO with the traditional K-means clustering algorithm to generate the 'K-GWO' algorithm, deriving a capacitated version of the K-GWO algorithm by incorporating a capacity constraint into the aforementioned algorithm, and finally, developing 2 new clustering heuristics. The resulting algorithm is used in the clustering phase of the cluster-first route-second method to solve the CVR problem. The algorithm is tested on a number of benchmark problems with encouraging results.

1. Introduction

The Capacitated Vehicle Routing Problem (CVRP) is one of the variants of the classical Vehicle Routing Problem (VRP). It was first introduced in[1]. In the CVRP, a fleet of homogeneous vehicles with fixed capacity depart from a central depot. These vehicles must satisfy the demands of a number of spatially separated customers at minimum cost. This must be done without exceeding any vehicle's capacity. The cost considered in the problem is either the total distance travelled or the total travelling time of all the vehicles. Several methods for solving CVRP have been reported in the literature. These methods are generally categorized into exact methods, heuristics, metaheuristics or a combination of two or more of them. For a detailed review of CVRP solution methods, interested readers are referred to [2]. Cluster-first route-second method is one of the heuristics employed to solve CVRP, where clustering represents the core step within the heuristic [3]. In the cluster-first route-second method, customers are first grouped into feasible clusters that maintain the vehicle capacity constraint. Hence, the clustering phase can be viewed as a capacitated clustering problem (CCP). One vehicle is assigned to serve the customers within each cluster. Efficient routes are then established (route second). Route establishing requires determining the sequence to visit customers, by each vehicle, within its assigned cluster.

Grey Wolf Optimizer (GWO) is a new metaheuristic inspired by the leadership hierarchy, as well as the hunting mechanism observed in grey wolves. It was first introduced by [4] in 2014. In GWO, four types of wolves: alpha, beta, delta, and omega are used to simulate the chain of command as illustrated in figure 1, taken from reference [4]. As can be seen from figure 1, the dominance is decreased from top to bottom. Alpha wolves dominate the beta wolves, which in turn dominate the delta wolves. Finally, the omega wolves are dominated by all the higher ranked wolves. The grey wolf hunting behaviour is mainly divided into three steps: tracking, encircling and attacking the prey. The wolves estimate the prey location, through an iterative process. The GWO algorithm starts by a population of randomly generated wolves. The population of randomly generated wolves represent solutions. Similar to the grey wolves' behaviour, the population of wolves (solutions) estimate the optimum solution (the prey) through an iterative process. The three best search agents: alpha, beta and delta are used to update the rest of the population's location. In GWO, tracking the prey represents the exploration phase. The exploitation phase is represented by attacking the prey. The balance between the exploration phase and the exploitation phase represents a key issue in any metaheuristic. Results obtained on uni-modal and multimodal functions verified, respectively, the exploitation and exploration abilities of GWO. Being a new algorithm, modest applications of the GWO algorithm were reported in the literature. In [5], the GWO algorithm was employed for optimum feature subset selection. When GWO is Compared against particle swarm optimization (PSO) and genetic algorithms (GA), GWO showed better accuracy. In [6], an improved version of Grey Wolf Optimizer (GWO) is employed for training q-Gaussian Radial Basis Functional-link nets neural networks. Competitive results were obtained when compared to other metaheuristic methods.

The K-means clustering algorithm is one of the widely used traditional partitioning approaches. It is simple and easy to implement. However, it suffers from the possibility of falling into local minima. To overcome this drawback, the K-means algorithm is usually hybridized with another optimization algorithm. The merits of combining a number of bio-inspired algorithms to K-means have been studied in [7] and [8]. Experimental validation showed the benefits of such integration.

In the current work, we integrate the GWO algorithm with K-means to introduce the (K-GWO), which stands for clustering with GWO. We further incorporate a capacity constraint to render the K-GWO suitable for solving capacitated clustering problems. We refer to this algorithm in this work as (capacitated K-GWO). We use the capacitated K-GWO to solve CVRP. The proposed capacitated K-GWO is used in the clustering phase of the cluster-first route-second technique to solve CVRP. To the best of our knowledge, this is the first time to employ such integration for solving CVRP.

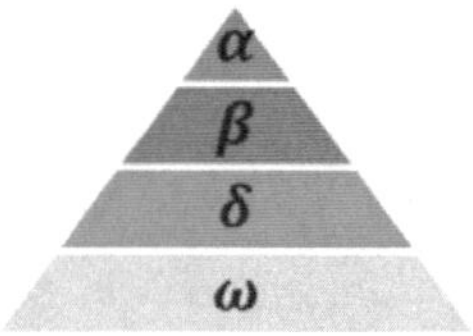

Figure 1. Hierarchy of grey wolf [4]

The rest of this work is organized as follows: in Section 2 we introduce the problem definition. Details of the solution methodology are given in Section 3, where grey wolf optimization algorithm is first introduced, followed by the K-GWO algorithm, and finally the capacitated K-GWO algorithm. Experimental computations and results are presented in section 4. Finally, concluding remarks are included in section 5.

2. Problem definition

The vehicle routing problem (VRP) is one of the most studied combinatorial optimization problems [9]. The classical VRP requires determining the optimal routes used by a fleet of vehicles to satisfy the demands of a number of spatially separated customers. Many variants of VRP exist in the literature and in practical applications. These variants usually depend on the different constraint(s) imposed on the problem. The CVRP considered in this work considers distribution of a single commodity from a central depot to a set of customers. A given number of identical vehicles are available. Each vehicle has a fixed capacity. The total demand assigned to each vehicle must not exceed the vehicle's capacity. CVRPs may consider deterministic or stochastic demands. In this work, we consider predetermined deterministic customers' demands. This assumption is based on the fact that many suppliers depend on long term contracts with their customers, and hence they can determine their customers' demands. Practical experience shows that customers prefer to be served once. Therefore, for the convenience of customers, demands may not be split between more than one vehicle. Accordingly, each customer should be visited exactly once by only one vehicle. The objective is to minimize the total distance traveled by all the vehicles to serve all the customers. Each vehicle starts at the depot, serves all its assigned customers, and ends the trip back at the depot.

According to [10], VRP is an NP-Hard problem. Heuristics and metaheuristics are therefore considered to solve the problem and its associated variants [11].

3. Proposed solution methodology

Our approach to solve the CVRP is based on the cluster-first route-second method. In the clustering phase, customers are grouped into feasible clusters. A feasible cluster requires that the total customers' demands within the cluster do not exceed the capacity of a single vehicle. In the routing phase, customers within each cluster are suitably ordered, which amounts this phase to solving a traveling salesman problem TSP [12].

In the present work, clustering will be performed using the proposed capacitated K-GWO algorithm. Details of the proposed algorithm are presented below.

3.1. Grey Wolf Optimizer (GWO)

Previous research shows that bio inspired metaheuristics are efficient in solving CVRPs. Such efficiency is recorded in terms of solution quality, as well as computational time. Therefore, many researchers are motivated to examine different bio inspired metaheuristics in solving the CVRP. The Grey Wolf Optimizer (GWO) algorithm is a new bio inspired metaheuristic that has been introduced in 2014 by [4] . It is inspired by both the social hierarchy of wolves, as well as their hunting behavior. In GWO, the search starts by a population of randomly generated wolves (solutions). During hunting (optimization), these wolves estimate the prey's (optimum) location through an iterative procedure. In order to formulate the hierarchy of wolves, the fittest solution is referred to as the alpha ( ). While the second and third best solutions are called beta ( ) and delta ( ), respectively. The wolves , , and are responsible for guiding the search (hunting), while other wolves follow. The hunting behaviour is mainly divided into three steps: tracking, encircling and attacking the prey. The encircling behaviour is mathematically modelled according to the following equations:

$$\vec{D} = |\vec{C}.\vec{X}_p(t) - \vec{X}(t)| \tag{1}$$

$$\vec{X}(t+1) = |\vec{X}_p(t) - \vec{A}.\vec{D}| \tag{2}$$

Where $\vec{D}$ represents the distance between the position vector of both the prey $\vec{X}_p$ and a wolf $\vec{X}$, and t represents the current iteration number. $\vec{A}$ and $\vec{C}$ are coefficient vectors, and are calculated as follows:

$$\vec{A} = 2\vec{a}.\vec{r}_1 - \vec{a} \tag{3}$$

$$\vec{C} = 2\vec{r}_2 \tag{4}$$

Where r_1 and r_2 are random vectors $\in [0,1]$, and the value of $\vec{a}$ is linearly decreased from 2 to 0 as iterations proceed.

According to (1) and (2), a wolf can randomly update its position in the area around the prey.

In order to model how the three best solutions , , and guide the other search agents, the following formulas are used:

$$\vec{D}_{\alpha} = |\vec{C}_1.\vec{X}_{\alpha} - \vec{X}|, \vec{D}_{\beta} = |\vec{C}_2.\vec{X}_{\beta} - \vec{X}|, \vec{D}_{\delta} = |\vec{C}_3.\vec{X}_{\delta} - \vec{X}| \quad (5)$$

$$\vec{X}_1 = \vec{X}_{\alpha} - \vec{A}_{\alpha}.(\vec{D}_{\alpha}), \vec{X}_2 = \vec{X}_{\beta} - \vec{A}_{\beta}.(\vec{D}_{\beta}), \vec{X}_3 = \vec{X}_{\delta} - \vec{A}_{\delta}.(\vec{D}_{\delta}) \quad (6)$$

$$\vec{X}(t+1) = \frac{\vec{X}_1 + \vec{X}_2 + \vec{X}_3}{3} \quad (7)$$

From (5), (6) and (7) it can be seen that the position $\vec{X}$of any wolf is determined by the position of the three best solutions. Position updating of grey wolf is shown in figure 2 taken from reference [4].

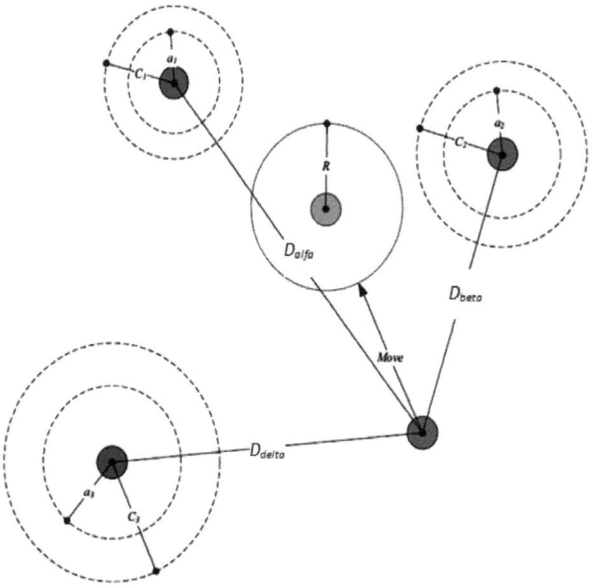

Figure 2. Position updating of grey wolf [4]

The hunting ends by attacking (approaching) the prey, this step represents the exploitation phase. It is performed by decreasing the value of $\vec{a}$, linearly from 2 to 0. This in turn decreases the value of $\vec{A}$. To avoid local stagnation, random values of $\vec{A}$ greater than 1 are employed to force the wolf away from the prey. This step represents the exploration phase. Using adaptive values of a and consequently A, which is depicted by linearly decreasing the values of a from 2 to 0, guarantees a balance between exploration and exploitation. This balance is achieved since half of the iterations are dedicated to exploration when |A| • 1, while the rest of the iterations is dedicated for exploitation when $|A| < 1$. This balance is one of the GWO's strengths. To sum up, values of $|A| < 1$ oblige the search agents to move towards the prey while values of |A| >1 oblige them to diverge from it.

The value of $\vec{C}$ represents another component that influences exploration, where $\vec{C} \in [0,2]$. From (1) we can conclude that $\vec{C}$ represents the weight of the prey in defining the distance; values of $C > 1$ emphasize its effect while $C < 1$ reduce it.

Integrating the GWO with K-means algorithm to overcome the drawbacks of the K-means is discussed in the next section.

3.2. Clustering with grey wolf optimize

The K-means clustering algorithm assigns points to k clusters based on their proximity to the cluster's centroid. Initially centroids are selected at random and then iteratively reallocated until a predetermined stopping criterion is met. The K-means algorithm may be divided into an initialization phase, a cluster assignment and centroids update phase, and finally an exploration and evaluation phase [5]. K-means algorithm suffers from two main drawbacks, namely: the dependence on initial centroid values, and its tendency to fall into local optima. Its integration with bio-inspired optimization algorithms to overcome the aforementioned drawbacks is studied in [7], [13].

In this work, we introduce for the first time a combination of the K-means algorithm and the GWO. In our proposed K-GWO algorithm, wolves represent solutions. Each wolf holds a set of Kccentroids that correspond to K clusters. Each centroid is a D dimensional vector. Consequently, each wolf is represented by a $(K \times \mathrm{D})$ vector. A population of N wolves collaboratively hunts for the best possible configuration of the clusters (prey). The best configuration of clusters is reflected by the optimal positions of the centroids. These centroids are subsequently used to grow clusters following the basic K-means principle: points are assigned to the cluster with nearest centroid.

The algorithm aims at minimizing an objective function defined as

$$F = \sum_{\mathrm{j}=1}^{\mathrm{k}} \sum_{\mathrm{i}=1}^{\mathrm{N}} \left\| x_{i,j} - cen_j \right\|^2 \qquad (8)$$

Where, k is the number of clusters, N is the number of points to be clustered, $x_{i,j}$ is the ith data point belonging to the jth cluster, and cen_j is the centroid of cluster j. As mentioned earlier, each search agent represents a set of k centroids $(cen_1, cen_2, \ldots cen_k)$ which, when used in equation (8), gives an indication on how fit this agent is. The fittest search agent is the one associated with the minimum value of F . Wolves use equations (3) to (7) for updating their positions depending on the positions of , , and wolves.

We further extend our proposed K-GWO, to incorporate a capacity constraint. Details are discussed in the next section.

3.3. Capacitated K-GWO

As mentioned earlier, the capacitated clustering problem (CCP) consists of imposing a capacity constraint on each cluster. Mainly, there exists three approaches to handle constraints: direct approach, Lagrange multipliers, and penalty method [14]. The penalty method is selected in our proposed method, since it is the simplest to implement in practice, without reducing the quality of the results. In our approach, constraints violations are incorporated in the objective function defining a penalized function. This function penalizes infeasible solutions relative to the amount of constraint's violation [15]. Accordingly, the new fitness function can be defined as

$$F_P(X) = F(X) + p \times (\Delta\, Capacity)^c \qquad (9)$$

Where: $F_P(X)$ is the penalized fitness function, $F(X)$ is the original fitness function, and $\Delta\, Capacity$ is the amount of capacity violation, p and c are user defined penalty parameters. In the current study, these parameters are determined experimentally. Incorporating the penalized fitness function to the K-GWO algorithm yields a capacitated K-GWO algorithm appropriate for capacitated clustering problems.

3.4. Cluster assignment

The cluster assignment phase is among the phases of K-means clustering algorithm where points are assigned to their corresponding clusters. The classical method for assignment is performed according to minimum distance to centroid as illustrated in figure 3. In the CVRP, when customers are assigned to clusters based only on minimum distance to centoid, infeasible clusters might be formed. In order to obtain feasible clusters, we develop two new heuristics. In both heuristics, Euclidean distances between every customer and all the centroids are calculated. In our first developed heuristic, shown in figure 4, points are assigned to the cluster based on minimum distance until capacity limit is reached. At this stage, the remaining non-clustered customers will violate capacity constraints if assigned to any cluster. For every non-clustered customer, we calculate the amount of capacity violation if the customer is assigned to each cluster. Every non-clustered customer is then assigned to the cluster having the minimum capacity violation.

```
for each search agent
    calculate the distance between every customer x_i and all centroids (cen_1, cen_2, ... cen_k)
    while x_i is not assigned
            Assign x_i to its nearest centroid
                    for every cluster K
                    Calculate capacity violation ( Δ Capacity )
                    end for
    end while
end for
```

Figure 3. Pseudo code of assignment procedure used in KmeansO

In our second developed heuristic, shown in figure 5, points are assigned to the nearest cluster based on minimum distance and maximum demand [16]. In this case a ratio is calculated using (10). Assignment is conducted according to this ratio until capacity limit is reached. At this stage, non-clustered customers are assigned to clusters according to the minimum capacity violation explained in the first heuristic above.

$$\text{ratio} = \mathrm{d_{ij}}/\mathrm{D} \tag{10}$$

Where $\mathrm{d_{ij}}$ is the distance between customer i and centroid j and D is the customer's demand.

```
for each search agent
    calculate the distance between every customer x_i and all centroids (cen_1, cen_2, ... cen_k)
    while x_i is not assigned
        if capacity is not violated
          Assign x_i to its nearest centroid
        else
          Assign x_i to centroid with minimum amount of violation
        end if
    end while
end for
```

Figure 4. Pseudo code of assignment procedure used in KmeansP

Based on the above mentioned heuristics, three versions of the capacitated K-GWO will be compared. The first version uses the classical assigning technique, where points are assigned to the nearest centroid, and, in this work, will be referred to as KmeansO. The second version will be referred to in this work as KmeansP, and uses our first assigning heuristic, where points are assigned to clusters with minimum distance and minimum violation. Finally, the third version, and in this work will be referred to as KmeansR, and uses our second assigning heuristic, where points are assigned to clusters with minimum ratio and minimum violation.

The capacitated K-GWO is utilized in the clustering phase of the cluster-first route-second technique to solve the CVRP. Experimental computations are carried out in order to investigate the efficiency of the proposed algorithm as will be seen in the next section.

```
for each search agent
    calculate the distance between every customer x_i and all centroids (cen_1, cen_2, ... cen_k)
    calculate the ratio between the distance and customer's demand according to (10)
    while x_i is not assigned
        if capacity is not violated
            Assign x_i to centroid with minimum ratio
        else
            Assign x_i to centroid with minimum amount of violation
        end if
    end while
end for
```

Figure 5. Pseudo code of assignment procedure used in KmeansR

4. Computational results and analysis

In order to verify the efficiency of the proposed capacitated K-GWO algorithm, a set of experiments are carried out. The proposed algorithm is tested using benchmark problems downloaded from the web http://www.branchandcut.org/. In this work, 20 instances from three different sets are tested. These sets are known in the literature as set A, set B, and set P. All three sets A, B and P are developed in [17]. For the instances in set A and set B, customers' locations and demands are randomly generated. Instances in set P are modified instances from the literature. Each test instance has the following nomenclature: the first letter refers to the set name followed by the number of customers and finally the minimum number of required clusters. The minimum number of required clusters corresponds to the minimum number of vehicles to employ. The minimum number of vehicles to employ is obtained by dividing the total customers' demands by the capacity per vehicle, rounding up fractions. For each instance, the depot and customers' coordinates, as well as customers' demands, are given.

Computational experiments are coded using MATLAB 2009a, on a PC with Pentium Dual Core T3200 Processor 2.0GHz and 2.0 GB memory. In the experiments, we run each instance for 10 independent runs and the best result is reported. The stopping criterion is the number of generations. The parameters of the proposed algorithm were set experimentally after a number of test runs. Chosen values of the parameters are as follows: the wolf population size is 20 and the maximum number of generations is 500. Table 1 summarizes the results of these runs, where the three versions of the capacitated K-GWO are compared. The values reported in Table 1 are those of the total distance travelled by all vehicles after optimally routing the vehicles within each cluster. The best value among the three versions is reported in the last column named "Best".

Table 1 shows that each of the three versions, KmeansFnO, KmeansFnP, KmeansFnP, are able to obtain the best value in 10, 9, and 11 instances out of the 20 test instances, respectively. This shows that the versions are comparable in terms of the solution quality each version can obtain. However, during the test runs, we noticed that the first version obtained many non-feasible solutions before being able to obtain a feasible solution. The second and third versions however, were able to reach feasible solutions faster than the first version. This is explained by the fact that the first heuristic minimizes the distance to centroids, and capacity constraints are only considered in the objective function penalty. Contrary to the other 2 versions, where capacity constraints are considered during cluster assignment, in addition to the penalty within the objective function. The computational time for solving the test problems is in the order of a few seconds for any of the three versions.

Table 1. Comparison of results for the three versions

Instance	KmeansFnO	KmeansFnP	KmeansFnR	Best
A-n33-k6	828	754*	754*	754
A-n34-k5	792*	822	839	792
A-n36-k5	814*	846	846	814
A-n39-k6	868	856	837*	837
A-n60-k9	1395*	1421	1423	1395
A-n62-k8	1366	1329*	1329*	1329
B-n31-k5	684	672*	672*	672
B-n34-k5	793*	817	817	793
B-n39-k5	571	564*	566	564
B-n41-k6	845*	920	877	845
B-n44-k7	940*	954	944	940
B-n50-k7	765*	783	774	765
B-n50-k8	1348*	1359	1361	1348
P-n20-k2	231	217*	217*	217
P-n21-k2	211*	211*	211*	211
P-n22-k2	216*	216*	216*	216
P-n45-k5	549	510*	510*	510
P-n55-k8	608	620	606*	606
P-n76-k4	622	614*	614*	614
P-n76-k5	660	650	635*	635

*Best value obtained among the three versions

In order to examine the solution quality obtained using the proposed methodology, we compare the best values obtained using the three versions against optimum solutions of the 20 test instances. The best values of the proposed capacitated K-GWO is obtained from the last column of Table 1. Table 2 shows the best solution obtained using the proposed method, the "optimum" solution according to http://www.branchandcut.org/, and the deviation of our best solution from the optimum. The "optimum" routes and the corresponding total cost are provided in the website.

Deviation from optimum is calculated using (11)

$$Deviation\% = \frac{optimum - modified}{optimum}\% \quad (11)$$

Table 2 shows that the proposed solution method was able to reach the optimum in 4 instances out of the 20 instances. The maximum deviation from optimum recorded was 3.5%, and the average deviation from optimum for the 20 test instances was 1.76%. It is noticed from Table 2 that the solution quality is independent of the problem sizes.

Table 2. Comparison of best obtained results to optimum

Instance	Best	Optimum	Deviation%
A-n33-k6	754	742	1.6
A-n34-k5	792	778	1.8

A-n36-k5	814	799	1.9
A-n39-k6	837	831	0.7
A-n60-k9	1395	1354	3.0
A-n62-k8	1329	1288	3.2
B-n31-k5	672	672	0
B-n34-k5	793	788	0.6
B-n39-k5	564	549	2.7
B-n41-k6	845	829	1.9
B-n44-k7	940	909	3.4
B-n50-k7	765	741	3.2
B-n50-k8	1348	1312	2.7
P-n20-k2	217	216	0.5
P-n21-k2	211	211	0
P-n22-k2	216	216	0
P-n45-k5	510	510	0
P-n55-k8	606	588	3.1
P-n76-k4	614	593	3.5
P-n76-k5	635	627	1.3

5. Conclusion

Grey wolf optimization algorithm is a new bio inspired algorithm. It has proved its efficiency for solving capacitated and non-capacitated optimization problems. K-means is one of the widely used clustering algorithms. In this paper, a capacitated K-GWO algorithm is developed and used to solve CVRP. Three main contributions are performed in this work: First integrating GWO with K-means algorithm. The K-means algorithm is used first as the classical K-means in the literature is presented. Our second contribution is concerned with enhancing the performance of the k-means algorithm by developing 2 other cluster assignment heuristics within the K-means algorithm. This resulted in having 3 different versions of the cluster assignment phase of solving the CVRP. Finally, our third contribution is penalizing the objective function of the CVRP to account for the capacity constraints of the problem. The 3 different versions of the proposed algorithm are compared against each other, and against optimum solutions on benchmark problems. Experimental analysis shows the ability of the algorithm to obtain optimum and near optimum solutions for benchmark problems. Applied to 20 benchmark instances, the maximum recorded deviation from optimum was 3.5%, with an average of 1.76%. The three versions obtain comparable solutions when compared against each other. However, our 2 developed versions obtain feasible solutions faster than the classical K-means version.

Future research includes enhancing the initialization phase of the proposed algorithm, apply statistical analysis and design of experiment methods to set tight ranges for the proposed algorithm parameters. The proposed solution method could be used to solve different variants of VRP and CVRP.

References

[1] Dantzig G B and Ramser J H 1959 The truck dispatching problem *Manag. Sc.* **6** 80–91.

[2] Takes F 2010 *Applying Monte Carlo Techniques to the Capacitated Vehicle Routing Problem* MSc Thesis Leiden University – The Netherlands.

[3] Simchi-Levi D, Chen X and Bramel J 2007 *The logic of logistics: theory, algorithms, and applications for logistics and supply chain management* (New York: Springer-Verlag).

[4] Mirjalili S, Mirjalili S M and Lewis A 2014 Grey wolf optimizer *Adv. Eng. Soft.* **69** 46–61.

[5] Emary E, Zawbaa H M, Grosan C and Hassenian A E 2015 Feature subset selection approach by gray-wolf optimization *Afro-European Conference for Industrial Advancement (Advances in Intelligent Systems and Computing* vol 334*)* ed A Abraham (Springer) pp1-13.

[6] Muangkote N, Sunat K and Chiewchanwattana S 2014 An improved grey wolf optimizer for training q-Gaussian Radial Basis Functional-link nets *Comp. Sc. Eng. Conf. (ICSEC)* pp 209-214.

[7] Tang R, Fong S, Yang X S and Deb S 2012 Integrating nature-inspired optimization algorithms to k-means clustering *Proc 7th Inter. Conf. Dig. Inf. Manag. (ICDIM '12)* pp.116-123.

[8] Fong S, Deb S, Yang X S and Zhuang Y 2014 Towards Enhancement of Performance of K-Means Clustering Using Nature-Inspired Optimization Algorithms *The Scientific World* **2014** 1-16.

[9] Toth P and Vigo D 2002 Models, relaxations and exact approaches for the capacitated vehicle routing problem *Discrete Applied Mathematics* **123** 487–512.

[10] Lenstra J K and Rinnooy-Kan A H G 1981 Complexity of vehicle routing and scheduling problems *Networks* **11** 221–8.

[11] Villegas J G 2010 *Vehicle routing problems with trailers* PHD Thesis Université de Technologie de Troyes – France.

[12] Cordeau J F, Laporte G, Savelsbergh M W P and Vigo D 2007 *Handbooks in Operations Research and Management Science* vol 14 ed C Barnhart and G Laporte pp. 367-428.

[13] Hashmi A, Gupta D, Upadhyay Y and Goel S 2013 Swarm intelligence based approach for data clustering *IJIRS* **2** 572-89.

[14] Yang X S 2010 *Nature-Inspired Metaheuristic Algorithms* 2nd ed. (United Kingdom: Luniver Press) pp. 29-41.

[15] Rocha A M A C and Fernandes E M G P 2009 Self adaptive penalties in the electromagnetism-like algorithm for constrained global optimization problems *Proc. 8th World Congress on Structural and Multidisciplinary Optimization (Portugal)* pp 1–10.

[16] Geetha S, Poonthalir G and Vanathi P T 2009 Improved k-means algorithm for Capacitated Clustering Problem *INFOCOMP* **8** pp 52-9.

[17] Augerat P 1995 *Approche polyèdrale du problème de tournées de véhicules* PHD Thesis Institut National Polytechnique de Grenoble - INPG.

IOP Conf. Series: Materials Science and Engineering **83** (2015) 012015 doi:10.1088/1757-899X/83/1/012015

Mediator Effect of TPM between TQM and Business Performance in Malaysia Automotive Industry

Ahmad, M.F[1,a], Zakuan, N[2,b], Raja Zuraidah RM Rasi [1,c] and MNN Hisyamudin[3,d]

1 Department of Production and Operation, Faculty of Technology Management and Business, Universiti Tun Hussein Onn Malaysia (UTHM), Parit Raja, Batu Pahat, Johor, Malaysia
2 Department of Management, Faculty of Management, Universiti Teknologi Malaysia (UTM), Skudai, Johor, Malaysia
3 Department of Materials & Design Engineering, Faculty of Mechanical Engineering Universiti Tun Hussein Onn Malaysia (UTHM), Parit Raja, Batu Pahat, Johor, Malaysia

E-mail: [a]mohdfauzi@uthm.edu.my, [b]norhayatimz@utm.my, [c]rzuraida@uthm.edu.my, [d]nhisham@uthm.edu.my

Abstract. Total Quality Management (TQM) is vital management tool in ensuring a company can success in the continuously growing competition in the global market. In order to survive in the global market with intense competition amongst regions and enterprises, the adoption of tools and techniques are essential in improving business performance. However, only few previous studies have examined the mediators and moderators between TQM and business performance. This present research proposed a TQM performance model with mediator effect of TPM with structural equation modelling, which is a more comprehensive model for developing countries, specifically for Malaysia. A questionnaire was prepared and sent to 1500 companies from automotive industry and the related vendors in Malaysia, giving a 21.3 per cent rate. The result concludes that TPM is partial mediation between and TQM and Business Performance with indirect effect (IE) is 0.25 which can be categorised as high mediator effect.

1.0 Introduction

The concept of total quality management (TQM) has been developed as a result of intense global competition [1][2]. Firms that manage the international trade in global competition have put emphasis on TQM philosophy, procedures, tools and techniques[3][4][5]. Juran [6] defines TQM as philosophy aimed at achieving business excellence through the application of tools and technique, as well as the management of soft aspects, such as human motivation in work. Furthermore, Demirbag et.al [7] defines TQM as a management philosophy aims to contribute continuous improvement in the organization with the participation of all employees to achieve customer satisfaction by producing better, cheaper, faster and safer than competitors. However, the study of mediators is neglected and is referred to less frequently in literature review [8][9]. The fundamental systems-interactive paradigm of organisational analysis features the continual stages of input, throughput (processing), and output, which demonstrate the concept of openness and closeness. Processing is the process of changing from one "look" to another, or one culture to another [10]. In this study, the author defines input as TQM; processing as application tools and techniques namely TPM; and output as business performance as shown in Figure 1. Thus, one of the objectives of this study is to empirically analyse the impact of TPM between TQM and business performance.

2.0 Literature Review

The purpose of TPM is to minimise breakdown and to maximise equipment availability of production systems at minimal cost [12]. TPM can be defined as an improvement programme that establishes a comprehensive productive-maintenance system throughout the entire life of the equipment with the participation of all employees through voluntary team-based activities [15]. However, only few previous studies have examined the mediators and moderators between TQM and business performance [16]. The inconsistent results between TQM and BP are because the mediators and moderators have been overlooked in research designs [17]. This present research proposed a TQM performance model with mediator effect of TPM with structural equation modelling, which is a more comprehensive model for developing countries, specifically for Malaysia.

Research Hypothesis
Relationship among TQM, TPM, and Business Performance

H1: Relationship between TQM and Business Performance
Most previous studies indicate a significant relationship between TQM practices and business performance [18]; [19]. In contrast, other studies show that TQM does not improve business performance [20]; [7]. Other findings show partial correlation between TQM practices and business performance [7]. Accordingly, the author proposes that:

H1: TQM practices are positively significant and have direct effects on business performance.

TPM is significantly supported by TQM for improving business performance [21]. Teeravaraprug et al., [22] suggest that TQM and TPM should be implemented before lean production. Two sets of factors are critical for the effectiveness of TQM and TPM: (1) universally significant factors for all three approaches such as leadership, process management, and strategic planning; and (2) approach-specific factors such as equipment management and focus on customer satisfaction [23]. Thus, TQM practices are positively correlated with TPM.
H2a: TQM practices are positively significance and direct effect on TPM.

Ahuja and Khamba [24] note the critical success factors of TPM, such as top management leadership and involvement, maintenance practices, as well as holistic TPM initiatives, enhance business performance in Indian industry. Konecny and Thun [21] indicate that TQM and TPM supported by human resource practices significantly improve business performance. Ahmad et al.,[1] and Nakajima [12] indicate that TPM practices have impact on business performance. Therefore, TPM practices are positively correlated with business performance. Accordingly, the author proposes that:
H2b: TPM practices are positively significance and direct effect on business performance.

There are lack of empirical evidences of TPM as mediator between TQM and business performance in previous work [16]. TQM provides soft and hard aspects such knowledge, skill, continuous improvement and work environment and culture to support adoption of TPM implementation [24],[25]. Meanwhile, TPM has positively impacts on business performance [6], [27]. Thus, TQM will be helpful for effective and efficient in TPM adoption, in turn increasing business performance. ***H2: TPM is a mediator between TQM and business performance.***

2.0 Methodology

A total of 1500 questionnaires were mailed to these companies. A total 327 questionnaires were returned which represented about 21.8% response rate. The number of returned questionnaires that were found to be usable in this study was 319, which represented about 21.3% response rate.

3.0 Result

SEM

TPM as the mediator variable was included into the model, as shown in Figure 1. Table 1 shows that the relationship between TQM and BP was reduced when TPM was included in the model, but the relationship was still significant with rc from 0.81 (CR=13.177, p<0.01) to 0.55 (CR=7.449, p<0.01) based on SEM result. The result also showed that TQM had a significant and direct effect on TPM with rc=0.80 (CR=14.346, p<0.01). In addition, TPM had a significant and direct effect on BP with rc=0.31 (CR=4.334, p<0.01). The goodness-of-fit indices for the structural model (χ^2/df=2.220, GFI=0.920, AGFI=0.902, TLI=0.937, CFI=0.952, and RMSEA=0.044) were well within the generally accepted limits, indicating a good fit to the data. Thus, it can be concluded that TPM partially mediated the relationship between TQM and BP. Standard indirect effect (IE) of TQM to BP was 0.250, which can be categorised as high effect of mediation [27].

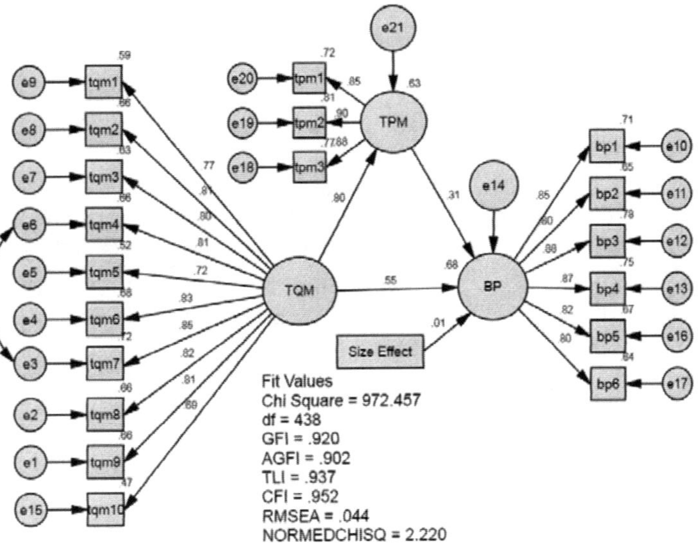

Figure 1 Mediator testing for TPM between TQM and BP

Table 1: Mediator testing for TPM between TQM and BP

No.	Hypotheses	Links in the model	Standardised Estimate (rc)	CR	p-value	Result	Remark
1.	H1	TQM → BP	0.81	13.177**	0.000	Supported	Without TPM
2.	H1	TQM → BP	0.55	7.449**	0.000	Supported	With TPM
	H2a	TQM → TPM	0.80	14.346**	0.000	Supported	With TPM
	H2b	TPM → BP	0.31	4.334**	0.000	Supported	With TPM

Note: *p<0.05; **p<0.01 (one-tailed test)

4.0 Discussion

It was expected that TPM was a mediator between TQM and business performance. This finding is supported by Seth & Tripathi [23] and Konecny & Thun [21]. Konecny & Thun [21] prove that TPM is a mediator between TQM and business performance in order to improve quality, flexibility, cost, and time in the context of manufacturing industry in Germany. The results of this present study were also in agreement with Nakajima [12] who states that the objectives of TPM are zero breakdown and zero defects to improve productivity and quality, to reduce inventory and delivery cycle time, and to improve safety and employee morale.

5.0 Conclusion

The result concludes that TPM is partial correlation between and TQM and BP with indirect effect (IE) is 0.25 which can be categorised as high moderator effect based on SEM result as shown Table 1. It has proved that the impact of TPM as mediator based on system theory and the importance of TPM in TQM implementation.

Acknowledgement
This research was supported by Malaysian Ministry of Higher Education (MOHE) under Exploratory Research Grant Scheme (ERGS) to the Office of Research, Innovation, Commercialization and Consultancy (ORRIC) at the Universiti Tun Hussein Onn Malaysia (vot:E044).

References:

[1] M.F Ahmad, N. Zakuan, A. Jusoh, and J. Takala, "Relationship of TQM and Business Performance with Mediators of SPC, Lean Production and TPM," *Procedia - Social and Behavioral Sciences*, vol. 65, no. 2012, pp. 186-191, Dec. 2012.

[2] M.F.B Ahmad and S. Yusof, "Comparative study of TQM practices between Japanese and non-Japanese electrical and electronics companies in Malaysia: Survey results," *Total Quality Management & Business Excellence*, vol. 21, no. 1, pp. 11-20, Jan. 2010.

[3] M.F.B Ahmad, S. Yusof, and N. Yusof, "Comparative study of quality practices between Japanese and non-Japanese base electrical and electronic companies in Malaysia: A Survey," *Jurnal teknologi*, vol. 47, no. A, pp. 75-89, 2008.

[4] M. F. Ahmad, N. Zakuan, A. Jusoh, and J. Takala, "Relationship of TQM and Business Performance with Mediators of SPC, Lean Production and TPM," *Procedia - Social and Behavioral Sciences*, vol. 65, no. 2012, pp. 186-191, Dec. 2012.

[5] M. F. Ahmad, N. Zakuan, a. Jusoh, S. M. Yusof, and J. Takala, "Moderating Effect of Asean Free Trade Agreement between Total Quality Management and Business Performance," *Procedia - Social and Behavioral Sciences*, vol. 129, pp. 244-249, May 2014.

[6] J. M. Juran, *Juran's Quality Handbook*. Singapore: McGRAW-HILL, 1998.

[7] M. Demirbag, E. Tatoglu, M. Tekinkus, and S. Zaim, "An analysis of the relationship between TQM implementation and organizational performance: Evidence from Turkish SMEs," *Journal of Manufacturing Technology Management*, vol. 17, no. 6, pp. 829-847, 2006.

[8] M.F. Ahmad, M. Arif, N. Zakuan, S. Rahman, T. Abdullah, and Fadzil, "The Effect of Demographics on Customer Satisfaction amongst Malaysia Hajj Pilgrims: Survey Result," *Applied Mechanics and Materials*, vol. 660, pp. 1000-1004, 2014.

[9] M. F. Ahmad, N. Zakuan, A. Jusoh, and J. Takala, "Review of Relationship between TQM and Business Performance," *Applied Mechanics and Materials*, vol. 315, no. 2013, pp. 166-170, Apr. 2013.

[10] W. B. Rouse, "A theory of enterprise transformation," *Systems Engineering*, vol. 8, no. 4, pp. 279-295, 2005.

[11] L. Hans, "Measuring maintenance performance- in search for a maintenance productivity index," *Int. J. Production Economics*, vol. 63, no. 8, p. 47, 2000.

[12] S. Nakajima, *Introduction to Total Productive Maintenance (TPM)*. Cambridge: Productivity Press, 1988.

[13] A. B. Abdallah, "The Influence of 'Soft' and 'Hard' Total Quality Management (TQM) Practices on Total Productive Maintenance (TPM) in Jordanian Manufacturing Companies," *International Journal of Business and Management*, vol. 8, no. 21, pp. 1-13, Oct. 2013.

[14] A. B. Abdallah and Y. Matsui, "JIT and TPM•: Their relationship and impact on JIT and competitive performances," *TPM*, vol. 1, no. 9, pp. 1-18, 2007.

[15] B. Dal, P. Tugwell, and R. Greatbanks, "Overall equipment effectiveness as a measure for operational improvement.," *Journal of Operations and Production Management*, vol. 20, no. 12, pp. 1488–1502, 2000.

[16] E. Sadikoglu and C. Zehir, "Investigating the effects of innovation and employee performance on the relationship between total quality management practices and firm performance: An empirical study of Turkish firms," *International Journal of Production Economics*, vol. 127, no. 1, pp. 13-26, Sep. 2010.

[17] R. M. Baron and D. A. Kenny, "The Moderator-Mediator Variable Distinction in Social Psychological Research•: Conceptual , Strategic , and Statistical Considerations," *Social Psychology*, vol. 51, no. 6, pp. 1173-1182, 1986.

[18] M. Jun, S. Cai, and H. Shin, "TQM practice in maquiladora: Antecedents of employee satisfaction and loyalty," *Journal of Operations Management*, vol. 24, no. 6, pp. 791-812, Dec. 2006.

[19] J. C. Bou and I. Beltrán, "Total Quality Management, human resource strategy and firm performance•: an empirical study," *Total Quality Management*, vol. 16, no. 1, pp. 37-41, 2005.

[20] P. Corredor and S. Goñi, "TQM and performance: Is the relationship so obvious?," *Journal of Business Research*, vol. 64, no. 8, pp. 830-838, Aug. 2011.

[21] P. A. Konecny and J. H. Thun, "Do it separately or simultaneously—An empirical analysis of a conjoint implementation of TQM and TPM on plant performance," *International Journal of Production Economics*, vol. 133, no. 2, pp. 496-507, Oct. 2011.

[22] J. Teeravaraprug, K. Kitiwanwong, and N. SaeTong, "Relationship model and supporting activities of JIT , TQM and TPM," *Songklanakarin Journal of Science and Technology*, vol. 33, no. 1, pp. 101-106, 2011.

[23] D. Seth and D. Tripathi, "Relationship between TQM and TPM implementation factors and business performance of manufacturing industry in Indian context," *International Journal of Quality & Reliability Management*, vol. 22, no. 3, pp. 256-277, 2005.

[24] I. P. S. Ahuja and J. S. Khamba, "An evaluation of TPM initiatives in Indian industry for enhanced manufacturing performance," *International Journal of Quality & Reliability Management*, vol. 25, no. 2, pp. 147-172, 2008.

[25] S. A. Brah and W. K. Chong, "Relationship between total productive maintenance and performance," *Journal of Production Research*, vol. 42, no. 12, pp. 2383-2401, 2004.

[26] I. P. S. Ahuja and J. S. Khamba, "An evaluation of TPM initiatives in Indian industry for enhanced manufacturing performance," *International Journal of Quality & Reliability Management*, vol. 25, no. 2, pp. 147-172, 2008.

[27] J. Cohen, "Statistical Power Analysis for the Behavioral Sciences." NJ: Lawrence Erlbaum, Hillside, 1988.

[28] D. Seth and D. Tripathi, "Total Quality Management & Business A critical study of TQM and TPM approaches on business performance of Indian manufacturing industry A Critical Study of TQM and TPM Approaches on Business Performance of Indian Manufacturing Industry," *Total Quality Management*, vol. 17, no. 7, pp. 37-41, 2006.

Relationship Between Lean Production and Operational Performance in the Manufacturing Industry

Raja Zuraidah RM Rasi[1], **Umol Syamsyul Rakiman**[2], **Md Fauzi Bin Ahmad**[3]
Department of Production and Operations Management,
Faculty of Technology Management and Business,
Universiti Tun Hussein Onn Malaysia,
Johor, Malaysia

E-mail: [1]rzuraida@uthm.edu.my, [2]avid8787@yahoo.com, [3]mohdfauzi@uthm.edu.my

Abstract. Nowadays, more and more manufacturing firms have started to implement lean production system in their operations. Lean production viewed as one of the mechanism to maintain the organisation's position and to compete globally. However, many fail to apply the lean concepts successfully in their operations. Based on previous studies, implementation of lean production in the manufacturing industry is more focused on the relationship between Lean and Operational Performance of one dimension only. Therefore, this study attempted to examine the relationship between Lean Production (LP) and Operational Performance in 4 dimensions which are quality, delivery, cost and flexibility. This study employed quantitative study using questionnaires. Data was collected from 50 manufacturing industries. The data was analysed using Statistical Package for Social Science (SPSS) 22.0. This study is hoped to shed new understanding on the concept of Lean Production (LP) in regards of Operational Performance covering the 4 dimensions.

1. Introduction

In the era of globalisation, organisations focus on maintaining its competitiveness at the global level by offering quality products and good services. Lean production is one of the mechanism used by many organisations that is said can improve operational performance. Many benefits of lean production are now well recognised, such as improved productivity and reduced costs through the removal of waste in operations.

Many firms implement lean production to enhance competitiveness, but the majority of them were disappointed with results [1]. This happened to be in the developed country which has long implemented it which might reduce the implementation in the developing country as the results are discouraging. In order to avoid that, with still flourishing hope as Malaysia also implement Lean Production, a positive relation between LP and Operational Performance might stop that from happening as LP promises many improvements towards Operational Performance. The aim of this study is to fill the gap by reviewing the relationships between Lean Production (LP) and operational performance in the manufacturing industry in Malaysia.

2. Literature Review

2.1 Lean Production

Lean is a form of continuous improvement with the main aim to identify and eliminate waste. Waste refers to the activities that does not affect value aspect of a product or service based on customer's perception [2]. According to [2], waste in this context refers to excess in production, waiting time, inventory, process scrap, transportation, damage or defect of the product and movement.

Excess in production can be defined as waste took place when excess order from the customer happened or producing the product earlier, prior to the customer's order. The next type waste is the waiting time, it would be considered as waste if idle time, storage time and waiting time for delivering the product to customer or transfer time for semi-finished product from department to department for assembly occurred. While inventory can be considered as waste, if the raw material is not used in the production, work-in-process, finished materials and operational supply excess.

Besides that, excess in production took place when the work done in producing the product does not contribute towards any benefits or value to the company. Transportation is the allocation of an item from one spot to the other and movement of product more than once is considered as waste. Product defect is such as warranty claim and returned product. Movement on the other hand refers to the unnecessary tools or workers movement in executing the given task.

Lean has been used broadly in most industries because this practice gives a lot benefits towards the organisation in increasing the work pace, reduce work capital, increase financial flow, increasing inventory turnover, market domination, increasing profit and finally fulfilling the demand from customers [3]

2.2 Lean Production

Lean Production originated from the Just-in-time (JIT) concept a few years back. There were a few companies in the world which failed to adopt this concept in the 1980s, even when all of the benefits is clear from the implementation of the JIT concept [5].

LP often implemented as proxy in Toyota Production System (TPS), where it flourished from the experiments and initiatives of Taiichi Ohno for three decades in the Toyota Motor company. Later, TPS officially being introduced in the United States of America in 1984 when New United Motor Manufacturing Inc. (NUMMI) was established and initially taken over as a partnership with the United States of America [6].

According to [7], LP can be defined as a system of business to organise and manage product development, operation, suppliers and customer relations with the aim of reducing the usage of human power, space capital and time in the production of product and reducing defects in satisfying the customer in comparison with previous mass production approach.

Aside from that, based on the research from the International Motored Vehicles Programme (IMVP) in the Massachusetts Institute of Technologies (MIT) showed that LP combined both the characteristics of mass production and craft production. From this combination, LP had the ability to reduce cost per unit and improving quality dramatically. At the same time, it also prepares a wider array of product and form of production system that is flexible which is essential in meeting the ever changing customer demand [8]. Many practitioners and researchers stated that investment in LP can reduce the cost greatly, productive work force, reducing lead time and even better quality [9] & [6].

2.3 Operational Performance

Operational performance conventionally discussed from the aspect of priorities of strategic operational competition in the manufacturing industry [10]. Competition priorties is a critical operational dimension for any process or supply chain in fulfilling internal or external customer's satisfaction, both or future customer. Competition priorities is planned for processing and and creating supply chain [11]. Based on previous study, most of the researcher adopt four (4) main elements in measuring operational performance in the manufacturing industry which covers quality, delivery, flexibility and cost [12],[13],[14].

2.4 Quality

Quality is the ability of a product or service to meet the customer's demand and satisfaction [3] (Heizer dan Barry, 2011). In measuring operational performance, quality were discussed in the form of product performance, product endurance and product's acceptance within the limit of design specifications [13].

In the manufacturing industry, product durability usually being measured by the time consumed to produce output. The longer it takes for a product to last or the longer it takes for it to be malfunctioned, the higher the durability of the product. Therefore, companies always measure product durability through the average time in between production [15].

2.5 Delivery

Product delivery refers to the time taken for a product to be delivered to the customer. Reduction of the lead time, faster delivery than the competitors and on time product delivery to the customer had been used to measure operational performance in the manufacturing industry [13] & [14].

2.6 Flexibility

Flexibility of production can be defined as how far the manufacturing operation react to to the constant change of number of order, type and characteristics of the produced product. Flexibility allows the company to react promptly to the change in the market such as in reacting to the competitors, customer and reducing the waiting time between the product's order and delivery [15]. Based on [3], flexibility refers to the ability of a company to react withn penalty in term of time, cost and customer's value.

2.7 Cost

Cost refers to the needed payment to produce product. Workers productivity, production cost and reduction in inventory is used in measuring the operational performance in themanufacturing industry [14]. While according to [15], workers productivity is the cost or the amount of working hours needed to produce per unit of output. In other word, the lower the labour cost used to produce an output, the higher the workers productivity.

2.8 Lean Production and Operational Performance

In order to increase operational performance, lean production essentially need to be practiced by the higher management in the organisation. As an example, Just in Time (JIT) is one of the LP elements which can eliminate waste in the manufacturing process [3]. A few studies also found that LP has a significant relation with the operational performance in the manufacturing industry [14] & [16].

2.9 Lean Production and Quality

Based on the study by [14] & [16], lean production implementation has a positive relation with quality in affecting operational performance. As an example, [17] found that LP helped to increase product's quality and durability [16]. While according to [3], LP affects work force, inventory reduction, quality, and flexibility.

2.10 Lean Production and Delivery

Lean production also has a positive relation in regard to the operational performance and the delivery performance in the manufacturing industry [14] & [16]. This is because one of the objective of LP implementation is to reduce variation and overcoming activities that does not contribute to value development in the manufacturing process which will increase the processing time. As an example, LP will positively affect the delivery time punctuality, average cycle time, reducing waiting time in the manufacturing process and reducing the time taken to deliver the product to the customer [16].

2.11 Lean Production and Flexibility

Lean Production also has a positive relation in regards to the flexibility performance. As an example, LP can increase speed, reduce the manufacturing cost, ease introducing new product and at the same time helping the company to produce output in accordance to the real market demand and not based on forecast. This situation can increase variation of product, amount of flexibility and the ability to assimilate product to the market. Reduction in the waiting time is an important element in increasing productivity [16].

2.12 Lean Production and Cost

Eventhough the relation is not that apparent, there is positive relation between LP and cost in manufacturing company. The level of relation between LP and cost is slightly good. In practice, LP will reduce inventory level [16]. Besides that, according to [18], he found that LP practise can reduce the waiting time and production cost effectiveness aside from increasing the quality of product and productivity.

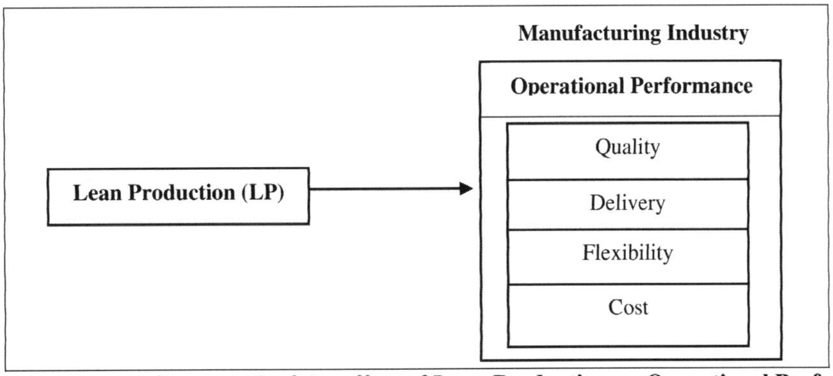

Figure 1: Conceptual framework of the affect of Lean Production on Operational Performance

Based on the past literatures, theoretically LP should have a positive relation with Operational Performance as it positively affect quality which is the first dimension of Operational Performance. Figure 1 shows the relation between LP and the other three dimensions of Operational Performance aside from quality which are delivery, flexibility and cost.

3.0 Methodology

This research adopted a quantitative approach as a way to measure the relation between Lean Production and operational performance. In alliance with that approach, a sample need to be selected from a population. According to [19], sampling framework is a methodology being used in selecting the sample from a population. In this study, purposive sampling was chosen to collect the precise data. According to [20], sampling is a procedure where a group of subject with a certain characteristic was chosen as the respondent. In this study, only manufacturing companies that practise Lean Production were chosen in answering the questionnaire.

3.1 Population and Sample

Population for this study consist operation manager, director of a company, production manager and knowledgable individual in Lean Production (LP) in the manufacturing industry. Based on the data base from Federation of Malaysian Manufacturers (FMM) 2012 there are 206 manufacturing company from the selected sample area. From the population, the sample size of the study was determined based on [21]. As much as 136 were purposively selected. The selected standard error for this research is 5 percent which brings towards the degree of freedom of 95 percent. Thus, the selected chi-square value equals to 3.84. Below is the formula showing the method to calculate the sample size based on [21]:

$$s = X^2NP(1-P) + d^2(N-1) + X^2P(1-P) \quad (3.1)$$

s= Needed sample size.
X^2= Chi-square value for the level of freedom towards the desired freedom (3.841).
N= Population size.
P= Proportionality of population (assumed as 0.50 since the sample size is not maximum)
d= Level of accuracy is expressed as a proportion (0.50) .

3.2 Data Collection Instrument

Questionnaire was used as the study instruments in collecting the data.

3.3 Data Collection Procedure

Data were collected from primary and secondary data source. The primary data was from the survey and the secondary data was from the literature review. This will be used in the analysis part.

4.0 Data Analysis

Statistical Package for Social Science (SPSS) 22.0 version was used in analysing the data. After the data has been collected, normality test was conducted to justify whether the data is normal or not. If the data is normal, then Pearson Regression method will used in explaining correlation between Lean Production and operational performance. If the value exceeds 0.01 then there is positive correlation betwwen the tested variables which in this case is between LP and operational performance.

4.1 Normality Test

According to [20], normality test is to identify whether a data set for a study is normally distributed or not. In this study skewness and kurtosis is used in identifying the data normality. In SPSS, the measurement of skewness and kurtosis should not far from zero. Z value is used in testing the normality. The data is normal if the Z value is in between -1.96 and +1.96 [22], [23], [24]. The statistical value is divided with the standard error to obtain the the Z value in the skewness and kurtosis.

Table 1: Normality test result (Skewness and Kurtosis)

Details	N	Min	Standard deviation	Skewness			Kurtosis		
	Statistic	Statistic	Statistic	Statistic	Standard error	Z Value	Statistic	Standard error	Z Value
Lean Production (LP)	51	3.782	0.589	-0.346	0.333	-1.039	0.358	0.656	0.546
Quality	51	4.0327	0.681	-0.387	0.333	-1.161	-0.348	0.656	-0.53
Delivery	51	4.000	0.686	-0.186	0.333	-0.558	-1.132	0.656	-1.725
Flexibility	51	3.719	0.734	-0.581	0.333	-1.742	0.772	0.656	1.176
Cost	51	3.510	0.563	-0.092	0.333	-0.275	-0.747	0.656	-1.139

Based on Table 1, Z value on skewness shows no elements exceeding the Z value as dictated by [22], [23], and [24]. The Z value for Lean Production in term of skewness, is -1.039 and Kurtosis, Z = 0.546, which confirm that the data is normal.

The Z value for quality is -1.161, while the Z value for kurtosis is -0.530. This indicates that data for quality is normal. The next one is delivery where the Z value is, -0.558 for skewness and Z = -1.725 for kurtosis which is normal. While the Z value for flexibility for skewness and kurtosis each is -1.742 and 1.176 which confirm that the data is normal. The last one in the normality test is cost. The Z is -0.275 at skewness and for kurtosis, Z is -1.139, which is the indication that the data is normal. As a conclusison all of the collected data is normal and can be proceeded for analysation.

4.2 Relation Betwwen Lean Production (LP) and Operational Performance

Table 2: Pearson correlation between Lean Production (LP) and Operational Performance

Variable		Lean Production (LP)	Quality	Delivery	Flexibility	Cost
Lean Production (LP)	Pearson Corelation	1	.608**	.387**	.439**	.486**
	Sig. (2-tailed)		.000	.005	.001	.000
	N	51	51	51	51	51
Quality	Pearson Corelation	.608**	1	.628**	.592**	.530**
	Sig. (2-tailed)	.000		.000	.000	.000
	N	51	51	51	51	51
Delivery	Pearson Corelation	.387**	.628**	1	.450**	.420**
	Sig. (2-tailed)	.005	.000		.001	.002
	N	51	51	51	51	51
Flexibility	Pearson Corelation	.439**	.592**	.450**	1	.547**
	Sig. (2-tailed)	.001	.000	.001		.000
	N	51	51	51	51	51
Cost	Pearson Corelation	.486**	.530**	.420**	.547**	1
	Sig. (2-tailed)	.000	.000	.002	.000	
	N	51	51	51	51	51

**. Correlation is significant at level 0.01 (2-tailed).

Pearson Regression method was used in analysing the data from the respondents in stating whether there is significant correlation between LP and Operational Performance. Based from the result in Table 2, there is a positive correlation between LP and Operational Performance which can be seen as follows. The correlation value which is (r) between LP and quality is 0.608, delivery ($r = 0.387$), flexibility ($r = 0.439$) and cost ($r = 0.486$). Hence, all of the correlation values are more than 0.01 which indicates there is a significant positive correlation between LP and Operational Performance.

4.2.1 Lean Production (LP) has a positive relation with quality in affecting Operational Performance in manufacturing industry.

Positively, from the result it shows that the implementation of LP had a positive relation with quality in affecting Operational Performance in the manufacturing industry. This result is in line with the study done by [14], [16].

4.2.2 Lean Production (LP) had a positive relation with delivery in affecting the Operational Performance in the manufacturing industry.

In term of delivery, the result from the survey indicated a positive relation between LP and Operational Performance in the manufacturing industry which is in conformity with study conducted by [14], [16].

4.2.3 Production (LP) had a positive relation with flexibility in affecting the Operational Performance in the manufacturing industry.

Lean Production had been confirmed to have a positive relation with flexibility in the manufacturing industry as being apposed by [16].

4.2.4 Lean Production (LP) had a positive relation in reducing cost in affecting the Operational Performance in the manufacturing industry.

It has been confirmed by implementing Lean Production, it will positively reduce cost in affecting the Operational Performance. This finding is in line with ither previous study from [3], [16] & [20].

4.3 Discussion

In the first dimension, LP is positively related with quality. In LP, product conformity with the specifications begins since the early stage in order to avoid waste and increase the quality. This is because the worker must strictly follow the guidelines, procedure and specifications in producing a product as a way to eliminate the possibility of waste and subsequently the quality of a product will be kept assured. As an example, if a products being produced from aluminium with the cutting dimension of 30cm X 21cm, the procurement department must order the right metal that is aluminium. The subsequent department which is production department must cut the aluminium based on the specifications to avoid problem in the next process since the aluminium is not fit for processing since it is not the meant size. These are some of the examples of how LP practise in a manufacturing company can affect quality since the initial stage.

This practise will then affect the delivery (second dimension) as when the production runs smoothly with quality, there will be no delay in delivering the product to the customers. The combination of quality product and on time delivery will then affect costs (third dimension). When there is no issue of quality such as defects and delivery is on time, there will be no extra cost of reproduction to replace defect product and send it back again to the customer which will save cost tremendously.

When there is no complaint from the customer on the quality and delivery, the manufacturing company can create new product since they have more time to improve what they are producing instead of wasting time repairing mistakes that is avoidable. This is exactly as being said in [25], when operational performance is increased new product development will come next. They also can produce more products as everything is as planned. This is the flexibility (fourth dimension) of a manufacturing company which is affected positively from the LP practise.

5.0 Conclusion

As a conclusion, the four dimensions of Operational Performance is positively related with the Lean Production practise in the manufacturing industry. This thus conform the finding to the previous studies on LP and operational performance. It is important to confirm this finding as an encouragement for other industries in other nations whether developed or soon to be developed as LP carries multiple dimensions benefits which cover quality, delivery, flexibility and cost.

Practically the four dimensions of Operational Performance will be positively affected by the LP practise. Nevertheless nothing is perfect, problems might occur here and there in the manufacturing company. This is where LP comes in handy as it avoids problems and increase Operational Performance since it promotes conformity with specifications to deliver result and avoid waste both in the work force and the products.

6.0 Acknowledgements

The author is grateful to the Department of Production and Operation Management for providing the facilities to carry out the research. This research was supported by Malaysian Ministry of Higher Education (MOHE) under Exploratory Research Grant Scheme (ERGS) to the Office of Research, Innovation, Commercialization and Consultancy (ORRIC) at the Universiti Tun Hussein Onn Malaysia (vot:E044).

7.0 References

[1] Anand, G., Ward, P.T., Tatikonda, M.V. & Schilling, D.A. 2009, "Dynamic capabilities through continuous improvement infrastructure", Journal of Operations Management, vol. 27, no. 6, pp. 444-461.

[2] Paul Myerson. (2012). Lean Supply Chain& Logistics management. The McGraw-Hill Companies, Inc. [10] Narasimhan, R., & Das, A. (2001). The impact of purchasing integration and practices on manufacturing performance. Journal of Operations Management,19(5), 593-609.

[3] Heizer J. & Render B. (2011). Operation Management. (10th edition). Pearson Education Limited.

[4] Kerber B. & Dreckshage B. J. (2011). Lean Supply Chain Management: A Framework for Material Managers. Taylor and Francis Group LLC.

[5] Arnold J. R. T., Chapman S. N. & Clive L. M. (2008). Introduction to Materials Management. Pearson Education, Inc., Upper Saddle River, New Jersey 07458.

[6] Shah, R., & Ward, P. T. (2007). Defining and developing measures of lean production. Journal of operations management, 25(4), 785-805.

[7] Kerber B. & Dreckshage Brian J. (2011). Taylor and Francis Group, an Informa business.

[8] Wong, Y. C., & Wong, K. Y. (2011). Approaches and practices of lean manufacturing: The case of electrical and electronics companies. African Journal of Business Management, 5(6), 2164-2174.

[9] Al Samadi, S. (2009). Kaizen strategy and the drive for competitiveness: challenges and opportunities. Competitiveness Review, 19(3), 203-211.

[11] Krajewski L.J., Ritzman L. P. & Malhotra M. K. (2010) Operations Management: process & supply chain. (9th edition).Prentice Hall Upper Saddle River, New Jersey 07458.

[12] Hallgren, M., & Olhager, J. (2009). Lean and agile manufacturing: external and internal drivers and performance outcomes. International Journal of Operations & Production Management, 29(10), 976-999.

[13] Rahman, S., Laosirihongthong, T., & Sohal, A. S. (2010). Impact of lean strategy on operational performance: a study of Thai manufacturing companies.Journal of manufacturing technology management, 21(7), 839-852.

[14] Chavez, R., Gimenez, C., Fynes, B., Wiengarten, F., & Yu, W. (2013). Internal lean practices and operational performance: The contingency perspective of industry clockspeed. International Journal of Operations & Production Management, 33(5), 562-588.

[15] Williams, C. (2013). Management (7th edition). United State of America

[16] Laugen, B. T., Acur, N., Boer, H., & Frick, J. (2005). Best manufacturing practices: what do the best-performing companies do?. International Journal of Operations & Production Management, 25(2), 131-150.

[17] Kannan, V. R., & Tan, K. C. (2005). Just in time, total quality management, and supply chain management: understanding their linkages and impact on business performance. Omega, 33(2), 153-162.

[18] Cua, K. O., McKone, K. E., & Schroeder, R. G. (2001). Relationships between implementation of TQM, JIT, and TPM and manufacturing performance. Journal of Operations Management, 19(6), 675-694.

[19] Levy, P. S., & Lemeshow, S. (2013). Sampling of populations: methods and applications. John Wiley & Sons.

[20] Chua, Y. P. Kaedah Penyelidikan. Malaysia: Mc Graw Hill. 2006.

[21] Krejcie, R. V., & Morgan, D. W. (1970). Determining sample size for research activities. Educational and psychological measurement, 30(3), 607-610.

[22] Cramer, D. (1998). Fundamental statistics for social research. Step-by-step calculations and computer techniques using SPSS for Windows. London and New York: Routledge.

[23] Cramer, D., & Howitt, D. (2004). The sage dictionary of statistics: A practical resource for students in the social sciences (pp. 168).

[24] Doane, D. P., & Seward, L. E. (2011). Measuring skewness: a forgotten statistic. Journal of Statistics Education, 19(2), 1-18.

[25] Jürgens, U. (2000). *New product development and production networks.* Berlin: Springer. (pp. pp446)

Implementation of RS-485 Communication between PLC and PC of Distributed Control System Based on VB

Chuan Lian Zhang[1], Zhi Da Huang[2], Gui Qing Zhou[2] and Kil To Chong[1, 3]

[1]Chonbuk National University, Jeonju, South Korea

[2] Huaihai Institute of Technology, Lianyungang, China

[3]Corresponding author, Advanced Electronics and Information Research Centre, CBNU, Jeonju, Korea

E-mail:kitchong@jbnu.ac.kr

Abstract. This paper focuses on achieving RS-485 communication between programmable logical controller (PLC) and PC based on visual basic 6.0 (VB6.0) on an experimental automatic production line. Mitsubishi FX_{2N} PLCs and a PC are chosen as slave stations and main station, respectively. Monitoring software is developed using VB6.0 for data input/output, flow control and online parameters setting. As a result, all functions are fulfilled with robust performance. It is concluded from results that one PC can monitor several PLCs using RS-485 communication.

1. Introduction

PLC as a high-performance control device has been widely used in distributed control systems [1]. There usually are three platforms to monitor PLC, which are touch panels, configuration software and universal software platform such as VB, C, and VC [2]. Each platform has advantages and disadvantages. Touch panel does not need to consider communication problems but has limited functions; configuration software has varieties of functions but is too expensive for medium production system; the software developed on universal software platform can customize for some small and medium-scale production systems [3] but need to consider communication problems. In these systems, people prefer to use senior program platforms to develop a simpler, more applicable and economical communication program to monitor production system. Therefore, it has great practical significance to find a highly efficient and cost-effective communication method.

There are many works that use different platforms to monitor different brand PLCs with different communication methods. For example, a combination of VB6.0, FX_{2N} PLCs and RS-232 was come up with by a group [4], another group's combination using RS-422 [5]. Besides that, some papers also use the combination of Siemens PLCs (S7-200), VB and RS-232 [6] or RS-485 [7].

This project is mainly about using RS-485 communication to implement communication between PLCs and PC and developing the monitoring software. RS-232, RS-422 and RS-485 are the most three common communication ways between PLCs and PC [5]. Only RS-485 can be used effectively over long distance and in electrically noisy environments. Multiple receivers can be connected to a network in a parallel way. So RS-485 series port communication is a good option to implement communication between PLC and PC, which also is the most difficult and key part of this paper. The monitoring software will be developed with VB 6.0, which has advantages about interface programing. After finishing those two parts, some experiments will be done on an experimental automatic production line only with one PLC. As a result, the monitoring software shows good performance on controlling the automatic produce line. Therefore, it is can control more PLCs and can be applied in practical automatic production line.

The system hardware are described in Section II. Section III gives an overview about implementation of communication between PC and PLC. The structure of program of PLC and VB6.0 is presented in Section IV. The results, presented in Section V, is a clear illustration of all functions such as monitoring all devices, controlling the automatic line and jogging and so on.

2. System hardware design

To fulfill these functions of system, the platform should be suitably integrated with hardware and software. When the requests are produced, the computer sends signals to the PLCs through the converter called USB to RS-485. Each PLC is connected with a 485ADP adapter, which transform signals in RS-485 to PLCs' format signal [8]. All PLCs are connected with each other in a parallel way. Each PLC can achieve different controlling functions according to system requirements. Figure 1 shows basic structure of the system.

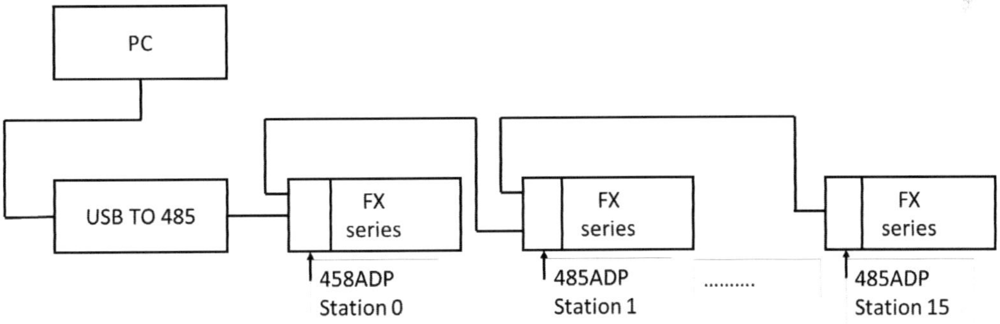

Figure 1. System structure.

3. Implementation of communication between PC and PLC

3.1. Initialization communication format

The communication format decides communication setting (data length, parity, and baud rate, etc.) between computer and PLCs. The communication format can be set using the special data register D8120 in PLCs. Setting D8120 value makes it in accordance with external equipment, which refers to VB's MSCOMM Control in this system. Figure 2 shows parameters that chose and set in the PLCs. Figure 3 shows communication parameters set with VB6.0 [8, 9]. After setting same communication format, the system can send and receive data according to communication protocol.

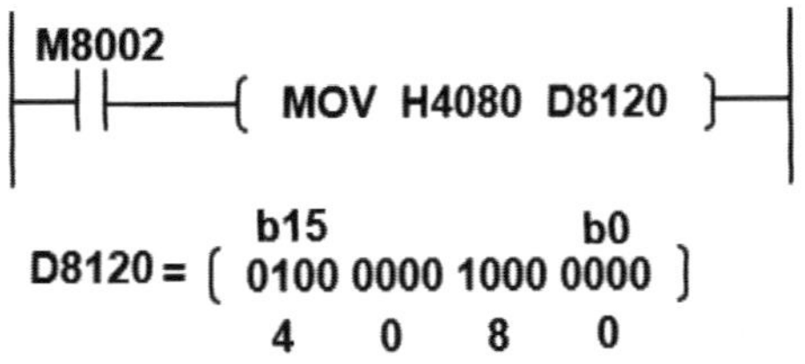

Figure 2. Setting parameters in PLCs.

MSComm1.Settings="9600,N,7,1"

Figure 3. Setting parameters in VB6.0.

3.2. Introduction to communication protocol

FX series PLC provides two basic two formats for dedicated protocol, which can be selected by setting special data register D8120. This system chose format 1. Figure 4 shows specific communication protocol about reading and writing data. According to communication protocol, it can use different operation command to get a complete command string, which is corresponding string sent from VB6.0. For example, the monitoring software needs state of device X1 on, then it get the string "0500FFBRAX00101" according to communication protocol. For more details, please refer to Mitsubishi PLC communion Manual [8].

Table 2. control protocol format

<table>
<tr><th>Description</th><th>Control protocol</th></tr>
<tr><td>To read data from PLC to computer</td><td>Command format:
<table><tr><td>ENQ</td><td>Station No.</td><td>PC No.</td><td>Command</td><td>Message wait time</td><td>Character area A</td><td>(sum check code)</td><td>(CR/CF)</td></tr></table>
Return code format:
<table><tr><td>STX</td><td>Station No.</td><td>PC No.</td><td>Character area B</td><td>ETX</td><td>Sum check code</td></tr></table>
</td></tr>
<tr><td>To write data from computer to PLC</td><td>Command format:
<table><tr><td>ENQ</td><td>Station No.</td><td>PC No.</td><td>Command</td><td>Message wait time</td><td>Character area A</td><td>(sum check code)</td><td>(CR/CF)</td></tr></table>
Return code format:
<table><tr><td>ACK</td><td>Station No.</td><td>PC No.</td></tr></table>
</td></tr>
</table>

4. Software program design

This section can be divide into two parts: PLC's program structure and VB6.0's program structure, which use USB to RS485 converter to connect to each other and has same communication parameters.

4.1. PLC's program structure

There are five PLC programming languages: Ladder diagram (LD), Sequential Function Charts (SFC), Function Block Diagram (FBD), Structured Text (ST), and Instruction List (IL). This allows program developers to select the language best suited to each particular task depending on their advantages and disadvantages. This paper adopted SFC method because of multiple branches requirements. PLC's program structure is selection branch structure and the selection command is given from main station (PC) by users. One FX_{2N} PLC has about 1000 state devices, which is the basic component of SFC program. According to FX_{2N} PLC manual, State devices (S0~S9) are used as initial step of SFC and middle steps use state devices (S20~S499). For more details, please refer to programming manual of SFC of FX_{2N} PLC.

PLC's program structure has two main parts: initialization communication parameters and selection branches. As figure 5 shows, communication parameters was firstly set. And then there are four branches for different functions coded according to production line requirements. Based on different commands from users, different branches was selected to run to fulfill different functions. At last, all these branches return to main branch.

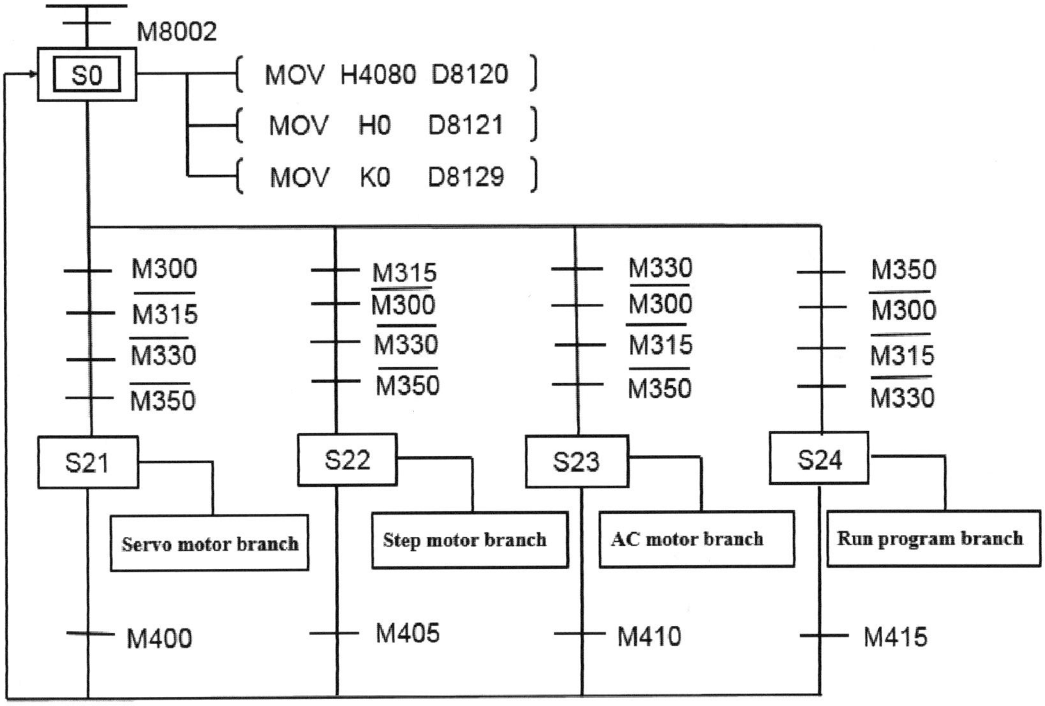

Figure 4. PLC's program structure.

4.2. VB6.0's program structure

Depending on automatic production line requirements and programmer developer's experiences, the monitoring software was designed to have 6 windows, which are main window, communication parameters setting and test window, run window, jog window, online devices' parameters setting and monitoring window, and help window. Every window has different functions and program structure.

Main window provides access to other 5 windows, which needs account and password and can prevent monitoring software operating by other people. Communication parameters, such as communication ports, can be changed with communication parameters setting and test window, which also can check if communication between PLC and PC works. In terms of online devices' parameters setting and monitoring window and run window, the main program's structure is a loop to scan state of PLCs' devices and there is a program interrupt to be used for devices' parameters setting. Besides, online devices' parameters window gives users access to read and write any PLC devices and workers can control and get real-time state information of automatic production line with run window. Jog window makes it easy for technics to tune production line after it was installed, while help window just shows some basic hints to solve some common errors.

5. RESULTS

In order to evaluate the performance of monitoring software and RS-485 communication, kinds of experiments are done on this system, such as online parameters setting and tuning. This part chooses three windows to show main functions of monitoring software.

5.1. Online devices' parameters setting and monitoring

As figure 6 shows, the online devices' parameters setting and monitoring window has two main parts, the left part is used to set and monitor all devices while the right part only can show value or state of some common devices such as input devices X0~X7 and timer devices T0~T7. The left part has four small parts: communication test section, output devices Y0~Y7 setting and resetting section, state devices operation section and numerical devices operation section.

This window demonstrates the ability of monitoring software to achieve online parameters setting and monitoring. For example, workers want to set timer T0's value as 100s. Firstly, operator needs to choose 'T' at numerical devices operation section and inputs timer device's address value '0000' to address textbox of numerical devices operation section. After that T0's value also should be input to the corresponding textbox of same section. Finally, click 'write' button. Referring to FX_{2N} PLC communication user manual, timer devices' value has four bits, which explains that device T0's address value is '0000' instead of '0'.

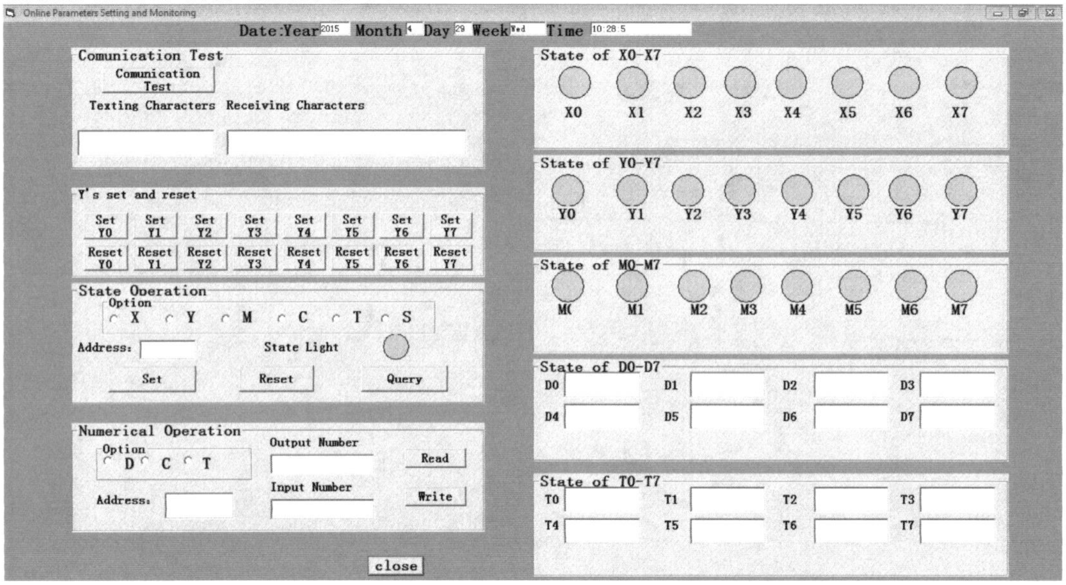

Figure 5. Online devices' parameters setting and monitoring window.

5.2. Jog control

Performance of jog control is an important evaluation, which is very useful for production system tuning and maintaining. Figure 7 shows that this window has three main sections. And it was utilized to control three different kind motors: step motors, AC motors and servo motors. Each section has same structure, which has three buttons from up to down to achieve motors' clockwise rotation, counter clockwise rotation and stop function. The interlock of motors motion control was considered by PLC program instead of this window's program since PLC has better reliability.

5.3. Controlling and monitoring automatic production line

Run window is a custom design window depending on practical control subject's requirements and it can show real-time state of automatic production line such as state of sensors and position of motors. The function of this production line is to select different color balls and put them into different places. With monitoring all PLC devices used by this production line and some VB6.0's functions, this window can show operating state of production line in a animation way. For example, servo motor's image of this window can move depending on production line sensor's state and setting velocity. Though there is a delay, it is acceptable for most automatic production line. With this function, one worker can control many machines in offices.

As figure 8 shows, the main content of this window is an animation picture of production line and four buttons can be found in the bottom of window, which shows different functions: reset, start, stop and quit. Those small blue rectangles indicate all sensors' position in production line and their color can become to red when sensors' state is on. The biggest rectangle stands for conveyor belt while step motor is shown using a big blue circle.

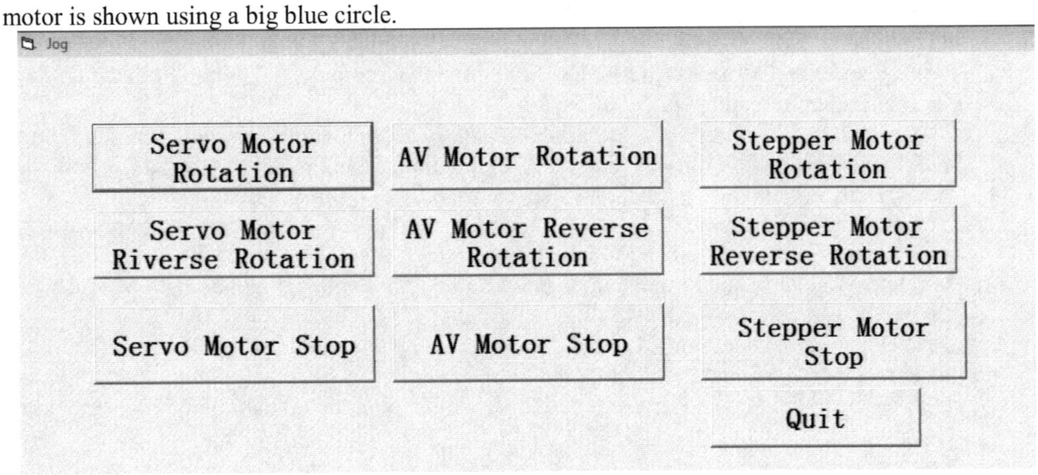

Figure 6. Log window.

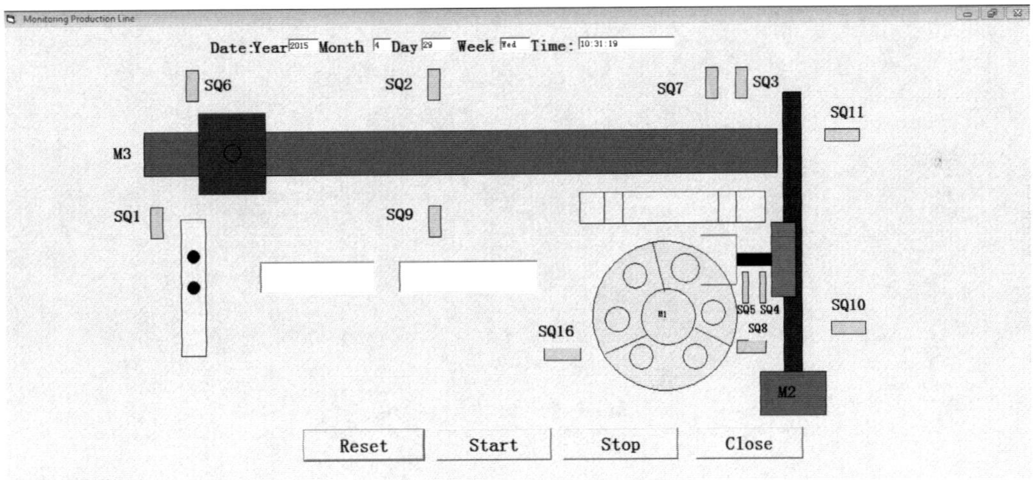

Figure 7. Run window.

6. Conclusion

In conclusion, the monitoring software shows good performance on controlling the experimental automatic production line. Besides that, Experimental results also shows that monitoring software interface is friendly, that has good maneuverability, and that shows reliable performance. Comparing to other works, the innovation of this paper is the original combination of VB6.0, PLC, and RS-485, which other works maybe use different brand PLCs and kinds of software.

Acknowledgement

The research is supported by the National Research Foundation of Korea (NRF) grant funded by the Korea government (MEST) (No. 2013R1A2A2A01068127) and (No.2013R1A1A2A10009458).

References

[1] Z. Chuan-wei and W. Yun-rui, "Implementation of Serial Communication Between PLC and PC," Control and Sensor, pp. 66-70, 2012
[2] Y. Song, Z. Z. Tang, and G. Chuan, "Communication betwenn Computer and PLC by Using VB Programming," Electrical Transmission, pp. 41-43, 1999.
[3] C.-g. MA, X.-q. LIU, D.-a. ZHAO, and T.-h. PAN, "Design of Communication Software of VB6. 0 and PLC with Distributed Control System Based on VB," Journal of Agricultural Mechanization Research, vol. 3, pp. 96-99, 2004.
[4] Z. Zi-sheng, Z. Bo, Y. Jie, L. Ya-fei, and D. Qian, "The communication between the PC and PLC based on VB6. 0 in the ESP," in Computing, Control and Industrial Engineering (CCIE), 2011 IEEE 2nd International Conference on, 2011, pp. 306-309.
[5] S. Jian, L. Peng, F. Yaqiong, L. Ruixiang, and C. Le, "The implementation and application of programming port communication between industry PC and Mitsubishi FX series PLC," in Intelligent System and Knowledge Engineering, 2008. ISKE 2008. 3rd International Conference on, 2008, pp. 1324-1327.
[6] M. ZHOU, F.-t. LI, and W.-m. WU, "Research on Communication Between PC and Multi-PLC Based on VB [J]," Computer Engineering, vol. 4, pp. 38-40, 2009.
[7] L. Pengfei and N. Luhua, "Remote control Laboratory based on LabVIEW," in Intelligent Computation Technology and Automation, 2009. ICICTA'09. Second International Conference on, 2009, pp. 84-87.
[8] "<Mitsubishi,FX communication (RS-232C, RS485) User's Manual, March 2000>."
[9] Z. Chong-zhi, "The realization of PC and Mitsubishi FX2N PLC serial port communication," Computer & PLC Application, pp. 37-39, 2005.03.

IOP Conf. Series: Materials Science and Engineering **83** (2015) 012018 doi:10.1088/1757-899X/83/1/012018

Autonomous Mobile Robot Navigation Using Harmonic Potential Field

Subbash Panati[1], Bayanjargal Baasandorj[2], Kil To Chong[3,*]

[1,2,3]Chonbuk National University, Jeonju-si, South Korea

E-mail: subbash.19@jbnu.ac.kr[1],to_bayanaa@jbnu.ac.kr[2], kitchong@jbnu.ac.kr[3,*]

Abstract. Mobile robot navigation has been an area of robotics which has gained massive attention among the researchers of robotics community. Path planning and obstacle avoidance are the key aspects of mobile robot navigation. This paper presents harmonic potential field based navigation algorithm for mobile robots. Harmonic potential field method overcomes the issue of local minima which was a major bottleneck in the case of artificial potential field method. The harmonic potential field is calculated using harmonic functions and Dirichlet boundary conditions are used for the obstacles, goal and initial position. The simulation results shows that the proposed method is able to overcome the local minima issue and navigate successfully from initial position to the goal without colliding into obstacles in static environment.

Keywords : Autonomous mobile robot, harmonic potential field, obstacle avoidance, path planning

1. Introduction

Mobile robot navigation has been an area of robotics which has been researched extensively during the past three decades. This has led to development of different algorithms focusing on solving the issues related to mobile robot navigation namely path planning and obstacle avoidance. Artificial potential field, Vector field histogram method, bug algorithms etc., are some of the most commonly used schemes in mobile robot navigation. Artificial potential field method for mobile robot navigation proposed by J Borenstein et, al [1]. has been one of the most used navigation scheme. The core principle of this scheme is based on attractive and repulsive potential field concepts. The goal point is assumed to be negatively charged and rest other elements in the working environment of the robot like the obstacles and the initial point of the robot are assumed to be positively charged. The magnitude of the attractive or repulsive potential field is calculated based on the distance between the robot and the goal or the obstacles respectively. The attractive feature of the artificial potential field method is its simplicity and elegance with respect to mathematical analysis. The main bottleneck of the artificial potential field methods lies in its inability to counter the issue of local minima in the working environment. Local minima is a condition where in the magnitude of the attractive potential field and repulsive potential field is equal. This situation leads to the robot getting stuck in a position and being not able to compute its next move. In order to overcome the issue of artificial potential field, Conolly et al [2] proposed harmonic potential field method. Harmonic potential filed method is a novel artificial function based on harmonic functions, which overcomes the limitations of potential field methods. Harmonic functions are solution to Laplaces equations. The most important property of harmonic functions is that they are free from local minima[3-5]. The core idea of this method lies in creation of only one minimum in the working environment i.e, the global minimum which is represented by the goal. If the goal is represented by a global minimum and no other minimum exists in the environment then the robot will arrive at the goal location always. Harmonic potential field algorithm has gained a lot of interest in mobile robot navigation due to its distinct feature of avoiding local minima and it also produces a smooth path for mobile robot navigation [6] . Unlike potential field methods, harmonic potential field method not only performs well in static environment but also works efficiently in dynamic environment and can cope the rapid changes in the environment. Some of the research articles related to the use of harmonic potential field for mobile robot navigation are mentioned as follows, Kalavsky et al [7] have explained the mechanism to generate potential field using harmonic functions and to formulate a algorithm for computing a safe path for mobile robot in the environment with created potential field. Keymeulen et al [8] have presented the effect of boundary conditions for the obstacles. The application of potential field method for autonomous ground vehicles has been presented in [9-11]. This paper presents a path planning of a differential wheel drive mobile robot using harmonic potential field approach, Dirichlet boundary conditions are used to specify the boundary values for the obstacles. Simulation is carried out to evaluate the proposed method, Matlab software has been used for setting the boundary conditions and generating the environment for implementation of harmonic potential field method. The simulation results shows the effectiveness of the proposed approach.

The organization of the paper is as follows. Kinematics modelling of the robot is discussed in section 2. The description of harmonic potential field is discussed in section 3. Simulation results are discussed in section 4. Discussion and conclusion are presented in section 5.

2. Kinematic Modeling of Wheeled Mobile Robot

The geometrical description of a mobile robot is shown in fig.1. Assuming that the mobile robot wheels are non-deforming, and moves without slip on the surface, two wheels are independently driven by two actuators to achieve the motion and orientation, the kinematic model of the wheeled mobile robot is given by following equation eq.1,

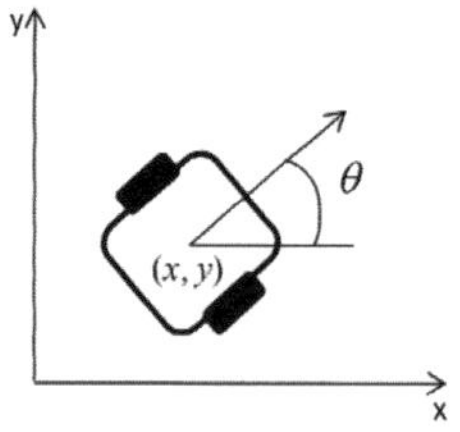

Figure 1. A two wheeled non-holonomic mobile robot

$$\begin{pmatrix} \dot{x} \\ \dot{y} \\ \dot{\theta} \end{pmatrix} = \begin{pmatrix} V cos\theta \\ V sin\theta \\ \omega \end{pmatrix} \quad (1)$$

3. Harmonic potential field

Harmonic Potential filed method is a novel method based on harmonic functions, it overcomes the limitations of Potential field methods. The most important trait of harmonic potential field method is that they are free from local minima. Harmonic potential field method uses harmonic functions and boundary value conditions to overcome the local minima problem. To build an artificial potential, we use harmonic function. A harmonic function should satisfy Laplaces equation, it should not have local extrema in a space free from singularities, it should have second order derivatives. Solution of laplace equation is also known as the mobile robot velocity potential:

$$V = -\nabla\Psi_{i,j} = \begin{pmatrix} V_x \\ V_y \end{pmatrix} = -\begin{pmatrix} \frac{\partial\Psi_{i,j}}{\partial x} \\ \frac{\partial\Psi_{i,j}}{\partial y} \end{pmatrix} \tag{2}$$

A harmonic function should also satisfy principle of superposition and principle of maxima and minima. These principles indicate that the harmonic function has its extremes only on the boundary, so it does not have local maxima/minima inside the boundary. Hence, it is convenient for us to define boundary conditions for boundary of all obstacles and boundary of goal. In this research Dirichlet boundary conditions have been used. The Dirichlet boundary condition states that the boundary of all obstacles will be assigned with the maximum value in the region and the boundary of goal position has the minimum value in the region. By defining the boundary conditions in this format the potential field is harmonic field with only global minimum represented by goal position.

3.1. Finite Difference Approximation Method

The finite difference approximation has been used in the proposed research to find the numerical solution of Laplace equation. Finite element method and finite volume method are some of the other methods used to find the numerical solutions to Laplace equations. The main advantage of using finite difference approximation method lies in its simplicity to implement and can be used to obtain numerical solution of partial differential equation(PDE)[12]. By using segmentation of environment, we can obtain potential distribution of the environment as shown in fig.2.

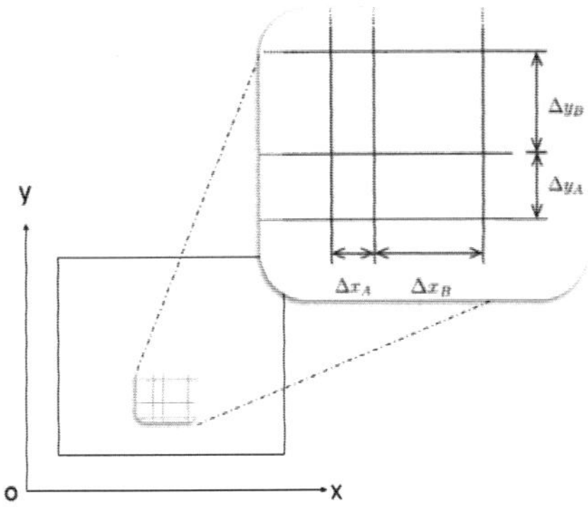

Figure 2. Sub-region of potential distribution in the environment

In order to solve finite difference approximation for the derivative ψ_{xx} we use the following equations:

$$\Psi_{i-1,j} = \Psi_{i,j} - T_x\Delta x_A + \frac{1}{2}\Psi_{xx}\Delta x_A^2 \tag{3}$$

$$\Psi_{i+1,j} = \Psi_{i,j} - T_x\Delta x_B + \frac{1}{2}\Psi_{xx}\Delta x_B^2 \tag{4}$$

Subtracting equation eq.3 and eq.4, and solving for $\Psi_x = \frac{\partial\Psi}{\partial x}$ while neglecting the terms involving $\Psi_{xx} = \frac{\partial^2\Psi}{\partial x^2}$, results in

$$\Psi_x = \frac{\Psi_{i+1,j} - \Psi_{i-1,j}}{\Delta x_A + \Delta x_B} \tag{5}$$

Adding equations eq.3 and eq.4, and solving for $\Psi_x = \frac{\partial \Psi}{\partial x}$, after replacing the expression for Ψ_x, from eq.5, results in the following expression:

$$\Psi_{xx} = \frac{2(\Psi_{i-1,j} + \Psi_{i+1,j} - 2\Psi_{i,j} + \frac{(\Psi_{i+1,j}-\Psi_{i-1,j})(\Delta x_A - \Delta x_B)}{(\Delta x_A + \Delta x_B)})}{\Delta x_A^2 + \Delta x_B^2} \tag{6}$$

To simplify the expression we introduce the following definitions:

$$\alpha_x = \Delta x_A - \Delta x_B,\ \beta_x = \Delta x_A + \Delta x_B,\ r_x = \frac{\alpha_x}{\beta_x},\ \gamma_x^2 = \Delta x_A^2 + \Delta x_B^2, \tag{7}$$

Thus

$$\Psi_{xx} = \frac{2(\Psi_{i-1,j} + \Psi_{i+1,j} - 2\Psi_{i,j} + r_x\left(\Psi_{i+1,j} - \Psi_{i-1,j}\right))}{\gamma_x^2} \tag{8}$$

Similarly, we can obtain the following equation Ψ_{yy} for the following derivatives in y:

$$\Psi_{yy} = \frac{2(\Psi_{i,j-1} + \Psi_{i,j+1} - 2\Psi_{i,j} + r_x\left(\Psi_{i,j+1} - \Psi_{i,j-1}\right))}{\gamma_y^2} \tag{9}$$

If we now replace the results of eq.8 and eq.9 into the Laplace equation $\Psi_{xx} + \Psi_{yy} = 0$, results in the following finite-difference approximation:

$$\frac{2(\Psi_{i-1,j} + \Psi_{i+1,j} - 2\Psi_{i,j} + r_x\left(\Psi_{i+1,j} - \Psi_{i-1,j}\right))}{\gamma_x^2} + \frac{2(\Psi_{i,j-1} + \Psi_{i,j+1} - 2\Psi_{i,j} + r_x\left(\Psi_{i,j+1} - \Psi_{i,j-1}\right))}{\gamma_y^2} = 0$$

An explicit solution for the value of the unknown $\Psi_{i,j}$ at the center of the computational cell can be obtained from above equation:

$$\begin{aligned} \Psi_{i,j} &= \frac{\gamma_y^2(\Psi_{i-1,j}+\Psi_{i+1,j}+r_x(\Psi_{i+1,j}-\Psi_{i-1,j}))}{2(\gamma_x^2+\gamma_y^2)} \\ &+ \frac{\gamma_x^2(\Psi_{i,j-1}+\Psi_{i,j+1}+r_y(\Psi_{i,j+1}-\Psi_{i,-1}))}{2(\gamma_x^2+\gamma_y^2)} \end{aligned} \tag{10}$$

If we consider a rectangular region where the increments in both x and y are uniform which means

$$\begin{aligned} \Delta x_A = \Delta x_B = \Delta x,\ r_x = \alpha_x = 0 \\ \Delta y_A = \Delta y_B = \Delta y,\ r_y = \alpha_y = 0 \\ \beta_x = 2\Delta x, \gamma_x^2 = 2\Delta x^2 \\ \beta_y = 2\Delta y, \gamma_y^2 = 2\Delta y^2 \end{aligned}$$

and eq.10 simplifies to

$$\Psi_{i,j} = \frac{\Psi_{i-1,j} + \Psi_{i+1,j} + \Psi_{i,j-1} + \Psi_{i,j+1}}{4} \tag{11}$$

IOP Conf. Series: Materials Science and Engineering **83** (2015) 012018 doi:10.1088/1757-899X/83/1/012018

4. Simulation

The simulation environment is modelled with static obstacles (walls), the mobile robot is required to traverse through these obstacles from initial position to reach the goal position (see fig.10). The area of the environment is 100×100 metres. Fig.5 and fig.6 illustrate the path which the robot followed to reach the target and displays the result of the proposed method in this paper. It can be observed that the mobile robot could safely avoid obstacles and successfully navigate towards the goal position.

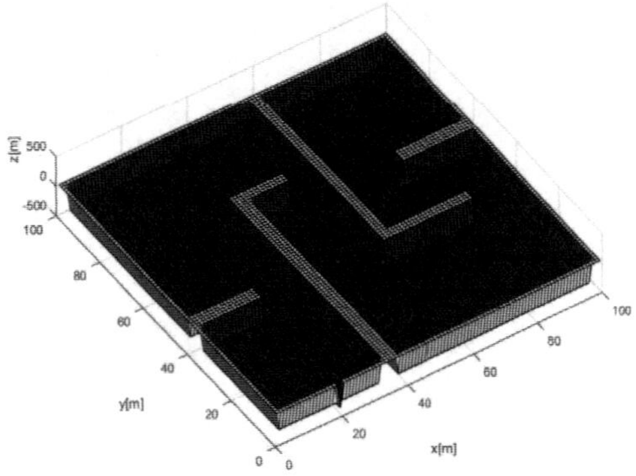

Figure 3. Simulation environment

Harmonic potential field method has the ability to generate path for any given initial position and any given target point of the mobile robot. The vector field generated is shown in fig.4 below

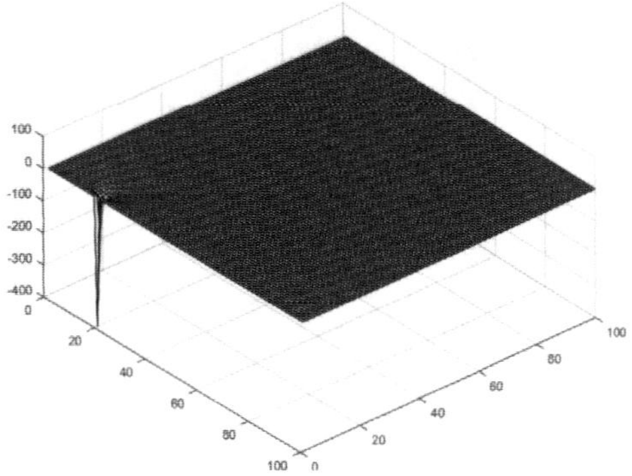

Figure 4. Potential field

In order to evaluate the proposed path planning scheme, two scenarios have been chosen. In the first scenario, the initial position and target position of the robot are assigned to be (90,90) and(2,20) respectively. The simulation result for the first scenario is shown in fig.5

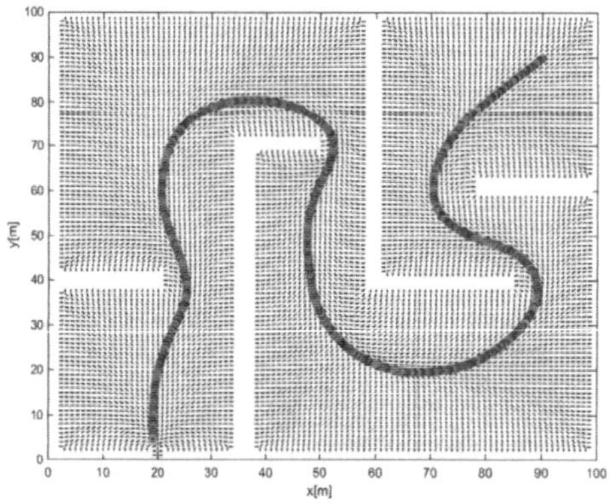

Figure 5. Mobile robot navigation: Scenario-1

The simulation results of the second scenario are shown in fig.6, in this scenario the initial position is considered to be (90,10) and goal position is considered to be (4,80). The simulation results for the second scenario is shown in fig.6.

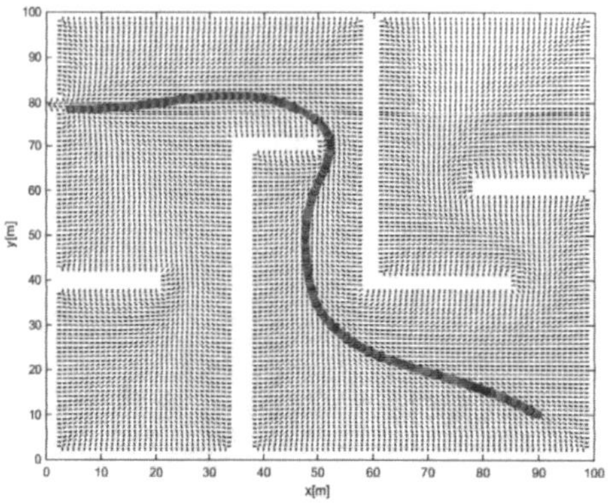

Figure 6. Mobile robot navigation: Scenario-2

5. Discussion

This paper presents a harmonic potential field based navigation scheme for mobile robots. The potential fields are generated using the Laplaces equation, Dirichlets conditions are used to specify the boundary conditions for the obstacles, Initial and goal positions. Finite difference approximation method is used to find the numerical solution of laplace equation. The proposed method has been evaluated using matlab simulation. Two types of simulation scenarios with different initial and goal points have been considered. It can be observed from the simulation results that the streamlines generated by the harmonic potential field method is smooth and it enables the robot to completely avoid the obstacles and efficiently reach the goal position in static environment.

6. Conclusion

The mobile robot navigation in static environment based on Harmonic potential field method has been discussed in this paper. Two scenarios with different initial position and goal position were considered and mobile robot navigation based on harmonic potential field was simulated in matlab environment. The simulation results show that harmonic potential field can be used to efficiently navigate the mobile robot in static environment.

Acknowledgments
The research is supported by the National Research Foundation of Korea (NRF) grant funded by the Korea government (MEST) (No. 2013R1A2A2A01068127) and (No. 2013R1A1A2A10009458) and BK21 PLUS.

References
[1] Borenstein, J. Koren, Y. (1989), *Real-time obstacle avoidance for fast mobile robots*,IEEE Transactions on Systems Man and Cybernetics, Vol. 19, pp. 1179-1187.
[2] C.J. Conolly J. B. Burns R. Weiss, *Path Planning Using Laplace's Equation*, International Conference of Robotics and Automation., Cincinnati, USA. pp. 2102- 2106, 1990.
[3] K. Sato, *Deadlock-Free Motion Planning Using the Laplace Potential Field*, Advanced Robotics, Vol. 3, No. 5, pp 449-461, 1993.
[4] Akishita, S. Kawamura, and K. Hayashi, *New Navigation Function Utilizing Hydrodynamic Potential for Mobile Robot*, Proc. 1990 IEEE Int. Workshop on Intelligent Motion Control, pp 413-417, August 1990.
[5] S. Waydo and R. M. Murray, *Vehicle Motion Planning Using Stream Functions*, Proc. 2003 IEEE Int. Conf. Robotics and Automation, pp 2484-2491, September 2003.
[6] C.I. Connolly, R. Gruppen, *On the applications of harmonic functions to robotics*, J. (Robotic Syst. 10 (7)) (1993) 931946.
[7] Kaavsk, Michal, and elmra Ferkov. *Harmonic Potential Field Method for Path Planning of Mobile Robot*, (2012).
[8] D. Keymeulen and J. Decuyper, *The Fluid Dynamics applied to Mobile Robot Motion: the Stream Field Method*, Proc. 1994 IEEE Int. Conf. Robotics and Automation, pp 378-385, May 1994.
[9] E. Rossetter, *A Potential Field Framework for Active Vehicle Lanekeeping Assistance*, Ph.D. dissertation, Dept. Mech. Eng., Stanford Univ., Palo Alto, CA, 2003.
[10] S. Shimoda, Y. Kuroda, and K. Iagnemma, *Potential Field Navigation of High Speed Unmanned Ground Vehicles on Uneven Terrain*, Proc. 2005 IEEE Int. Conf. Robotics and Automation, pp 2839-2844, April 2005.
[11] T. Hesse and T. Sattel, *Path-Planning with Virtual Beams*, Proc. 2007 American Controls Conference, pp 3904-3905, July 2007.
[12] Numerical partial differential equations, Available at: `http://en.wikipedia.org/wiki/Numerical_partial_differential_equations`, (Accessed: 10th Dec 2014)

Using dynamic programming technique for solving a class of bi-criteria multistage transportation problem

E Ellaimony[1], H Abdelwali[1], J Al-Rajhi[1] and M Al-Ardhi[1]
[1] Assistant Professor, Automotive and Marine Engineering Technology Department, College of Technological Studies, PAAET, Kuwait.
Emails: ee.ellaimony@paaet.edu.kw , ha.mohamed@paaet.edu.kw , jm.alrajhi@paaet.edu.kw , alardhi@paaet.edu.kw

ABSTRACT. This paper presents a solution algorithm to solve one class of bicriteria multi-stage transportation problem. This class represents bi-criteria multistage transportation problem without any restrictions on the intermediate stages. A way to find the non-dominated extreme points in the criteria space is developed. This method involves a parametric search in the criteria space. The presented algorithm is based also on fruitful application of the methods for treating transportation problems and the dynamic programming technique to obtain the shortest route for transportation networks and methods for solving bicriteria linear programming. The solution algorithm as well as an illustrative example are also included in this paper.

Keywords: Transportation Problems, Multi-stage Transportation Problem. Bi-Criteria Transportation Problem, Dynamic Programming.

1. INTRODUCTION

The main purpose of a traditional transportation problem is to determine the optimal transportation pattern of a certain good from supplies to demand customers so that the total transportation cost become minimum [1, 2, 3]. It is investigated by many researches and known as Hitchcook Koopman transportation problem. In the traditional or classical transportation problem, the cost of transportation is directly proportional to the number of units transported. A lot of efficient algorithms had been developed for solving the transportation problems when the cost coefficients and the supply and demand quantities are known. The real world problems are multi-objective in nature. Most of the practical transportation problems appear with two objectives known as bi-criteria transportation problems. There are two objectives, minimization of total transportation cost and minimization of transportation time, or minimization of total deterioration.

Deterioration is relevant in the case of certain perishable or decaying items. The degree of deterioration may depend on the route, mode and time of transportation. Clearly, this problem can be solved using any of the multi-objective linear programming techniques.

Intensive investigations on multi-objective linear transportation problems have been made by several researchers. Aneja and Nair [4] presented an algorithm for solving bicriteria transportation problem using weighted sum method. A duality study for a linear multi-objective transportation problem is presented by Isermann [5]. Solving multi-objective transportation problem and an adjacent search is prepared by Diaz [6, 7]. Current et al [8] prepared a review of multi-objective design of transportation networks. Ringuest J.L. and Rinks D.B. [9] proposed the interactive solutions for linear multi-objective transportation problems.

Practically, the transportation processes are done in many stages. The optimal solution for these problems depends on the nature of the problem and its network. Ivan Brezina and Adriana Istvanikova [10] presented a way of solving two-stage transportation problem. Al-Ellaimony E.E.M. [11] presented a mathematical model for all types of bicriteria multistage transportation problems. He presented both BMTP1, BMTP2, BMTP3, and BMTP4, and introduced a decomposition algorithm for solving BMTP3.

In this paper we introduce an algorithm for solving BMTP1 which is a bicriteria multi-stage transportation problem without any transportation restrictions on the intermediate stages. The presented algorithm consists of three phases. Phases 1 and 2, determine both the minimum transportation cost and deterioration between the first stage sources and the last stage destinations. The dynamic programming technique is applied on these two phases using the shortest route procedure. Phase 3, determines the points of the non-dominated set in the objective space of the bicriteria transportation problem presented. The algorithm is presented by Aneja and Nair [4].

2. DIFFERENT FORMS OF BICRITERIA MULTISTAGE TRANSPORTATION PROBLEMS[11]

The formulation of different bicriteria multistage transportation problems which are presented in this paper almost cover several real situation. The proposed formulations enable the decision maker discussing such transportation problems to find his goal.

2.1 Bicriteria Multistage Transportation Problems of the First Kind (BMTP 1):

This case represents bicriteria multistage transportation problems without any transportation restrictions on the intermediate stages. In order to obtain the mathematical formulation of the problems representing this case let us assume that:
c_{ij} = minimum transportation costs between the first stage sources and the last stage destinations.
d_{ij} = minimum deterioration of a unit while transporting from first stage sources to last stage destinations.
a_i = Availabilities of the first stage sources.
b_j = Requirements of the last stages destinations.
x_{ij} = Amounts transported from i to j.

Then the problem takes the form:

Minimize
$$z_1 = \sum_{i=1}^{m} \sum_{j=1}^{n} c_{ij}\, x_{ij}, \qquad (1)$$
$$z_2 = \sum_{i=1}^{m} \sum_{j=1}^{n} d_{ij}\, x_{ij},$$

Subject to
$$\sum_{j=1}^{n} x_{ij} = a_i\,, \qquad i = 1,2,\ldots,m, \qquad (2)$$
$$\sum_{j=1}^{m} x_{ij} = b_j\,, \qquad j = 1,2,\ldots,n, \qquad (3)$$
$$x_{ij} \ge o \qquad for\ \ all\ \ i, j.$$

2.2 Bicriteria Multistage Transportation Problems of the Second Kind (BMTP 2):

This case represents bicriteria multistage transportation problems in which the transportation at any stage is independent of transportation at the other stages.
To obtain the mathematical formulation of the problems representing this case let us assume that, for K^{th} stage, K=1,2,…,N; the availabilities of the sources are ($a_{i_k}^{k}$), i_k=1,2,…,m_k; the requirements at

the destinations are ($b^k_{j_k}$), j_k = 1,2,..,n_k; the cost of transporting one unit from source (i_k) to destination (j_k) is ($c^k_{i_k j_k}$); and the deterioration of one unit while transporting from (i_k) to (j_k) is ($d^k_{i_k j_k}$).

The mathematical formulation for this case can be written as (N) single stage bicriteria transportation problems in the form:

Min.
$$z_1^k = \sum_{i_k=1}^{m_k} \sum_{j_k=1}^{n_k} c^k_{i_k j_k} \; x^k_{i_k j_k} \tag{4}$$

$$z_2^k = \sum_{i_k=1}^{m_k} \sum_{j_k=1}^{n_k} c^k_{i_k j_k} \; x^k_{i_k j_k} \tag{5}$$

Subject to:

$$\sum_{j_k=1}^{n_k} x^k_{i_k j_k} = a^k_{i_k}, \; i_k = 1,2,\dots,m_k, \tag{6}$$

$$\sum_{j_k=1}^{m_k} x^k_{i_k j_k} = b^k_{i_k}, \; j_k = 1,2,\dots,n_k, \tag{7}$$

$$x^k_{i_k j_k} \geq o \quad for\ all\ i_k,\ j_k \quad k = 1,2,\dots,N; \tag{8}$$

The minimum transportation cost and deterioration are given as:

$$z_1^* = \sum_{k=1}^{N} z_1^k \qquad , \qquad z_2^* = \sum_{k=1}^{N} z_2^k \tag{9}$$

2.3 Bicriteria Multistage Transportation Problems of the Third Kind (BMTP 3):

This case represents bicriteria multistage transportation problems with some additional transportation restrictions on the intermediate stages which does not affect the bicriteria transportation problem nature at each stage.

The mathematical formulation of the problem representing this case is given us:

Min.
$$z_1 = \sum_{K=1}^{N} \sum_{i_k=1}^{m_k} \sum_{j_k=1}^{n_k} c^k_{i_k j_k} \; x^k_{i_k j_k} \tag{10}$$

$$z_2 = \sum_{K=1}^{N} \sum_{i_k=1}^{m_k} \sum_{j_k=1}^{n_k} d^k_{i_k j_k} \; x^k_{i_k j_k} \tag{11}$$

Subject to:

$$\sum_{j_k=1}^{n_k} x^k_{i_k j_k} = a^k_{i_k} \;, \; i_k = 1,2,\dots,m_k; \tag{12}$$

$$\sum_{j_k=1}^{m_k} x^k_{i_k j_k} = b^k_{i_k} \;, \; j_k = 1,2,\dots,n_k; \tag{13}$$

$$F_{r_k}\,(x^{k-1}_{i_{k-1}}\ j_{k-1} \quad , \; x^k_{i_k j_k} \;, \; x^{k+1}_{i_{k+1}} \quad j_{K+1}) = b^o, and \tag{14}$$

$$x^k_{i_k j_k} \geq o \quad for\ all\ i_k,\ j_k \quad k = 1,2,\dots,N; \tag{15}$$

Where:

F_{r_k}, k=1,2,…,N are linear functions representing the additional transportation restrictions at the (N) stages and, r is the umber of this linear functions at the kth stage.

2.4 Bicteria Multistage Transportation Problems of the Fourth Kind (BMTP4):
This case represents bicriteria multistage transportation problems in which the subtraction between the input and the output transportation commodity is known at the sources (destinations) of each intermediate stage. The assumed transportation restrictions in this case affect the transportation problem formulation of each individual stage.

The mathematical formulation of the problem representing this case is given as:

Min.
$$z_1 = \sum_{K=1}^{N} \sum_{i_k=1}^{m_k} \sum_{j_k=1}^{n_k} c_{i_k j_k}^{k} \; x_{i_k j_k}^{k} \tag{16}$$

$$z_2 = \sum_{K=1}^{N} \sum_{i_k=1}^{m_k} \sum_{j_k=1}^{n_k} d_{i_k j_k}^{k} \; x_{i_k j_k}^{k} \tag{17}$$

Subject to:

$$\sum_{j_1=1}^{n_k} x_{i_1 j_1}^{k} = a_{i_1}^{k} \;,\; i_1 = 1,2,\ldots,m_1; \tag{18}$$

$$\sum_{i_{k-1}=1}^{m_{k-1}} x_{i_{k-1} j_{k-1}}^{k-1} - \sum_{j_k=1}^{n_k} x_{i_k j_k}^{k} = b_{j_{k-1}}^{k-1} \;,\; j_{k-1} = 1,2,\ldots,n_{k-1}; i_k = i_k = 1,2,..,m_k; \quad k = 1,2,\ldots,N; \tag{19}$$

$$\sum_{i_N=1}^{m_N} x_{i_N j_N}^{N} = b_{j_N}^{N} \;,\; j_N = 1,2,\ldots,n_N; \tag{20}$$

$$x_{i_k j_k}^{k} \geq o \quad for\ all\ i_k, j_k \quad k = 1,2,\ldots,N; \tag{21}$$

(BMTP1) Can be solved as a bi-criteria single stage transportation problem. This should be done after obtaining the shortest path between first stage sources and last stage destination. These routes represent the minimum cost coefficient and minimum deterioration coefficient between first stage sources and last stage destinations. So, the problem will be transformed into a bicriteria single stage transportation problem.

(BMTP2) Can be solved as N bi-criteria single stage transportation problems. Then we can obtain the non-dominated solutions for each individual stage.

(BMTP3) The decomposition technique can be used to solve this model to utilize the special nature of the transportation problems to determine the points of the non-dominated set in the objective space.

(BMTP4) Can be solved using any algorithm for solving multiobjective linear programming problems. In this paper, we introduce an algorithm for solving bicriteria multi-stage transportation problem of the first kind (BMTP1). This algorithm uses the dynamic programming to find the shortest routes to transform the problem from multi-stage into a single stage transportation problem. The Aneja and Nair algorithm which is presented in [3] is used to obtain all of the non-dominated extreme points in the objective space for the studied case.

3. Dynamic Programming

Dynamic programming (D.P.) is a technique that can be used to solve many optimization problems. It is a useful mathematical technique for making a sequence of interrelated decisions. It provides a systematic procedure for determining the optimal combination of decisions. In most applications, D.P. obtains solutions by working backward from the end of a problem towards the beginning, thus breaking up a large unwieldy problem into a series of smaller, more tractable problems. D.P. divides problems into sub-problems called stages. After solving every sub-problem, the original problem solution can be achieved easily by using the state variables. State variables represent the links between stages which allow one to make optimum decisions for the remaining stages without having to check the effect of future decisions or decisions previously made. In other words, D.P. is based on a multistage decision process where a decision at one stage will affect the decision at subsequent stages. In contrast to other mathematical programming techniques, there does not exist a standard mathematical formulation for D.P. problems. But it is a general strategy for optimization rather than a specific set of rules. Consequently, the particular equation used must be developed to fit each problem. Abdelwali H. A. [12] introduced parametric multi-objective dynamic programming to solve some of automotive problems.

The D.P. technique is applied in this paper to find the shortest route between all main sources in the first stage to all destinations in the last stage of the multistage transportation problem for all objectives. The shortest route here means the minimum transportation cost, and the minimum deteriorations from all sources in the first stage to all destinations in the last stage of the transportation network. With a transportation network of m-sources at the first stage and n-destinations at the last stage, we must find (m*n) shortest routes. In other words, we should have (m*n) shortest route problems we need to solve!

The reason behind using D.P. to find these shortest routes and not using any other shortest route method comes from the fact that D.P. generates inherent parametric study for the problem solution. This inherent parametric study helps us to find the shortest route from any source to the last destinations. Another best advantage of these inherent results is reducing the number of computation required to find the shortest routes from all main sources to all last destinations from (m*n) problems into only n problems. For example, assume we have 12 main sources at the first stage, and 10 destinations at the last stage. To find the shortest routes from all main sources to all last destinations we need to solve 12*10= 120 problems. But only 10 problems (which equal the number of last destinations) can be solved using backward D.P. This reduces the time and effort required to find the shortest routes from all main sources to all last destinations.

4. Dynamic Programming Procedure Recursive Equation:

4.1 Recursive Equation

As mentioned above, the dynamic programming technique will be used to find the shortest route (where represent the transportations cost and deterioration) between all sources in the first stage to just one destination in the last stage. Then we need to repeat this procedure for all other destinations in the last stage. Of course this should be done for every single objective. So, the recursive equation for the backward dynamic program for the shortest route problem is illustrated in equations (22) to (25). Equations (22) calculate the minimum transportation cost to a certain destination from all the sources at the last stage. Equations (23) calculate the minimum transportation cost to all destinations from all sources at any other stage except the last stage. Equations (24) calculate the minimum transportation deterioration to a certain destination from all the sources at the last stage. Equations (25) calculate the minimum transportation deterioration to all destinations from all sources at any other stage except the last stage.

$\boldsymbol{For\ all\ (j_N = 1,2,3,\dots,n_N):\ calculate}$

$$F^N(i_N) = c^N_{i_N j_N},\ i_N = 1,2,\dots,m_N \tag{22}$$
$$F^K(i_k) = min.\{c^k_{i_k j_k} + F^{k+1}(i_k)\};\ j_k = 1,2,\dots,n_k;\ i_k = 1,2,\dots,m_k;\ k = N-1, N-2,\dots,1 \tag{23}$$

$For\ all\ (j_N = 1,2,3,\dots,n_N):\ calculate$
$$F^N(i_N) = d^N_{i_N j_N},\ i_k = 1,2,\dots,m_k \tag{24}$$
$$F^K(i_k) = min.\{d^k_{i_k j_k} + F^{k+1}(i_k)\};\ j_k = 1,2,\dots,n_k;\ i_k = 1,2,\dots,m_k; k = N-1, N-2,\dots,1 \tag{25}$$

Where:
$F^N(i_N)$: represents the optimum value of the studied objective for each source at the last stage (N).
$F^k(i_k)$: represents the optimum value of the studied objective for each source at any stage except the last stage.
K: stage number, k= 1,2,3,..., K
$c^N_{i_N j_N}$: transportation cost at the last stage (N) from its sources i_N to its destinations j_N.
$c^k_{i_k j_k}$: transportation cost at any stage (k) from its source i_k to its destinations j_k.
$d^N_{i_N j_N}$: transportation deterioration at the last stage (N) from its sources i_N to its destinations j_N.
$d^k_{i_k j_k}$: transportation deteriorations at any stage (k) from its source i_k to its destinations j_k.
i_N: the source number at the last stage (N).
j_N: the destination number at the last stage (N).
i_k: the source number at any stage (k) except the last stage.
j_k: the destination number at any stage (k) except the last stage.
m_N: the total number of sources at the last stage (N).
n_N: the total number of destinations at the last stage (N).
m_k: the total number of sources at stage (k).
n_k: the total number of destinations stage (k).

4.2 Solution Algorithm:
There are two algorithms exist in this paper. The first solution algorithm includes two phases (phase 1 and phase 2). This algorithm is used to find the minimum transportation costs and the minimum deteriorations between the first stage sources and the last stage destinations. This transforms the problem from bicriteria multi-stage into bicriteria single stage transportation problem. The second algorithm is summarized in (phase 3) to obtain the set of nondominated extreme points for the single stage bicriteria transportation problem which is generated from algorithm number 1.

4.2.1 Phase 1:
Step 1: Set $j_N = 1$.
Step 2: Use the recursive equations (22) to obtain the transportation cost ($F^N(i_N)$) between all sources ($i_N = 1,2,\dots m_N$) to destination number (j_N) at the last stage.
Step 3: Use the recursive equations (23) to obtain the minimum transportation cost ($F^k(i_k)$) between all sources ($i_k = 1,2,\dots m_k$) to all destinations for all stages (K= N-1, N-2, … 1).
Step 4: If ($j_N = n_N$), go to phase 2. Otherwise, set ($j_N = j_N+1$) then go to step 2 of this phase 1.

4.2.2 Phase 2:
Step 1: Set $j_N = 1$.
Step 2: Use the recursive equations (24) to obtain the transportation deterioration ($F^N(i_N)$) between all sources ($i_N = 1,2,\dots m_N$) to destination number (j_N) at the last stage.
Step 3: Use the recursive equations (25) to obtain the minimum transportation deterioration ($F^k(i_k)$) between all sources ($i_k = 1,2,\dots m_k$) to all destinations for all stages (K= N-1, N-2, … 1).
Step 4: If ($j_N = n_N$), go to phase 3. Otherwise, set ($j_N = j_N+1$) then go to step 2 of this phase 2.

4.2.3 Phase 3:

This phase determines the nondominated extreme points in the objective space for a bicriteria single stage transportation problem obtained from phases (1) and (2). The algorithm is validated by the following theorem [4].

4.2.3.1 Theorem:

A point $z^{(q)} = \left(z_1^{(q)}, z_2^{(q)}\right)$ is a nondominated extreme point is the objective space if and only if z(q) is recorded by the algorithm.

4.2.3.2 Step 1: For bicriteria single stage transportation problem, find

$z_1^{(1)} = Min.\ (z_1 / x \varepsilon M)$

And find

$z_2^{(1)} = Min.\ (z_2 / z_1 = z_1^{(1)}\ and\ x \varepsilon M)$

Record ($z_1^{(1)}$, $z_2^{(1)}$) and set q = 1.

Similarly, find

$z_2^{(2)} = Min.\ (z_2 / x \varepsilon M)$

And find

$z_1^{(2)} = Min.\ (z_1 / z_2 = z_2^{(2)}\ and\ x \varepsilon M)$

If $(z_1^{(2)}, z_2^{(2)}) = (z_1^{(1)}, z_2^{(1)})$, *stop.*

Otherwise record ($z_1^{(2)}$, $z_2^{(2)}$) and set q = q+1

Defines sets L = {(1,2)} and E = ϕ, and go to step 2.

4.2.3.3 Step 2: Choose an element (r, s) ∈L and set

$a_1^{(r,s)} = \left|z_2^{(s)} - z_2^{(r)}\right|\ and$

$a_2^{(r,s)} = \left|z_1^{(s)} - z_1^{(r)}\right|\ and$

Obtain the optimal solution ($\bar{x}$) to the transportation problem.

$$Minimize \sum_{i=1}^{m} \sum_{j=1}^{n} (a_1^{(r,s)} c_{ij} + a_2^{(r,s)} d_{ij})\ x_{ij}$$

Subject to: $x\ \varepsilon M, x \geq o$

If there are alternative optima, choose an optimal solution $\bar{x}$, for which

$$\sum_{i=1}^{m} \sum_{j=1}^{n} (c_{ij}\ \bar{x}_{ij}\ \min imum)$$

Let: $\overline{Z_1} = \sum_{i=1}^{m} \sum_{j=1}^{n} c_{ij}\ \bar{x}_{ij}$ and $\overline{Z_2} = \sum_{i=1}^{m} \sum_{j=1}^{n} d_{ij}\ \bar{x}_{ij}$

If ($\overline{z_1}$, $\overline{z_2}$) is equal to $(z_1^{(r)}, z_2^{(r)})\ or\ (z_1^{(s)}, z_2^{(s)})$

Set E = E ∪ {(r,s)} and go to step 3.

Otherwise record ($z_1^{(q)}$, $z_2^{(q)}$) such that

$z_1^{(q)} = \overline{z_1}$, $z_2^{(q)} = \overline{z_2}\ and\ set\ q = q+1,$

L = L ∪ {(r,q)}, (q,s)} and go to step 3.

4.2.3.4 Step 3: Set L = L - {(r-s)}. If L = ϕ, stop.

Otherwise go to step 2.

5. Illustrative Example

5.1 The Problem

We applied the presented algorithm to solve a simple 3-stage bicriteria transportation problem of the first kind (BMTP 1). The chosen 2-objectives are transportation cost and deteriorations. Figure 1 below represents this simple network. The availabilities of the main sources at stage 1 equals are those values of (a_1, a_2 and a_3). The requirements of the destinations at the last stage are those values of (b_1, b_2 and b_3). There are no transportation restrictions on the middle stages availabilities or requirements. The values of both the transportation unit costs and the deterioration amount while transporting the product from any source (i_k) to any destination (j_k) through a certain stage (k) is illustrated on the arcs as ($c^k_{i_kj_k}$, $d^k_{i_kj_k}$). It is required to find the non-dominated set of extreme points for this problem. All the problem data are included in Tables (1, 2 and 3) below. All these data are summarized in Table (1).

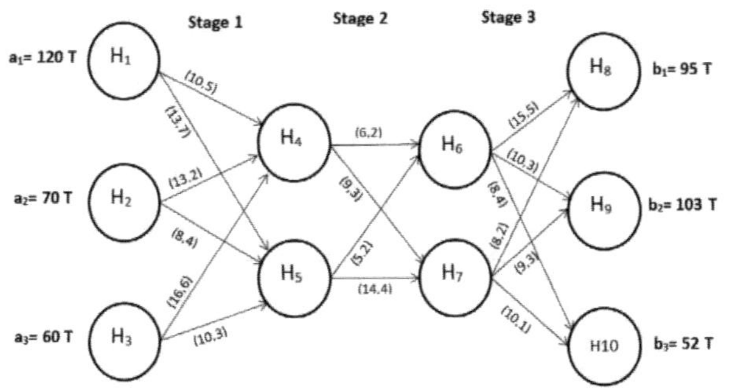

Figure 1, The illustrative example network.

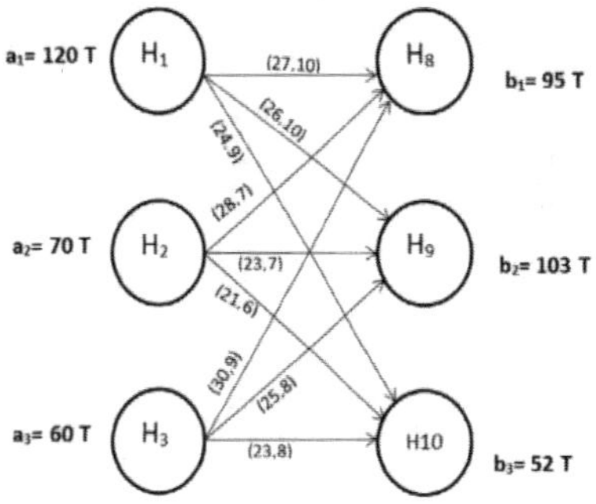

Figure 2, the equivalent network.

Table 1. Cost and deterioration of transportation & availabilities and requirements for stages (1, 2 and 3) resp.

Sources / Destinations	H4	H5	Availabilities
H1	(10,5)	(13,7)	120
H2	(13,2)	(8,4)	70
H3	(16,6)	(10, 3)	60
Requirements	No limit	No limit	

Sources / Destinations	H6	H7	Availabilities
H4	(6, 2)	(9,3)	No limit
H5	(5,2)	(14,4)	No limit
Requirements	No limit	No limit	

Sources / Destinations	H8	H9	H10	Availabilities
H6	(15,5)	(10,3)	(8,4)	No limit
H7	(8,2)	(9,3)	(10,1)	No limit
Requirements	95	103	52	

5.2 Dynamic Programming Results

Since we have 3 destinations in the last stage of this example, therefore, we need to solve 3 D.P. problems for each objective. So, the first step to solve this problem is to apply phase (1) and phase (2) in the solution algorithm to find the minimum transportation costs and the minimum deteriorations from all sources of the first stage to all destinations of the last stage. Then we use these results to convert the 3-stage network into a single stage problem. Table (2) below shows the D.P. stages to find the shortest route, for just problem, for the first objective (the transportation cost) from all sources of the first stage (nodes: H_1, H_2 and H_3) to node (H_8) only at the last stage. This same procedure is applied to find the shortest route from (H_1, H_2, and H_3) to both (H_8, H_9 and H_{10}) for both objectives.

Table (3) summarizes the shortest route for all other problems for both objectives. Figure 2 illustrates the equivalent single stage transportation problem according to D.P. results. Both c_{ij} and d_{ij} values are in dollars/ton.kilometer. The values of X_{ij} are in ton.kilometer.

Table 2. Dynamic Programming Stages To Find The Shortest Routes From (H_1, H_2 and H_3) To (H_8).

Stage (3):

State (S_3)	Alternatives	Optimal Solution	
	H_8	$F_3(X_3)$	D^*_3
H_6	15	15	H_8
H_7	8	8	H_8

Stage (2):

State (S_2)	Alternatives		Optimal Solution	
	H_6	H_7	$F_2(X_2)$	D^*_2
H_4	6+15= 21	9+8= 17	17	H_7
H_5	5+15= 20	14+8= 22	20	H_6

Stage (1)

State (S_1)	Alternatives		Optimal Solution	
	H_4	H_5	$F_1(X_1)$	D^*_1
H_1	10+17= 27	13+20= 23	27	H_4
H_2	13+17= 30	8+20= 28	28	H_5
H_3	16+17= 33	10+20= 30	30	H_5

Table 3. D.P. results for all problems of the 2 objectives.

Destination	For Objective 1		For Objective 2	
	Shortest Route	Min. Cost	Shortest Route	Min. Deterioration
H_8	$H_1 - H_4 - H_7 - H_8$	27	$H_1 - H_4 - H_7 - H_8$	10
	$H_2 - H_5 - H_6 - H_8$	28	$H_2 - H_4 - H_7 - H_8$	7
	$H_3 - H_5 - H_6 - H_8$	30	$H_3 - H_5 - H_7 - H_8$	9
H_9	$H_1 - H_4 - H_6 - H_9$	26	$H_1 - H_4 - H_6 - H_9$	10
	$H_2 - H_5 - H_6 - H_9$	23	$H_2 - H_4 - H_6 - H_9$	7
	$H_3 - H_5 - H_6 - H_9$	25	$H_3 - H_5 - H_6 - H_9$	8
H_8	$H_1 - H_4 - H_6 - H_{10}$	24	$H_1 - H_4 - H_7 - H_{10}$	9
	$H_2 - H_5 - H_6 - H_{10}$	21	$H_2 - H_4 - H_7 - H_{10}$	6
	$H_3 - H_5 - H_6 - H_{10}$	23	$H_3 - H_5 - H_7 - H_{10}$	8

5.3 Bicriteria Phase (3) Results

Phase 3 is then applied on the new equivalent problem to determine the efficient distribution and the non-dominated extreme points accompanying each iteration, the minimum transportation cost and minimum deterioration as single objective problems. The results are summarizes in Tables (4) and (5). The decision maker should Choose a non-dominated point and its related distribution from Tables (4), and (5) according to his/her point of view.

Table 4. Solution of the illustrative example in the objective space.

Iteration	L	E	Recorded Points
1			$Z^{(1)} = (6221,2118)$, $Z^{(2)} = (6293,2118)$
2	{(1,2)}	Φ	$Z^{(3)} = (6293,2118)$
3	Φ	{(1,2)}	$Z^{(3)} = (6293,2118)$

Table 5. Non dominated solutions and the related efficient points.

Non Dominated Points	Efficient Distribution	The Shortest Route
1	$x_{18}= 95$, $x_{19}= 25$, $x_{29}=$ 18, $x_{210}=52$, $x_{39}= 60$	$(H_1-H_4-H_7-H_8)$, $(H_1-H_4-H_6-H_9)$, $(H_2-H_5-H_6-H_9)$, $(H_2-H_5-H_6-H_{10})$, $(H_3-H_5-H_6-H_9)$
2	$x_{18}= 77$, $x_{19}= 43$, $x_{28}=$ 18, $x_{210}=52$, $x_{39}= 60$	$(H_1-H_4-H_7-H_8)$, $(H_1-H_4-H_6-H_9)$, $(H_2-H_4-H_7-H_8)$, $(H_2-H_4-H_7-H_{10})$, $(H_3-H_5-H_6-H_9)$
3	$x_{18}= 77$, $x_{19}= 43$, $x_{28}=$ 18, $x_{210}=52$, $x_{39}= 60$	$(H_1-H_4-H_7-H_8)$, $(H_1-H_4-H_6-H_9)$, $(H_2-H_4-H_7-H_8)$, $(H_2-H_4-H_7-H_{10})$, $(H_3-H_5-H_6-H_9)$

6. Conclusion

In this paper, we introduced 2 solution algorithms in 3 phases to solve the bicriteria multistage transportation problem of the first kind (BMTP1). The presented algorithms enable solving such problems more realistically to find all efficient extreme points. The bi-criteria multi stage transportation problem can be solved using the standard technique of a transportation problem.

Dynamic programming is a very powerful technique to reduce the multi-stage into a single stage transportation problem, when there are no restrictions in the middle stages sources availabilities and destinations requirements point of view. The decision maker then, will get all non dominated solutions with their efficient points and their related distributions. Therefore, s/he can choose any point from the non dominated solution which meets his/her requirement.

7. References

[1] Frederick Hillier, *Introduction to Operations Research, with Student Access Card*, Edition 10th, McGraw-Hill, 2014. ISBN-13: 978-1259162985 ISBN-10: 1259162982.

[2] Taha H.A., *Operations Research, An Introduction*, 9th Edition, Prentice Hall, 2010. ISBN-13: 978-0132555937 ISBN-10: 013255593X.

[3] Sang M. Lee and David L. Olson, *Introduction to Management Science*, 3e Paperback, Edition 3rd, Cengage Learning, 2005. ISBN-13: 978-0324415995 ISBN-10: 0324415990.

[4] Aneja Y.P., Nair K.P.K., Bi-criteria transportation Problems, Management Science, 1979, vol. 25, pp 1 – 11.

[5] Isermann H., The enumeration of all efficient solutions, Naval Research Logistic Quarterly 1979, vol. 26, pp 123 – 139.

[6] Diaz J.A., Solving multi-objective transportation problem, Ekonomicko- Matematicky Obzor, 1978, vol. 14, pp 267 – 274.

[7] Diaz J.A., Finding a complete description of all efficient solutions to a multi-objective transportation problem, Ekonomicko- Matematicky Obzor, 1979, vol. 15, pp 62 – 73.

[8] Current J., Min H., Multi-objective design of transportation networks: Taxonomy and annotation, European Jornal of Operations Research, 1986, vol. 26.

[9] Ringuest, J.L., Rinks, D.B., Interactive solutions for the linear multiobjective transportation problem, Eur. J. Oper. Res. 32, 1987, 96 – 106.

[10] Berzina, Istranikova, The way of solving two-stage transportation problems, Mathematical Methods in Economics 1999, 39 – 44.

[11] Ellaimony E.E.M., Different *Algorithms for Solving Multi-Stage Transportation Problems*, Ph.D., Helwan University, Egypt, 1985.

[12] Abdelwali H.A., *On Parametric Multi-Objective Dynamic Programming With Applications To Automotive Problems*, Ph.D., El-Minia University, Egypt, 1997.

Generalized Calogero system for simple Lie algebra series

Alexander Zuevsky

Institute of Mathemtics, Academy of Sciences of the Czech Republic, Prague Czech Republic

E-mail: zuevsky@yahoo.com

Abstract. We derive equations of motions and solutions for the generalized spin–Calogero models associated for the classical Lie algebra series B_n, C_n BC_n, and D_n using the pole expansion.

1. Generalized Calogero system

The generalized Calogero systems have been introduced in [3]. Corresponding Hamiltonian has the form:

$$H = \frac{1}{2m}\sum_{i=1}^{N} p_i^2 + \frac{1}{2}\alpha_1^2\sum_{i\neq j}^{N}\left(V(x_i - x_j) + V(x_i + x_j)\right) + \sum_{i=1}^{N}\left(\alpha_2^2 V(x_i) + \alpha_3^2 V(2x_i)\right), \quad (1)$$

where

$$V(x) = \wp(x),$$

the Weierstrass function [5]. When $\alpha_1 \neq 0$, to B_n-serie corresponds $\alpha_3 = 0$; to C_n-serie corresponds $\alpha_2 = 0$; to BC_n-series corresponds $\alpha_2 \neq 0$, $\alpha_3 \neq 0$; and finally to D_n-serie corresponds $\alpha_2 = \alpha_3 = 0$. These systems constitute relativistic limits of the generalized Ruijsenaars-Schneider system [1] with the Hamiltonian

$$H = mc^2\sum_{i=1}^{N}\cosh\left(\frac{\theta_i}{mc}\right)\prod_{j=1,i\neq j}^{N}\left(\sigma\left(\frac{ig}{mc}\right)\right)v(x_i, x_j), \quad (2)$$

where

$$v(x_i, x_j) = \sqrt{a + b\left(\alpha_1^2\left(\wp(x_i - x_j) + \wp(x_i + x_j)\right) + \alpha_2^2\wp(x_j) + \alpha_3^2(2x_j)\right)}, \quad (3)$$

for

$$a = \left(2\alpha_1^2 + \alpha_2^2 + \alpha_3^2\right)\wp(\eta)\sigma^2(\eta), \quad (4)$$

$$b = -\sigma^2(\eta), \quad (5)$$

$$\eta = \frac{ig}{mc}, \quad (6)$$

$$p = ms\sinh\theta, \quad (7)$$

where θ denotes rapidity.

2. Generalized non-abelian $2D$ Toda model

The generalized non-abelian Toda model chain equation has the form [2, 4]

$$\partial_+((\partial_- g_n)g_n^{-1}) = g_n g_{n-1}^{-1} - g_{n+1} g_n^{-1}, \tag{8}$$

The associated linear problem is then given by [2]

$$\partial_+ \psi_n(t_+, t_-) = \psi_{n+1}(t_+, t_-) + v_n(t_+, t_-)\psi_n(t_+, t_-), \tag{9}$$

$$\partial_- \psi_n(t_+, t_-) = c_n(t_+, t_-)\psi_{n-1}(t_+, t_-), \tag{10}$$

where

$$c_n = g_n g_{n-1}^{-1}, \tag{11}$$

$$v_n = (\partial_+ g_n) g_n^{-1}. \tag{12}$$

The constraints (according to [6]) for B_n

$$g_{-n} = g_n^{-1}, \tag{13}$$

$$v_{-n} = -v_n, \tag{14}$$

for C_n series are

$$g_{-n-1} = g_n^{-1}, \tag{15}$$

$$v_{-n} = -v_{n-1}, \tag{16}$$

2.1. Equations of motion

Similarly to Theorems 2.1 and 2.2 of [2] we find the following

Proposition. *The equations of motion for non-abelian 2D Toda system have*

$$\begin{aligned} \partial_t \Psi(x,t) &= \Psi(x+\eta, t) + \sum_{i=1}^{N} a_i(t) b_i^\dagger(t) \left(\alpha_1^2 \left(V(x - x_i(t)) + V(x + x_i(t))\right)\right. \\ &+ \left.\alpha_2^2 V(x) + \alpha_3^2 V(2x)\right)\Psi(x,t), \end{aligned} \tag{17}$$

$$\begin{aligned} -\partial_t \Psi^\dagger(x,t) &= \Psi^\dagger(x-\eta, t) + \Psi^\dagger(x,t) \sum_{i=1}^{N} a_i(t) b_i^\dagger(t) \left(\alpha_1^2 \left(V(x - x_i(t)) + V(x + x_i(t))\right)\right. \\ &+ \left.\alpha_2^2 V(x) + \alpha_3^2 V(2x)\right), \end{aligned} \tag{18}$$

have N pairs of linearly independent solutions with simple poles

$$\Psi = \sum_{i=1}^{N} s_i(t,k,z)\Phi(x - x_i, z) z^{x/\eta}, \tag{19}$$

$$\Psi^\dagger = \sum_{i=1}^{N} s_i^\dagger(t,k,z)\Phi(-x + x_i(t) - \eta, z) z^{-x/\eta}, \tag{20}$$

$$\Phi(x,z) = \frac{\sigma(z + x + \eta)}{\sigma(z+\eta)\sigma(x)} \left(\frac{\sigma(z-\eta)}{\sigma(z+\eta)}\right)^{x/2\eta}. \tag{21}$$

Note that here

$$\Phi(x,z) = \frac{1}{x} + A + \mathcal{O}(x), \tag{22}$$

where

$$A = \zeta(z+\eta) + \frac{1}{2\eta}\log\frac{\sigma(z-\eta)}{\sigma(z+\eta)}, \tag{23}$$

when $x \longrightarrow 0$.

Proof. By substituting $\Psi(x,t,z,k)$ into equations (17)–(18) and expanding over poles, we obtain (as before) for coefficients near $(x-x_i)^{-2}$:

$$\partial_t x_i s_i = \alpha_1^2 a_i (b_i^\dagger s_i),$$

$$s_{i,\alpha}(t,k,z) = c_i(t,k,z) a_{i,\alpha}(t),$$

$$\partial_t s_i x_i = (b_i^\dagger a_i).$$

For poles $(x-x_i)^{-1}$ we get

$$\partial s_i - \alpha_1 \left(\sum_{i\neq j, j=1}^{N} a_j(t) b_j^\dagger(t) V(x_i - x_j(t)) + \sum_{j=1}^{N} V(x_i + x_j) \right) s_i$$
$$+ \sum_{j=1}^{N} a_j b_j^\dagger \left(\alpha_2^2 V(x_i) + \alpha_3^2 V(2x_i)\right) s_i + \left(A - \zeta(\eta)\right) a_i(t) b_i^\dagger s_i$$
$$- a_i \alpha_1^2 \sum_{j\neq i} \left(b_i^\dagger s_j\right) \Phi(x_i - x_j, z) = 0,$$

$$\lambda_i(t) = \frac{\partial_t c_i}{c_i} + \left(\zeta(\eta) - A\right)\partial_t x_i - \sum_{j\neq i} \alpha_1^2 \left(b_i^\dagger \alpha_j\right) \Phi(x_i - x_j, z)\frac{c_j}{c_i} = 0.$$

In the matrix form we obtain:

$$\left(\partial_t + M(t,z)\right) C = 0,$$

where

$$M_{ij}(t,z) = \left(-\lambda_j + (\zeta(\eta) - A)\partial_t x_i\right)\delta_{ij} - (1-\delta_{ij})\alpha_1^2 b_i^\dagger a_j \Phi(x_i - x_j, z).$$

Canceling of poles $(x - x_i + \eta)^{-1}$ gives

$$-s_i k + \sum_{j\neq i} \alpha_1^2 a_i b_i^\dagger s_j \Phi(x_i - x_j - \eta, z) = 0,$$

$$\left(L_- - kI\right) C = 0, \qquad C = C(c_i),$$

where

$$L_{-,ij}(t,z) = \alpha_1^2 \left(b_i^\dagger a_j\right) \Phi(x_i - x_j - \eta, z).$$

For poles $(x + x_i + \eta)^{-1}$ we obtain:

$$-s_i k + \sum_{j\neq i} \alpha_1^2 a_i b_i^\dagger s_j \Phi(-x_i - x_j - \eta, z) = 0,$$

$$L_{+,ij}(t,z) = \alpha_1^2 \left(b_i^\dagger a_j\right) \Phi(-x_i - x_j - \eta, z.$$

For poles $(x+x_i)^{-1}$ it follows:

$$a_i \sum_{j=1}^{N} \alpha_1^2 \left(b_i^\dagger s_j\right) \Phi(-x_i - x_j, z) k^{x_j/\eta} = 0,$$

i.e.,

$$M_{+,ij}(t,z) = a_i \sum_{j=1}^{N} \alpha_1^2 \left(b_i^\dagger a_j\right) \Phi(-x_i - x_j, z) k^{x_j/\eta},$$

$$M_+ C = 0.$$

For B_n, C_n, BC_n series, near poles x^{-1} for $x \longrightarrow 0$ it follows:

$$\sum_{i \neq j}^{N} a_i \left(b_i^\dagger s_j\right) \Phi(-x_i, z) = 0,$$

i.e.,

$$M_{0,ij}(t,z) = \sum_{j \neq i}^{N} a_i \left(b_i^\dagger a_j\right) \Phi(-x_i, z) = 0,$$

$$M_0 C = 0.$$

For $(x+\eta)^{-1}$:

$$\sum_{i \neq j}^{N} a_i \left(b_i^\dagger s_j\right) \Phi(-\eta - x_i, z) \frac{1}{k} = 0,$$

i.e.,

$$L_{0,ij}(t,z) = \sum_{j \neq i}^{N} a_i \left(b_i^\dagger a_j\right) \Phi(-\eta - x_i, z) \frac{1}{k} = 0,$$

$$L_0 C = 0.$$

Similar results we obtain for equations on $\Psi^\dagger$.

□

2.2. Generalized spin Ruijsenaars models

Proposition. *With*

$$V(x) = \zeta(x) - \zeta(x+\eta),$$

we get for the generalized spin Ruijsenaars models we obtain

$$\begin{aligned}\partial_t^2 x_i = {} & \alpha_1^2 \sum_{i \neq j, j=1}^{N} (b_i^\dagger a_j)(b_j^\dagger a_i)(V(x_i - x_j) - V(x_j - x_i)) \\ & + \alpha_1^2 \sum_{j=1}^{N} (b_i^\dagger a_j)(b_j^\dagger a_i)\left(V(x_i + x_j) - V(-(x_i + x_j))\right) \\ & + \alpha_2^2 \left(V(x_i) - V(-x_i)\right) + \alpha_2^3 \left(V(2x_i) - V(-2x_i)\right)\right),\end{aligned}$$

$$\partial_t a_i = \alpha_1^2 \sum_{j \neq i}^{N} a_j (b_j^\dagger a_i) V(x_i - x_j) + \alpha_1^2 \sum_{j=1}^{N} a_j (b_j^\dagger a_i) V(x_i + x_j)$$

$$+\sum_{i\neq j}^{N}(b_i^\dagger a_j)(b_j^\dagger a_i)\left(\alpha_2^2 V(x_i)+\alpha_3^2 V(2x_i)\right)-\lambda_i a_i,$$

$$\partial_t b_i^\dagger=-\alpha_1^2\sum_{j\neq i}^{N}b_j^\dagger(b_i^\dagger a_j)V(x_j-x_i)-\alpha_1^2\sum_{j=1}^{N}b_j^\dagger(b_i^\dagger a_j)V(-(x_j+x_i))$$

$$+\sum_{j=1}^{N}(b_i^\dagger a_j)(b_j^\dagger a_i)\left(\alpha_2^2 V(-x_i)+\alpha_3^2 V(-2x_i)\right)+\lambda_i^\dagger b_i^\dagger.$$

Proof. Last three equations can be rewritten in the form:

$$\partial_t^2 x_i=\alpha_1^2\sum_{i\neq j,j=1}^{N}(b_i^\dagger a_j)(b_j^\dagger a_i)\left(V(x_i-x_j)+V(x_i+x_j)-V(x_j-x_i)+V(-(x_i+x_j)\right)$$

$$+\sum_{j=1}^{N}(b_i^\dagger a_j)(b_j^\dagger a_i)\left(\alpha_2^2\left(V(x_i)-V(-x_i)\right)+(\alpha_3^2+\alpha_1^2)\left(V(2x_i)-V(-2x_i)\right)\right),$$

$$\partial_t a_i=\alpha_1^2\sum_{j\neq i}^{N}a_j(b_j^\dagger a_i)V(x_i-x_j)+V(x_i+x_j))$$

$$\sum_{j=1}^{N}(b_i^\dagger a_j)(b_j^\dagger a_i)\left(\alpha_2^2 V(x_i)+(\alpha_3^2+\alpha_1^2)V(2x_i)\right)-\lambda_i a_i,$$

$$\partial_t b_i^\dagger=-\alpha_1^2\sum_{j\neq i}^{N}b_j^\dagger(b_i^\dagger a_j)\left(V(x_j-x_i)+V(-(x_j+x_i)\right)$$

$$\sum_{j=1}^{N}(b_i^\dagger a_j)(b_j^\dagger a_i)\left(\alpha_2^2 V(-x_i)+\alpha_3^2 V(-2x_i)\right)+\lambda_i^\dagger b_i^\dagger.$$

□

References

[1] K. Chen, B. Hou, W-L Yang, Integrability of the Cn and BCn Ruijsenaars-Schneider models. J. Math. Phys. 41 (2000), no. 12, 8132-8147.

[2] I. M. Krichever, A. Zabrodin, Spin generalization of the Ruijsenaars–Schneider model, the nonabelian two-dimensionalized Toda lattice, and representations of the Sklyanin algebra. (Russian) Uspekhi Mat. Nauk 50 (1995), no. 6(306), 3–56; translation in Russian Math. Surveys 50 (1995), no. 6, 1101–1150

[3] M. A. Olshanetsky, A.M. Perelomov. Completely integrable classical systems connected with semisimple Lie algebras. Lett. Math. Phys. 1 (1975/76), no. 3, 187–193.

[4] A. Razumov, M. Saveliev, A. Zuevsky. Non-abelian Toda equations associated with classical Lie groups. In: Symmetries and Integrable systems, Memorial volume dedicated to M.Saveliev, (JINR, Dubna, 2000, 190-203), math-ph/9909008;

[5] Serre, J-P.: *A course in arithmetic*, Springer-Verlag, (Berlin 1978).

[6] K. Ueno, K. Takasaki. Toda lattice hierarchy. Group representations and systems of differential equations (Tokyo, 1982), 1–95, Adv. Stud. Pure Math., 4, North-Holland (Amsterdam, 1984).

Remanufacturing Aided Added-Value Creation, Innovations Meeting to Deliver Sustainable Manufacturing

Ziyad Tariq Abullah[1,2] **, Shun Sheng Guo**[1] **and Sheng Bu Yun**[1]
1Wuhan University of Technology, Mechatronic School, No.205, Luoshi Road, Wuhan City 430070, Hubei Province, China
2 Institue of Technology –Baghdad, Mechanical school , Al-Za'franiya 10074, Baghdad, Iraq

E-mail: ziyad7tariq@yahoo.com

Abstract. End-of-life scrap steel such as vehicles bulks and bodies, steel wheel and shells are easily land filled at the end-of-life when treated in a developing country with non-industrial infrastructure. Research idea is about composite shape steel remanufacturing to be sheet steel for construction application through nested recovered pieces of scrap steel within new sheet steel base to meet innovation value creation of remanufactured steel and innovation eco-design of steel products to close supply chain through linkage developed and developing countries of non-industrial infrastructure economy. That can be satisfied through comprehensive business-education-training model conduction firstly at the developing countries to reduce costs and change the intensive labour remanufacturing paradigm collaboratively. Sustainable remanufacturing business model can be applied based on infrastructure of educational institutions such as institutes of technology to adopt environmental, economic, and social developments as triple bottom line sustainability. Such innovation value creation is driven by eco-design and eco-innovation enabling where the meet to deliver human development, employment, and education conscious environment and bench mark recommendations of development directions for upgrading to apply business that allows eco-societies to emerge, through cooperative steel scrap processing.

1. Introduction

Value creation-oriented remanufacturing copes with depletion of world resources which causes environmental and ecological burdens that lead to inevitable economic and social problems. Innovation should be focused to lengthen product use phase, sustain end-of-life processing and employ product upgradability, core availability and information flow to increase remanufacturing demand instead of to be as uncertainties that freeze remanufacturing feasibility within viability stage [1]. So dialogue and negotiation to fulfil information assessment systems and development strategies such as product service systems are required to be adopted nationally towards linked at a global level to be collaborative solutions that develop local-global sustainability [2, 3, 4]. Energy-saving and emission reduction by remanufacturing induce sustainable development which should be adopted by developing countries to substitute the lack of extended manufacturer responsibility. Manufacturer should be encouraged to supply common technical services system to know-how and learning to support design for remanufacturability and innovation. Also innovative technologies that disseminate economic and societal values to harvest high financial, societal, and environmental returns, reduce costs, increases jobs, and conserve scare resources are necessary to practise which required global collaboration. These can eliminate consumer dissatisfaction which is due to lack of activities to include customer within design and development processes, and also lack of funding of new innovative researches and

technologies to discover uncertainties of quantity and quality of remanufactured products as a result of lack of life cycle information. Eco-design and product-service system can together satisfy customers while degrade services to relieve environmental burdens. Such efficient paradigm removes consumer dissatisfaction through integrating users' satisfaction and environmental aspects to certain technical performances through application of eco-design. Such collaborative satisfaction and balancing of customer needs and environmental aspects can degrade technical performance and maintain user satisfaction to cut down environmental impacts by negotiation through service-oriented manufacturing/remanufacturing system to sell not only physical products but also accompanied services to certain product lifecycle feedbacks . [5, 6, 7]
Sustainability statement:-
Sustainable manufacturing is global closed-loop supply which starts at developed countries by eco-design and continues through developing countries by remanufacturing, where linkage in the cloud can certain flow of necessary information to deliver closed loop supply chain of flexible design to include business models, product services systems and customers as integral parts of such collaborative manufacturing. [3]

2. Eco-design and eco-innovation meeting

The following statements can show the delivering of innovations meeting through:-

1- Feedbacks of eco-design ideas can be extracted from remanufacturing of take back for remanufacturing of land filled end-of-life products.
2- Several ideas as eco-design requisitions which are mentioned in this study to keep continue steel remanufacturing.
3- Composite shape steel is eco-innovation since layers of new steel can be included recovered steel, and the sandwich configuration can be maintained through assembly with rivets and bolts.
4- Sustainable manufacturing of remanufactured steel is eco-innovation since structural steel section can be obtained.

3. Scrap steel remanufacturing motivations

- Internet of things quality management optimization

Eliminating uncertainty due to variety of types, conditions and remaining lives of components in end-of-life products can mitigate planning of remanufacturing operations. [8].

- Hybrid manufactured and remanufactured modules based new products

New products are manufactured based on remanufacturing of modules from old returned items. Remanufactured modules can be also used to remanufacture returned products to be like new [9].

- X- Grade as a service , mass customization and personalization

Physical products associated services bring more value to final customers than the pure physical objects without accessories and services. Cloud manufacturing systems result higher individualized and user-friendly product experiences and reflect intelligent and connected products throughout whole life-cycles. Product-service systems merit flexibility, adaptability, reactivity, expertise, and X-gradability. Such X-gradability approach enables sustainable products and satisfaction for users by delighting with intensive engagement in product design, functionalities, services, and individual experiences. Small series production of personalized products can be as integration and collaboration of virtual social network, internet of things, cloud manufacturing and product service system. [10]. Globalized manufacturing networks can encounter decreasing lifecycles, increased product complexity, mass customization and demand volatility. Cloud automated closed-loop control systems are necessary to deliver technology based business approaches globally and enable optimizing of enterprise performance in area of design and planning of manufacturing networks for mass customization and personalization. Through supporting and promoting of dynamic configuration and optimal routing of manufacturing networks based on facilities, time, complexity of environmental and constraints, and product services and personalization [11].

3- Configuration of society to conduct scrap steel remanufacturing

Reduce uncertainties, eliminate barriers and supply funds and technologies for remanufacturing should be symbiotic and solidarity on both international and national levels to include products updates, eco-design and eco-innovation. Further, societies should be responding to support open research, public regulations and policies, to substitute lack of extend manufacturer responsibility, and enact subsidies for remanufacturers, to affect positively the remanufacturing business and industry. The paper explains a proposed relation between three investigated enablers to help emerge scrap steel remanufacturing which include business model, remanufacturing barriers elimination scenario and sustainability initiation key factors.These enablers interact according to the following strategies:-

1-Institutes of technologies are the bases of emerging national scrap steel remanufacturing company.
2-National independent remanufacturers' are helped to emerge.
3-Recent private reveres logistics should be strengthened bifurcated and enlarged within national framework.
4-Design for remanufacturing is adopted as legislations which should be certain remanufacturability of new imported products.
5- Complex products remanufacturing can be conducted gradually.
6-National product services system should be established as consequences of remanufacturing maturity.
7-National cloud library and information system end-of-products are necessary to discover uncertainties and certain global remanufacturing.
8- National cloud library and information system can be linked with international cloud manufacturing systems.

3.1. Business model

Lack of strategic remanufacturing business models prevents developing countries from conduct sustainable manufacturing plans. Guidelines are proposed to help conducting of scrap steel remanufacturing business models in a developing country ,based on the literature of [1] and the website http://bragecrim2.wix.com/remanufacturing[12], to include :-

3.1.1. Business Strategy: New strategy can be emerged by exploiting educational institutions infrastructures to develop remanufacturing business and seek for value creation, environmental education, employment, skills accumulation and human development. End of life products are taken back to keep continue reverse supply chain as an end of life service instead of disposal. Current private reverse supply chains of Aluminum and Plastics collection for recycling can be integrated within national reverse logistics and take back for remanufacturing, by financial incentives and public media promotion, reverse supply chain can be strengthen through public discussion about sustainable development. Education disciplines can be enlarged to include remanufacturing and provide practical syllabus to be achieved by student to satisfy graduation requirements. Cost/labor ratio can be as low as possible due to exploit educational and social activities to substitute the labor intensive remanufacturing.

3.1.2. Customer segments: High environmental concern of customer generates high acceptability of remanufactured steel which is supported by low sensible price. Functionality oriented customers should be induced as national adoption of remanufacturing business model to prompt business-to-customer paradigm to reach customization and personalization market requirements.

3.1.3. Value proposition: Convenient remanufactured steel products can be customized and of cheap price and priority. Such value proposition can be certain by standard quality and like new warranty.

3.1.4. Customer relationship: Financial based voluntary return of end-of-life products are strategic involvement of customer in remanufacturing processes and activities to apply relationship based high exchanging.

3.1.5. *Network:* Hybrid centralized and decentralized return systems can be based on current reverse supply chain of take back for recycling which can also be enlarged as private suppliers. The main emerged reverse chain is a financial based disposal, where eco-design, design for multiple lifecycle continue provide remanufactured steel as a regular switching between national and independent remanufacturers and reverse logistics operators through launching direct and indirect distribution channels.
3.1.6. Resources: Educational institutions infrastructure and hybrid labor of skilled and educational trainers with sustainable students' contribution are required to perform remanufacturing of end-of-life steel into construction steel products. Such sustainable educational behavior leads to triple bottom sustainability.
3.1.7. Revenue: Remanufactured construction steel selling provides revenue with dynamic pricing to certain competition with imported products, so remanufactured steel should be of the lowest possible alternative price.
3.1.8. Cost structure: Value creation is to seek necessary minimum operating costs and drive successful continues sustainability. Symbiotic and solidarity of society are helpful to reduce labor cost while remanufactured steel selling should provide reasonable revenues to cover reverse logistics cost and rent , maintenance and buying of equipments.
3.1.9. Financial Business Plan: Public and educational institutions investments can start remanufacturing and profited selling of remanufactured steel can keep continue the business. While feedback form public sectors are necessary to support such adoption of human development activities. Financial assets owned by educational institution and banking leans for sustainability can drive reasonable profited remanufacturing business.

3.2. *Sustainability initiation key factors*
3.2.1. Financial incentives: Since Original manufacturers are foreigner companies and new products are imported, so primarily public opinion should be enabled to response as a community sustainable conscience which leads to prepare incubator to lunch steel remanufacturing and achieve reasonable profit for business self-continue. The creation of specific rules to take into account the specificities of reverse logistics operators and educational remanufacturers, are important to increase responsibility of end-of-life steel environmental conscious treatments. Without sectorial bifurcations to avoid heterogeneous and difference of business conductors, successful educational remanufacturers and other stakeholders can be directed towards central goal of sustainability. The government releases general laws that relate to the end-of-life treatment of products. Public instructions induced community rethinking to be applied by experts as strategic end-of-life steel remanufacturing possibilities. Goals and instruments instead of compulsory laws can navigate remanufacturer from different sectors to move towards sustainability with elastic regulations to certain monitoring and feedback.
3.2.2. Sustainable development encouraging: Financial support can promote sustainable development, where stimulating and rewarding of remanufacturing activities fosters environmental education and public discussion on sustainable development. Advertisement and green labeling of products increase contributes to environmental education and sustainability. Such central public discussion of educational institutions and remanufacturing conductors should be actively engaged in remanufacturing activities to promote triple bottoms development.
3.2.3. Strategies of remanufacturing: Growing demand of remanufactured steel for rebuilding and governmental financial incentives can substitute lack of foreigner original manufacturers to enter the remanufacturing market and extend their responsibility to remanufacturing end-of-life products. Original manufacturers can provide cloud based instructions, advising and supervision even there is no responsibility toward the end-of-life products so independent remanufacturers emerge and increase.
3.2.4. Users of remanufactured steel: Steel consumption behavior should be reviewed and measured to introduce sustainable using. Standardizing and unifying of sustainable eco-design should be widely circulated through metal working community. Cloud metal working and e-marketing accumulate eco-steel products.
3.2.5. Demand for remanufactured steel: Improved quality of remanufactured steel and obey certifications and new standards as enacted by associations of steel manufacturing can certain like new

and equivalent remanufactured steel and lead to increase demand. Introducing of technologies and methods of disassembly to lower costs for remanufacturing and decrease the price of remanufactured steel should source in this direction of demand increasing.
3.2.6. *Associations for remanufacturing:* Certifications, technical standards and legal regulations are prompted to increase quality of products and services and cause a positive influence on demand. The aim is, associations of steel manufacturing prove that remanufacturing of steel is successful through the reduction of the overall technological gap and environmental impacts.
3.2.7. Product development: Product service system and remanufacturing can be focused to establish national service oriented value creation. This can help establishing of information bank of lifecycle of products which can be linked with cloud manufacturing to provide stream of remanufactured parts to function within new products structure. Gradual transition from direct selling into accessories services accompany physical product can configure collaborative incubator which is very necessary with eco-efficiency to apply sustainability. Services in combination with products provide higher profits and product development prospective which motivate investment to remanufacture complex products. Product service system, reducing of consumption and eco-innovation approaches integrate end-of-life products as promising environmental protection strategies to be adapted for increasing the usage and productivity of resources by enabling multiple lifecycle in different applications.
3.2.8. Providers: National Company can be established to include technical institutions in one business model to remanufacture steel and provide metal working services to produce sustainable construction steel products to be used for rebuilding of infrastructures and services through whole the country. Step by step introduction of more complex products to be remanufactured to link national cloud of remanufacturing with cloud manufacturing of developed countries can be realized. Product service system and product lifecycle related information system construction can configure strength remanufacturer-customer relationships and deliver educational environment and pro-environmental using and consumption promotion. Complete substitution services system to be included detailed information about remanufactured modules and parts can be further crowdsourcing to get mature information to be exploited for remanufacturing of like new products.
3.2.9. Qualification: Environmental education and business entities realize enormous potentials that enable graduation and training requirements fulfillment and certification which can be exploited as sustainable efforts. Educational institution with remanufacturing teaching and trainings through comprehensive business model can link, at international level, associations, universities and companies collaboratively. In addition, special training and informal education can be increased to serve with public discussions contribute sustainable development of remanufacturing. Remanufacturing business can be delivered internationally as a cloud service and thus, international experts can be exploited for further maturity of information for remanufacturing. Public promotion and offering of trainings related to remanufacturing spread wide knowledge, so cloud of know-how services can be offered by original manufacturers to help developing countries conduct development and sustainability.
3.2.10. Reusability: environmental education, eco-design and eco-innovation realize eco-efficiency and consequently prompt reusability. Brainstorming is a powerful innovation tool which can be extracted and exploited through crowdsourcing to be focused through information system that can be accompanied remanufactured products. Stimulating and rewarding can encourage attraction, fondness and interesting of human to generate eco-ideas.

3.3. *Remanufacturing barriers elimination scenario*
Lack of standard quality of remanufactured products and absence of organizational remanufacturers are important immaturity reasons that lead to not emerge remanufacturing due to effort illegitimacy, uncertain availability of used products, and unguaranteed quality [13], which required cloud information system. Such cloud system can enable quality and quantity of returned steel to be consistent, so continues stream of end-of-life steel is certain. As high as suppliers of end-of-life steel, strengthening of remanufacturing can be obtained while no financial based selection is applied. Final customer should be included as supplier of end-of-life steel, since reverse logistic for recycling is door to door and thus, individual returns can be collected. Storage place is enough since there is no need for

IOP Conf. Series: Materials Science and Engineering **83** (2015) 012021 doi:10.1088/1757-899X/83/1/012021

special conditions to protect end-of-steel. Responsibility toward environment protection and financial incentives can encourage conscientious responding toward society and accept rewarding few incentives. Remanufacturing problems of scattering and dispersion can be treated through social activities and public media instructions to increase awareness of environment protection. Pre-test to discover remanufacturability is no need and automation can be introduced to disassembly and reassembly. Social activities can support emerging of disassembly and assembly workshops as a responsibility toward society and environment protection. No inspection costs and no need for special inspection devices, since accuracy of remanufactured steel are consistent and obtainable. High and stable demand of remanufactured steel can help emerge of national workshops of metal working to apply eco-design and eco-innovation. Such sustainable consumption and eco-purposing of steel prompts multiple lifecycle. High acceptance of service oriented remanufactured steel can be sold with new steel through the same distribution channels to eliminate customer suspicious since quality and attractiveness are comparable with new products. Collaborative working, with skilled labor, can remanufacture end-of-life steel according to structural steel standards. Remanufactured structural steel is durable and diversity of product can be manufactured by responding to standards without utilization of high materials. Design for remanufacturing can be prompt and appear in the next generation of structural steel finished products to certain continues remanufacturing business. Structural steel products should be as simple as possible to deliver disassembly. Information banking system can expand gradually to include all remanufactured and end-of-life modules and products. Such national digital library can contain detailed information such as CAD files, pricing, quantities, mechanical and material properties, product status and remanufacturing procedures. Exposure of such information, to cloud crowdsourcing, adaptive learning knowledge and expert systems, results into mature information about how to remanufacture end-of-life modules. A digital bank of eco-ideas can exploit remanufactured modules to function within eco-products through international cloud manufacturing system.

4-Case studies and experimental procedures

Various resources of end-of-life steel can be exploited such as vehicles hulks and bodies, steel wheel and different shells and cases. Any pieces of steel of uniform shape can be remanufactured into structural steel. Such steels are remanufactured into semi-finished products which contain hollow section of tubes and pipes. Doors, windows and decoration applications need further processing of semi-finished steel by metal working operations such cutting, shearing and welding. To keep sustainable development, recent methods of consumption should be changed and substituted by more environmental conscious methods. Shapes of doors and windows should be as simple as possible and can be covered by plastics or wood for decoration and engraving purposes. Figure (1) shows an algorithm of automotive steel body remanufacturing in an academic scale. Doors , engine cover , roof and back cover are disassembly , hammered and sheared into strips to be nested within new six meters length strips to form composite shape steel. Mechanical bolting or riveting can be used to nest the recovered steel within the structure of the new strips. Energy consumed is measured and accordingly Co2 standard emission value is selected, eco-audit can be offered by figure (1) also. Different example, car wheel is illustrated in figure (4), such end-of-life steel wheels can be disassembly by turning or milling processes to remove the welded areas. Transaction cutting by plasma can certain easiness of disassembly. Consequences of bench and hand fly cutting operations can change the rim into small pieces. Hammering and shearing are series to recovery regular steel shapes, eco-audit is also recorded in figure (4). The remanufacturing of steel oil filter shells contains removing of handles, bottoms and pipes and the rest shell is sheared at one side to form long sheet by unfolding. Shearing and fit to size can help obtaining of regular recovered steel to be nested within bases of new steel strips. Since the remanufactured end-of-life vehicle is belonged to the beginning of the decade of nineties as an application of take back for remanufacturing procedure. So it is suffered from lack of disassembly due to welding and high amount of forming to be difficult to be remanufactured. Legislations makers should onset design for remanufacturing through increasing disassembly and reduce amount of forming and increase the value of used steel. The doors interior guards, figure (2), as example, are hardened by cold forming and of several openings to simplify assembly and maintenance of door locking and windows auto open-close systems. The design makes them complex to be remanufactured and they are suitable for smelting. A feedback from the end-of-life doors can secrete an eco-design idea. Instead of one piece as in figure (3), two pieces with as low as possible forming, to avoid forming and related wrinkling which is encountered during remanufactured steel preparation, can be assembled by bolts. In this contrast, authors contacted several scrap steel processing equipments manufacturers and no ability to flatten the formed steel was discovered. Produce construction steel pipe of 22mm outer diameter and 3mm wall thickness required steel strip of 64mm width. To simulate this through the door guard of two pieces, eco-design states that the main dimensions can be roughly of 640mm each, figure(3), to produce ten pieces of 64mm width to be nested by remanufacturing within two bases of new steel of 64mm also. Finally coil of 64mm remanufactured steel can be obtained by repeating the nesting process, figure (7).

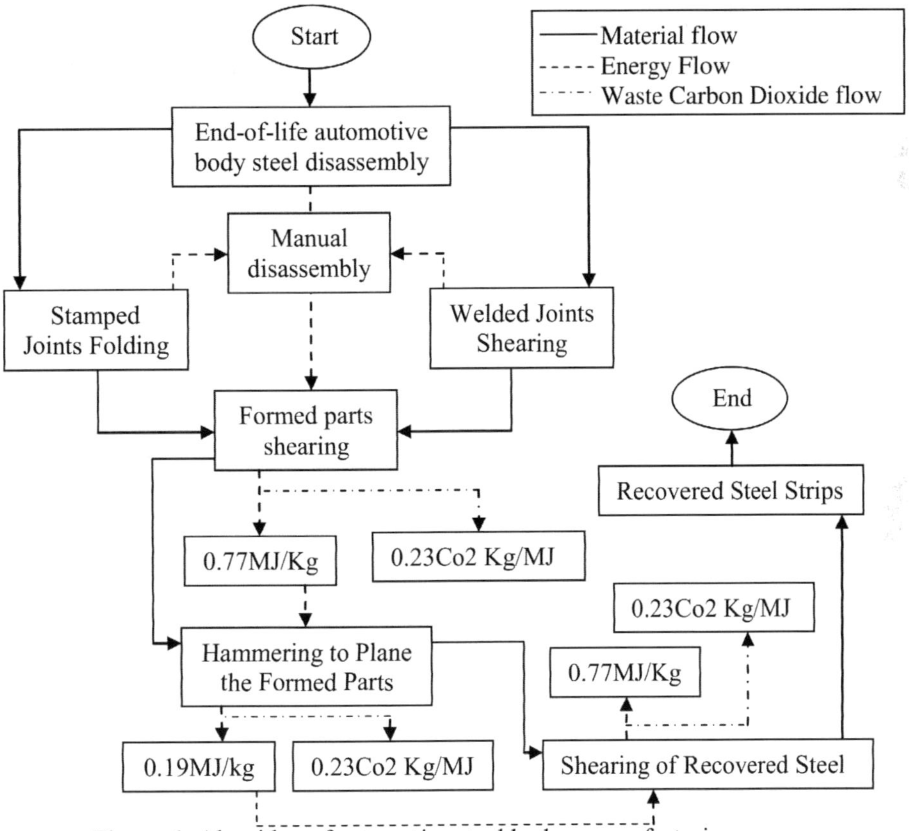

Figure 1. Algorithm of automotive steel body remanufacturing.

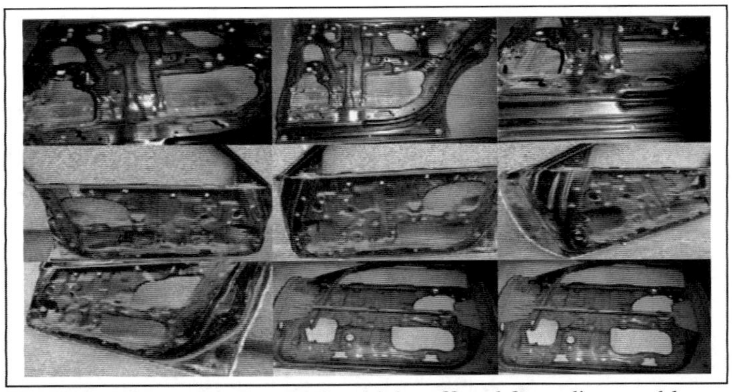

Figure 2. Car door's guards which suffered from disassembly and unnecessary forming.

IOP Conf. Series: Materials Science and Engineering **83** (2015) 012021 doi:10.1088/1757-899X/83/1/012021

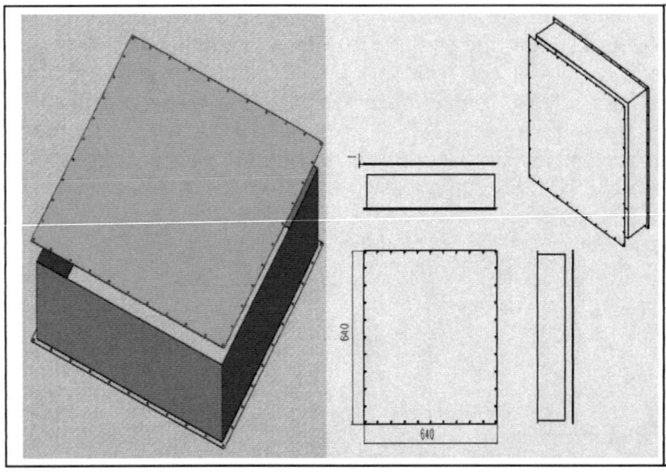

Figure 3. Eco-design idea to increase remanufacturing of car door's guard.

Figure 5. Eco-design idea to increase remanufacturability of car steel wheel.

Car wheels steel also suffer from non-design for remanufacturing by prompting disassembly. Eco-innovation states that two pieces, figure (5), assembly by using bolts can increase remanufacturability and value embodied by the steel wheel. This will save energy consumed and needed time for welded areas removing processes. The value can be raised by interchangeability attribute acquisition, both rim and hub can be designed to be modules of storage or pipe transportation systems. Also standardization of shape and size will increase the lifecycle of the wheel and reduce unnecessary metal forming. Design for disassembly strongly appears to weak application of remanufacturing of oil cases steel. Oil flow pipes and handles are welded to main body of the steel shell which is required elimination of welded areas by mechanical processing. So such non eco-design alters remanufacturing by consuming power and time. Academic scale processing of oil filter steel cases and corresponding eco-audit are illustrated in figure (6).

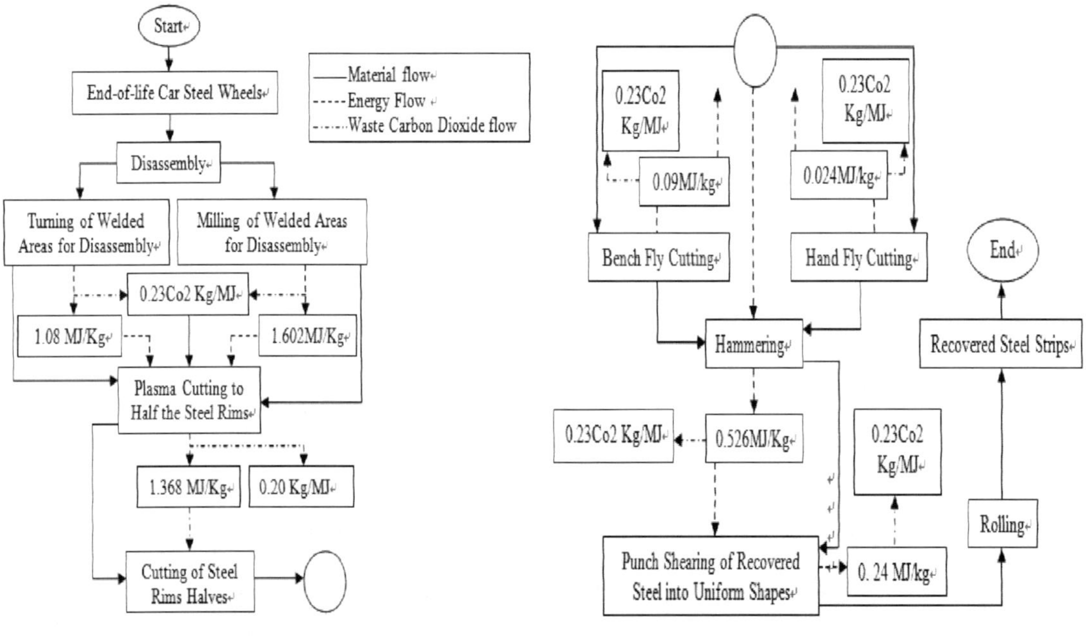

Figure4. Algorithm of steel car wheel remanufacturing.

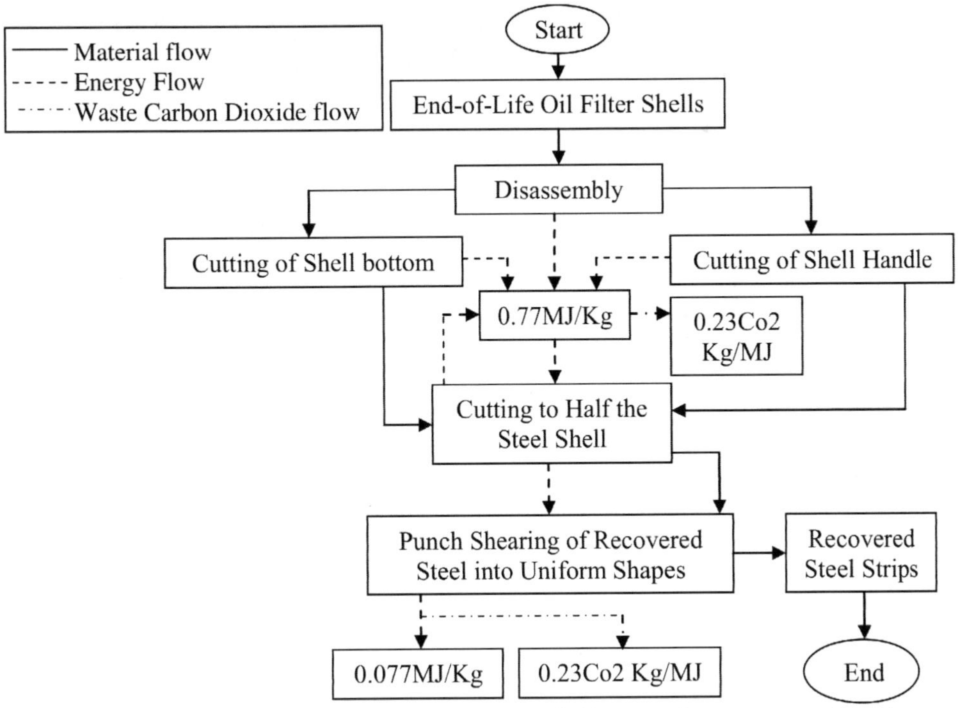

Figure 6. Algorithm of oil filters steel cases remanufacturing.

Recovered steel strips are mixed with new strips and since recovered steel is painted so its weldability is very low and mechanical bonding is more suitable. Aluminum rivets are strong enough to be used and self-locked pipes are another alternative. Arranging successive layers of recovered and new steels can certain new strips to be of new outer surfaces to deliver like new appearance and thus the required thickness can be obtained.

Table 1. Energies (MJ/Kg) of virgin recycled and remanufacture steels.

Virgin Steel	Recycled Steel	Remanufactured Steel
39-42	6.6-8	0.26 (vehicle body)
39-42	6.6-8	3.112-3.63 (car wheel)
39-42	6.6-8	0.198 (oil filter shell)

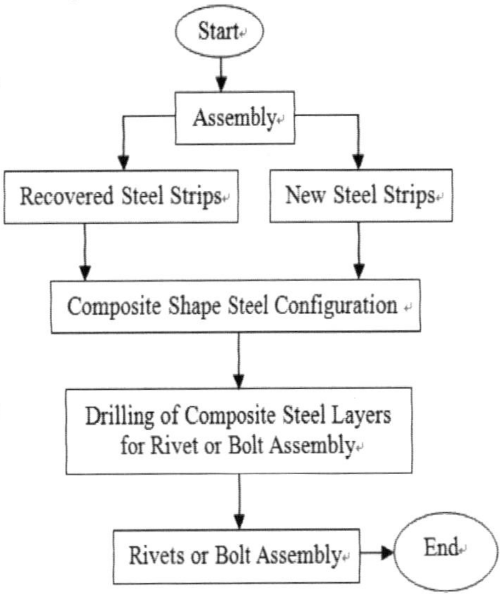

Figure 7. Composite shape steel strips remanufacturing procedure.

5-Conclusion

To be consistent with scientific literature which states that huge amount of vehicles are at the end-of-life each year and with directives of bigger cars manufacturers to reach 95% of recovery. This study represents a request to take in consideration aspirations of developing countries to conduct remanufacturing business. Industrial nations can crowdsourcing eco-design and eco-innovation to certain continues stream of remanufactured steel through design high value products to continue in service until the end-of-life. While remanufacturer societies are emerged based on educational institutions in the developing countries to close the supply chain as globally integrated sustainable manufacturing platform Remanufacturing of car body registered high required energy than the lowest which is registered by oil filter shells remanufacturing, but highly available resources of end-of-life vehicles prompts car body remanufacturing. Table (1) is a literature based comparison, [14], to show the level of embodied energy for virgin, recycled and remanufacturing steels. Such academic scale needs to be further searched to be applied in industrial scale to certain lower levels of remanufacturing energy and to apply automation at assembly level while intensive labor remanufacturing can be covered through education-business model.

Acknowledgements
Authors highly appreciate manufacturing infrastructure using facility and submit warm thanks to Wuhan University of Technology / Mechatronic School, Foreigners Office and Hu Bei Jin Da Li Chuang Ke Ji You Xian Manufacturing Company.

References

[1] Guidat, T., Barquet A.P., Widera, H., Rozenfeld, H. and Seliger G. 2014 Guidelines for the definition of innovative industrial product-service systems (PSS) business models for remanufacturing *Proceedings of the 6th CIRP Conference on Industrial Product-Service Systems.* **Procedia CIRP16** 193-198
[2] Wang-Chih Chena and Jahau Lewis Chena 2014 Eco-innovation by Integrating Biomimteic Design and ARIZ. *21st CIRP conference on Life Cycle Engineering*. **Procedia CIRP15** 401- 406
[3] Michelle Tenam-Zemach , Mandy Kirchgessner b, John Pecore c, Leanne Lai d and Steven Hecht 2014 Development of an innovative for analyzing the presence of environmental sustainability themes and an ecological paradigm in science content standards *Studies in Educational Evaluation* **41** 133–142
[4] L. Buys, K. Mengersen, S. Johnson, N. van Buuren and A. Chauvin 2014 Creating a Sustainability Scorecard as a predictive for measuring the complex social , economic and environmental impacts of industries, a case study :Assessing the viability of the dairy industry *Journal of Environmental Management***133** 184–192
[5] Guangdong Tian , Jiangwei Chu , Hesuan Hu and Hongliang Li2014 Technology innovation system and its integrated structure for automotive components remanufacturing industry development in China *Journal of Cleaner Production* **85** 419-432
[6] Xiqiang Xia , Kannan Govindan and Qinghua Zhu 2015 Analyzing internal barriers for automotive parts remanufacturers inChina using grey-DEMATEL approach *Journal of Cleaner Production* **87** 811-825
[7] C. Salazar, A. Lelah and D. Brissaud 2015 Eco-designing Product Service Systems by degrading functions while maintaining user satisfaction *Journal of Cleaner Production* **87** 452-462
[8] Jing Wang, Jun Zhao and Xun Wang 2011 Optimum policy in hybrid manufacturing/remanufacturing system *Computers & Industrial Engineering* **60** 411–419
[9] Afshin Mehrsai, BjØnar Henriksen, Carl C. RØtad, Karl A. Hriberni and Klaus-D. Thoben 2014 Make-to-x grade for the design and manufacturing of flexible, adaptive, and reactive products 24th *CIRP Design Conference* 199 – 205
[10] Mourtzis D. and Doukas M. 2014 Design and planning of manufacturing networks for mass customization and personalization: Challenges and Outlook *Robust Manufacturing Conference (RoMaC)* 1 – 13
[11] Dazhong Wu, David W. Rosen, Lihui Wang and Dirk Schaefer 2015 Cloud-based design and manufacturing: A new paradigm in digital manufacturing and design innovation *Computer-Aided Design* **59** 1–14
[12] http://bragecrim2.wix.com/remanufacturing.
[13] Qinghua Zhu , Joseph Sarkis and Kee-hung Lai 2014 Supply chain-based barriers for truck-engine remanufacturing in China*Transportation Research Part E* **68** 103–117
[14] Michael F. Ashby2013Materials and the Environment(UK: Elsevier Inc)

IOP Conf. Series: Materials Science and Engineering **83** (2015) 012022 doi:10.1088/1757-899X/83/1/012022

Projection Classification Based Iterative Algorithm

Ruiqiu Zhang[1], **Chen Li** [1] **and Wenhua Gao** [2]
[1]School of Design, South China University of Technology, Guangzhou 510640, Guangdong, China
[2]School of Mathematics, South China University of Technology, Guangzhou 510640, Guangdong, China

E-mail: whgao@scut.edu.cn

Abstract. Iterative algorithm has good performance as it does not need complete projection data in 3D image reconstruction area. It is possible to be applied in BGA based solder joints inspection but with low convergence speed which usually acts with x-ray Laminography that has a worse reconstruction image compared to the former one. This paper explores to apply one projection classification based method which tries to separate the object to three parts, i.e. solute, solution and air, and suppose that the reconstruction speed decrease from solution to two other parts on both side lineally. And then SART and CAV algorithms are improved under the proposed idea. Simulation experiment result with incomplete projection images indicates the fast convergence speed of the improved iterative algorithms and the effectiveness of the proposed method. Less the projection images, more the superiority is also founded.

BGA (Ball Grid Array) package as a high-density and high thermal performance package form has been widely used in CPU manufacturing, memory manufacturing and other manufacturing. Because of its contacts distributed in the packages below, it is very difficult to detect solder joint quality. Traditional artificial visual, automated optical inspection (AOI) and flying probe test technology are no longer applicable, although you can use X-rays penetrate the board to get the information of hidden solder joints distributed in the packages below, X-rays detection can only make a rough detection of solder joints if there is a bridge or weld defects ,but for little tin, viod ,weld and other smaller defects which impact the quality of the product ,is helpless because the projected image contains many overlapping information. Therefore, many scholars explore the use of hierarchical scanning method (Laminography) to reconstruct the joints [3][4], but the reconstructed image quality and is poor compared with CT (Computed Tomography) reconstructed image.

Although CT technology has been widely used in various industrial NDT field, its electronic detection for BGA chip solder joints was ineffective. The reason of high attenuation characteristics of the ray of flat shaped circuit board and solder joints makes the traditional algorithm of cone beam CT (such as FDK [1]) not play it more accurate and fast reconstruction of objects advantages, because Ray beam hardening [13] specific direction leads to the inaccurate projection data which can not reflect the actual material attenuation coefficient distribution. The application of CT technology in BGA solder joint inspection need to be considered: on the one hand, due to testing requirements of high efficiency in solder circuit board production line, a complete projection data algorithm must be abandoned, because it would waste a lot of time to acquire projection images. On the other hand, as far as possible to avoid the inaccurate projection data of the circuit board in the horizontal direction. Therefore, short scan system and less angle reconstruction seems to be a feasible solution, the former is to reduce the number of projection and consider short scan form analysis type of algorithm of complete projection data[10][11], the latter is more focused on the research for the incomplete projection data algorithm, such as iterative algorithm.

Iterative algorithm which is antithetical to analytic algorithm was introduced firstly into the field of image reconstruction by Gordon et al. The theory of reconstruction field has made great progress after a lot of scholars' efforts. Objects are usually discretized by iterative algorithm into a grid in accordance with the pixel or voxel (2D) (3D) size of objects. ART, EM, SIRT, SART, CAV are common iterative algorithm, the first is the basic iterative algorithm, the second is statistical iterative algorithm, the latter three are classified as synchronous iterative algorithm, they are the improved algorithm of ART algorithm in convergence speed and the quality of the reconstructed image. Usually the iterative algorithm is more suitable for incomplete reconstruction projection data, but these types of algorithm such as ART algorithm, SART algorithm, and CAV algorithm have large amount of calculation and the slow convergence speed, researchers have been exploring to accelerate the convergence speed of the iterative algorithm. For example, Mueller [5] studied the effect of access order to projection data on convergence speed, Herman [6] studied the effect of different relaxation factors on the convergence rate of the reconstruction, Wang studied that how OS technique can be applied to the SART algorithm to accelerate the convergence [12].

In conclusion, the analytical algorithm is not suitable for the detection of BGA solder joints on the circuit board. This paper is to improve the convergence speed of iterative algorithm and tells us that the objects of different materials have different backprojection speed, and then discusses the method of classification of projection data, the reconstructed objects are classified according to the solute, solution and air, the backprojection speed of iterative algorithm and material of objects are relevant. In addition, this paper describes the design of a virtual 3D cone beam circular orbit scan system and simulates the projection and reconstruction of BGA solder joint.

1. Iterative algorithm

The following discussion is focused on two representative iterative algorithm and the WP (Weighted Projection) type deformation based on projection data classification.

The SART algorithm proposed by Andersen in paper [7], the following shape type:

$$f_j^{k+1} = f_j^k + \frac{\lambda_k}{\sum_{i=1}^{M} w_{ij}} \sum_{i=1}^{M} \frac{w_{ij}}{\sum_{j=1}^{N} w_{ij}} (p_i - w^i f^k) \tag{1}$$

In the formula, f_j^k represents the image after k iterations, λ_k represents relaxation factor, w_{ij} is the weight coefficient, which reflects the contribution degree j voxel to article i ray, p_i represents projection, M represents pixel number in the total projection, N is the object voxel number.

CAV algorithm is developed from the synchronous Cimmino algorithm by Censor[8], Censor believes that if the weight coefficient matrix is sparse, then the weight coefficient should not be $\frac{1}{M}$ of Cimmino algorithm , it should be $\frac{1}{S_j}$, S_j is the number of non-zero elements in the j column in the weight coefficient matrix. So the CAV algorithm is obtained and the data update formula is as follows:

$$f_j^{k+1} = f_j^k + \lambda_k \sum_{i=1}^{M} \frac{p_i - \langle w^i, f^k \rangle}{\sum_{j=1}^{N} S_j w_{ij}^{\,2}} w_{ij} \tag{2}$$

Wherein, $\langle \bullet,\bullet \rangle$ represents point multiplication. The derivation of this formula can reference [7]

Iterative reconstruction is based on the difference between the projection data and the estimated projection to take backprojection to keep the process of reconstruction of objects constantly. It is generally believed that the weight coefficient matrix is determined by physical configuration of scanning system, effects of all voxels on pixels that arrive on the detector are same, of course, according to other a priori information, such as the density of a known object distribution can be used in the process of back-projection of the weights. Analysis of several common iterative algorithm which contains the following two:

A、 Projection deviation, expressed as $p_i - w^i f^k$

B、 The denominator of the average deviation.

Comparison of several algorithms shows that several iterative algorithms denominator are only concerned with the weight coefficient matrix, the paper [9] put the previous iteration solution into backprojection process to improve the efficiency of the Cimmino algorithm. It is only in a large number of iterations that reflects fast convergence. From the other point of view, this paper proposes that back-projection process should be linked with the projection data categories to accelerate convergence.

For quality detection of circuit board, testing personnel hope to analysis the quality of solder joints to determine whether there are defects weld, bridging and voids. Chip itself and the PCB substrate is not part of the testing personnel concerned, the utilization of fly needle of traditional test method is also based on this starting point, but such an electrical connection testing method for the detection of some defects such as voids, insufficient solder etc is helpless. If you take X ray scanning system, these parts will inevitably reflected in the projected image, a large parts of the amount of calculation will be used for the reconstruction that is not the testing personnel's concerned substances while in 3D backprojection reconstruction process. Inspired by this, this paper believes that the reconstruction of objects can be divided into three parts: the solute part (solder), solution (such as part of the chip and the substrate) and air part (peripheral reconstruction region), so we should give them different reconstruction speed to accelerate the convergence speed in the backprojection process. Here the specific approach is to solution, threshold for the recovery speed of vertices, respectively to the solute and air threshold linearly decreasing, as shown in Figure 1, to get the improved WP-SART, one of WP type projection algorithm, the form below:

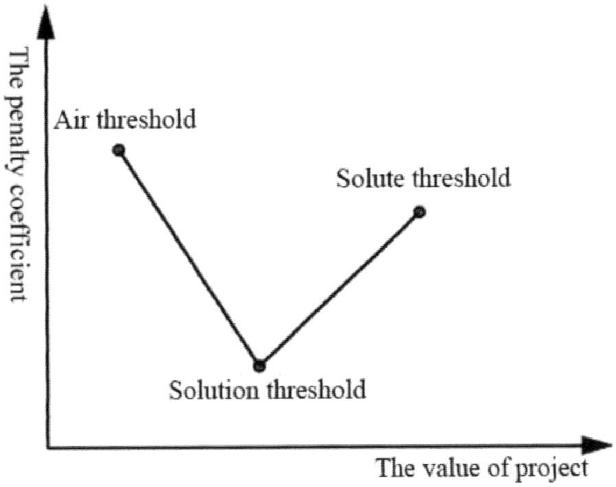

Figure 1. Penalty function

$$f_j^{k+1} = \frac{\lambda_k}{\sum_{i=1}^{M} w_{ij}} \sum_{i=1}^{M} \frac{w_{ij}}{\left(\left|\frac{\phi_{\eta_i} - \eta_i}{\phi_{\eta_i}}\right| + \mu\right)\sum_{j=1}^{N} w_{ij}} (p_i - w^i f^k) + f_j^k \tag{3}$$

In the formula, $\eta_i = \frac{p_i}{\max(p) - \min(p)}$, ϕ_{η_i} is a threshold value, μ is the correction factor, μ is a value of 0.1 in this paper. In the formula（3）, $\left|\frac{\phi_{\eta_i} - \eta_i}{\phi_{\eta_i}}\right| + \mu$ can be calculated and stored in advance according to the known projection data. Therefore, this improved method doesn't increase additional computation in the process of iteration.

Similarly, $\left|\frac{\phi_{\eta_i} - \eta_i}{\phi_{\eta_i}}\right| + \mu$ can be applied to the CAV algorithm and then WP-CAV algorithm can be obtained, here no longer say.

2 The simulation experiment

To compare the effect of reconstruction of traditional synchronous iterative algorithm with WP algorithm, this paper introduces the design of cone beam CT projection system of a virtual scanning circular orbit and virtual model of the BGA circuit board, setting up that its size is 24 x 64 x 84 (sections such as shown in figure [2]), voxel size is 1mm, the detector size is 128 x 80, pixels size is1.5mm, source to object rotation center axis distance is 100mm, the distance is 100mm from the ray source to the center of rotation axis of the object, 1 degree projection index.

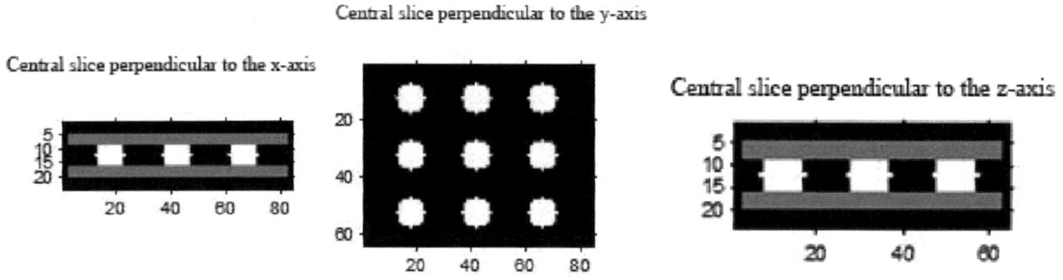

Figure 2. Slice of simulation BGA model

The two kinds of commonly method of reconstructed image evaluation is given to evaluate the convergence speed of the evaluation algorithm in this paper, one is to select key rows or columns of reconstructed image to compare with the original Sliced model, the other is reconstruction of objects and MSE (Mean Squared Error) [4]of the original model as follows:

$$MSE = \frac{1}{N}\sum_{j}^{N}(f_{jr} - f_j)^2 \tag{4}$$

In the formula, f_{jr} is the reconstruction of objects, f_j is the original model.

This experiment used SART, CAV, WP-SART, WP-CAV algorithm, set up iterative times as 50, coefficient of relaxation 0.3, penalty term correction factor 0.1, the reconstruction results of various data are as follows.

2.1 Reconstruction of 30 degrees to 150 degrees projection angles

The simulation results in Figure 4 obtained by the MSE of various algorithms, the convergence speed of WP algorithm is superior to the same type (WP-SART is better than SART, WP-CAV is better than CAV) , CAV algorithm outperforms the SART algorithm. Through nearly 50 iterations, SART and CAV algorithm are not convergent to a small MSE, but WP algorithm from fewer iterations have converged to a steady state. From the analysis of the central slice in Figure 3, we can see that WP algorithm has better detail resilience and the reconstruction effect of the center region of the image is better.

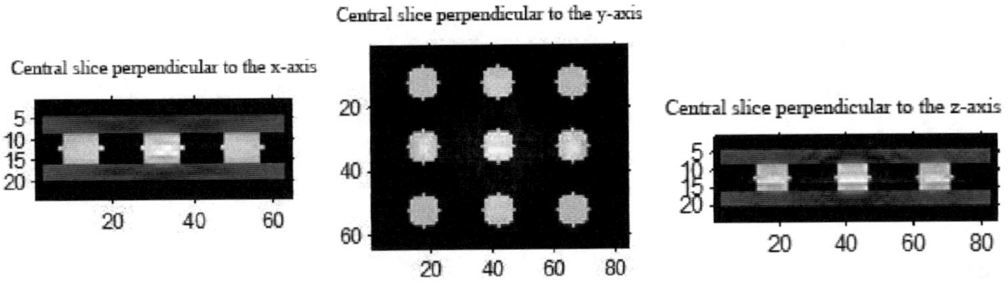

(a) SART

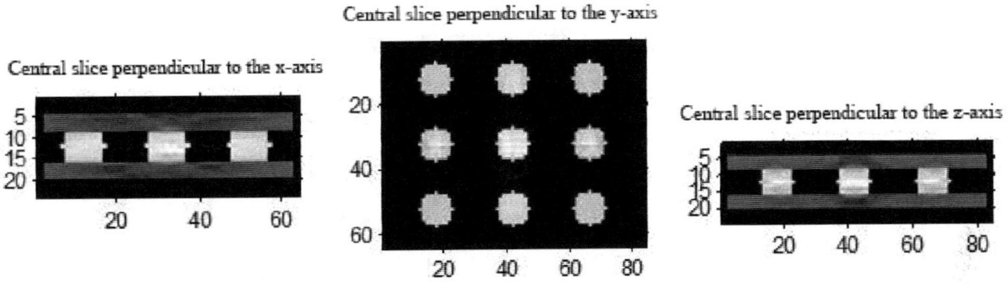

(b) WP-SART

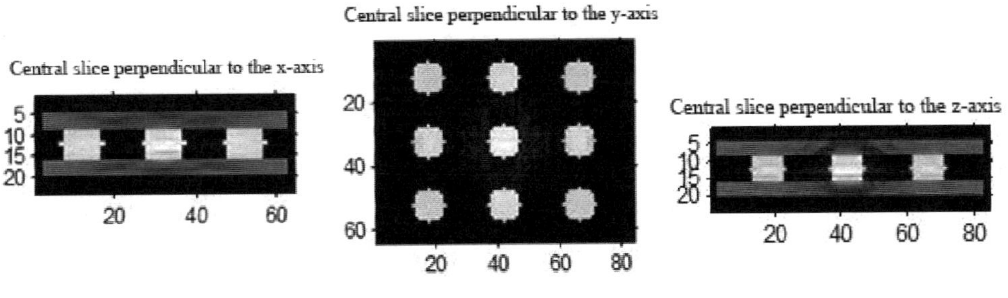

(c) CAV

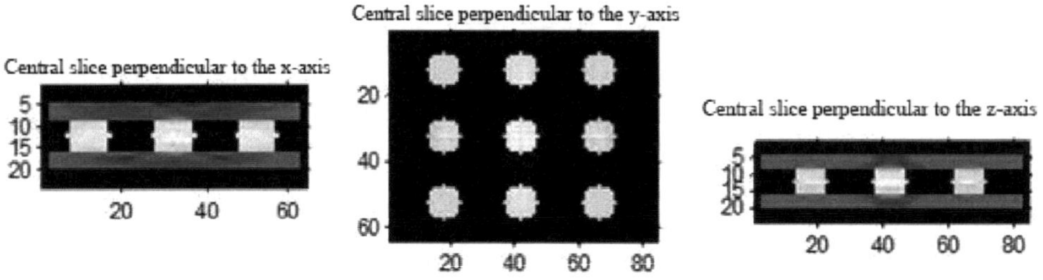

(d) WP-CAV

Figure 3. Slice of computed object

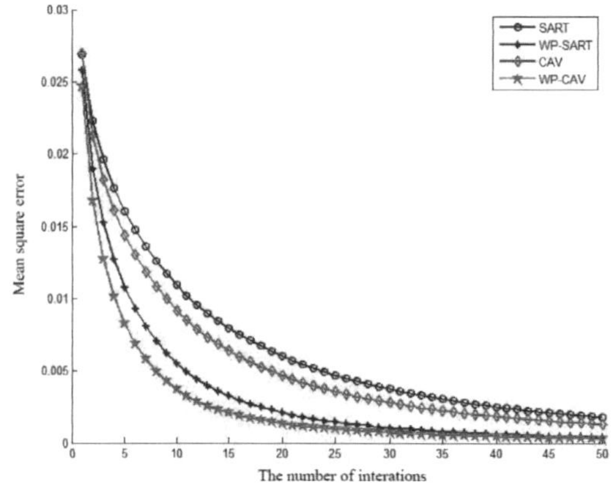

Figure 4. MSE under 30 to 150 degree projection

2.2 Reconstruction of 45 degrees to 135 degrees projection angles

Compared with the projection of 30 degrees to 150 degrees, due to reduced amount of data, the convergence speed of 45 degree to 135 degree projection is slow, but WP algorithm is still better than original algorithm, the convergence speed of WP-CAV algorithm is fastest shown in Figure 5.

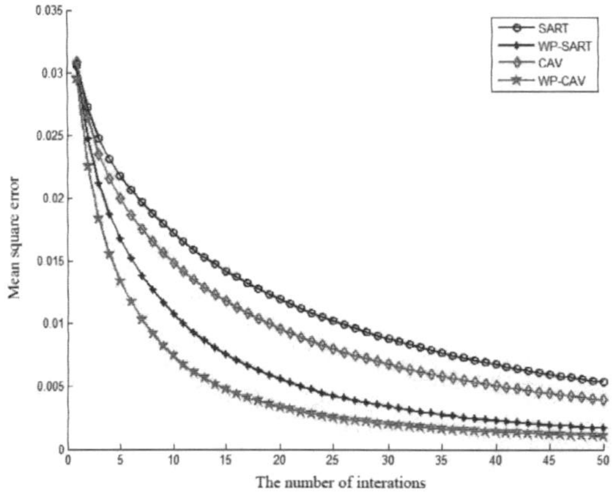

Figure 5. MSE under 45 to 135 degree projection

2.3 Reconstruction of 60 degrees to 120 degrees projection angles

Compared with the previous two kinds of amount of data, the amount of projection data based on 60 degree to 120 degree is loss seriously, various algorithms are non- adaptive, but WP algorithm is still better than the original algorithm. Figure 6 has shown that WP-SART algorithm convergence speed and CAV algorithm convergence speed are similar, but the WP-CAV algorithm still has a faster convergence speed.

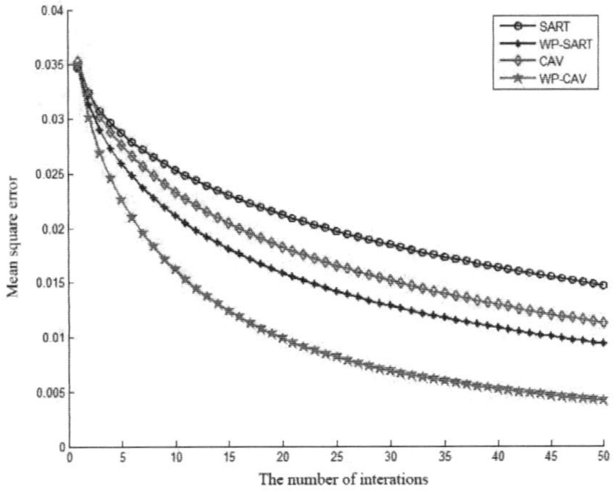

Figure 6. MSE under 60 to 120 degree projection

3 Conclusion

The iterative algorithm is well suited for this class of flat plate objects like circuit board, it does not need the ray on projection circuit board in the horizontal direction to converge. Thus it avoids the use of the inaccurate projection data. In addition, compared with the traditional analytic algorithm, iterative algorithm can avoid object rotation platform which can be replaced by the rotary imaging sector to meet the needs of production lines of the circuit board.

Because iterative algorithm convergence speed is slow, it has not been applied to BGA solder joint inspection which needs higher demands on the detection time. This paper proposes to deal with different parts of objects (solute, solution, air) to take a different back-projection algorithm to speed up the convergence speed, which is based on the classification of the detected object projection. Based on two typical synchronous iterative algorithm, simulation results show that the SART and CAV algorithm, the WP algorithm by using the algorithm of the same type improved by the method proposed in this paper, it's convergence speed is faster and can have better detail recovery capability. The less projection data used, the more obvious effect improved, thus the effectiveness of this method is proved.

The MSE chart shows that MSE of WP algorithm were significantly smaller if the iterations is same, the next step will research convergence of the algorithm qualitatively. In addition, the WP method will consider the object shape information to reduce the reconstruction region of objects to further accelerate the convergence speed of the algorithm.

References

[1] Feldkamp L A, Davis L C, Kress J W. Practical cone-beam algorithm[J]. JOSA A, 1984, 1(6): 612-619.

[2] ZHUANG Tian-yi. CT principles and algorithms [M] Shanghai: Shanghai Jiaotong University Press, 1992, 34-37.

[3] Roh Y J, Cho H S. Implementation of uniform and simultaneous ART for 3-D reconstruction in an x-ray imaging system[C]. Vision, Image and Signal Processing, IEE Proceedings-. IET, 2004, 151(5): 360-368.
[4] Moore T D, Vanderstraeten D, Forssell P M. Three-dimensional X-ray laminography as a tool for detection and characterization of BGA package defects[J]. Components and Packaging Technologies, IEEE Transactions on, 2002, 25(2): 224-229.
[5] Mueller K, Yagel R, Cornhill J F. The weighted-distance scheme: a globally optimizing projection ordering method for ART [J]. IEEE Transactions on Medical Imaging, 1997, 16(2): 223-230.
[6] Herman G T, Meyer L B. Algebraic reconstruction techniques can be made computationally efficient [positron emission tomography application][J]. Medical Imaging, IEEE Transactions on, 1993, 12(3): 600-609.
[7] Andersen A H, Kak A C. Simultaneous algebraic reconstruction technique (SART): a superior implementation of the ART algorithm[J]. Ultrasonic imaging, 1984, 6(1): 81-94.
[8] Censor Y, Elfving T. Block-iterative algorithms with diagonally scaled oblique projections for the linear feasibility problem[J]. SIAM Journal on Matrix Analysis and Applications, 2002, 24(1): 40-58.
[9] Guo W. Research on Reconstruction Algorithms of CT with Incomplete Projection Data[D]. JI LIN university, 2011.
[10] Lee S W, Wang G. A Grangeat-type half-scan algori- thm for cone-beam CT[J]. Medical physics, 2003, 30(4): 689-700.
[11] Wang G, Liu Y, Lin T H, et al. Half - scan cone - beam X - ray microtomography formula[J]. Scanning, 1994, 16(4): 216-220.
[12] Wang G, Jiang M. Ordered-subset simultaneous algebraic reconstruction techniques (OS-SART)[J]. Journal of X-ray Science and Technology, 2004, 12(3): 169-177.
[13] Hammersberg P, Mångård M. Correction for beam hardening artefacts in computerised tomography[J]. Journal of X-ray Science and Technology, 1998, 8(1): 75-93.

Improving University Ranking to Achieve University Competitiveness by Management Information System

M Dachyar, F Dewi

Industrial Engineering Department.
University of Indonesia, Depok 16424, Indonesia

E-mail: mdachyar@yahoo.com

Abstract. One way to increase university competitiveness is through information system management. A literature review was done to find information system factors that affect university performance in Quacquarelli Symonds (QS) University Ranking: Asia evaluation. Information system factors were then eliminated using Delphi method through consensus of 7 experts. Result from Delphi method was used as measured variables in PLS-SEM. Estimation with PLS-SEM method through 72 respondents shows that the latent variable academic reputation and citation per paper have significant correlation to university competitiveness. In University of Indonesia (UI) the priority to increase university competitiveness as follow: (i) network building in international conference, (ii) availability of research data to public, (iii) international conference information, (iv) information on achievements and accreditations of each major, (v) ease of employment for alumni.

Keywords – Competitiveness, Management Information System, University, Quality

1. Introduction

In 2015, ASEAN Economic Community will be implemented among ASEAN countries. ASEAN Economic Community (AEC) aims to make Southeast Asia a hub where products, services, investment, skilled workers, and capitals can move freely among ASEAN countries (ASEAN Secretariat, 2014).

University is not only as teaching establishment, but also as an organisation that creates new knowledge and support social communities (Numprasertchai & Yuen, 2006). University plays important role to increase economic competitiveness in local, provincial, and national scale (Lane, 2012). Based on Quacquarelli Symonds (QS) University Rankings: Asia report, University of Indonesia is ranked eight in ASEAN. This ranking is based on evaluation in academic reputation, employee reputation, faculty student ratio, paper per faculty, international students and academic staff, and inbound/outbound exchange students (QS University Rankings: Asia 2014, 2014). One way to increase organisation competitiveness is through strategic information system (Hemmatfar, Salehi, & Bayat, 2010).

Information system is a virtual system that helps management to organize operations in a company (Mc Leod & Schell, 2008). The basic goal of information system is to prepare organisation to adapt to changes (Salwe, Ahmed, Aloufi, & Kabir, 2010). Several usages of information system are as base for decision-making, virtual communication, e-commerce, knowledge management, etc. (Petter, DeLone,

& McLean, 2008). Information system implementation are consist of three steps; pre-implementation, implementation, and post-implementation.

Competitiveness is the ability to provide products and services in more efficient and effective manner compared to other firms in the same business (Blunck, 2006). Competitiveness is capacity to manage and access company resources efficiently (Rutkauskas, 2008). Competitive advantage is a part of strategic management study (Sigalas & Economou , 2013). There are two schools of teaching in competitive advantage literature. The first stated that competitive advantage is highly related to company performance (Sigalas & Economou , 2013).

Global university ranking serves to increase accountability and transparency, which helps universities in arranging strategy to encounter global competition (Jöns & Hoyler , 2013). There are many methods to rank universities, such as issued by Times Higher Education Supplement (THES), QS Top Universities, US News and World Report, and Shanghai Jiao Tong University.

Quacquarelli Symonds (QS) is a higher education consultancy service provider which was established in 1990. Annually since 2004, QS has issued QS World University Rankings. The evaluation is based on nine factors shown in Table 1.

Table 1. QS University Ranking: Asia Factors

Factors	Description	UI Score in 2014
Academic Reputation (30%)	Number of academic referees endorsing institution in QS Global Academic Survey	80.9
Employer Reputation (10%)	Number of employer referees endorsing institution in QS Global Employer Survey	88.6
Faculty Student Ratio (20%)	The ratio of faculty to students	73.8
Papers per Faculty (15%)	Research papers (Scopus) per faculty member	3.1
Citations per Paper (15%)	Citations per paper – adjusted for institutions with negligible activity in medicine, science and technology	27.1
International Faculty (2.5%)	Proportion of international faculty	75.7
International Students (2.5%)	Proportion of international students	12.4
Inbound Exchange Students (2.5%)	Proportion of inbound exchange students	25.8
Outbound Exchange Students (2.5%)	Proportion of outbound exchange students	15.7

Source: (QS University Rankings: Asia 2014, 2014)

The potential of information system in company management has widely been acknowledged (Díez et al, 2009). In strategic management, resource-based view (RBV) is a dominant concept used by in gaining competitive advantages (Peppard, Galliers, & Thorogood, 2014). The information system can increase company competitiveness if integrated with resources and business process (Louis, Hackney, & Braganza, 2008).

Strategic of Information system can change goals, process, products, organisational interaction with surroundings to increase competitiveness and minimize weaknesses (Turban, King, Viehland, & Lee, 2006). Strategic information system improves company differentiation, development, cost, innovation, and network, which lead to competitive advantages (Jaf, Xinping, & Sabr Jaf , 2012).

2. Method and Data

Delphi method is used to obtain expert feedback from a group of respondents through a series of controlled questionnaire (Hadaya, Cassivi, & Chalabi, 2012). Recommended number of respondent in Delphi ranged from 5 to 14 (Afshari, Yusuff, & Derayatifar, 2012). Delphi is also often used to select criteria for other method in research (Vidal, Marle-Jean, & Bocquet, 2010) (Afshari, Yusuff, & Derayatifar, 2012) (Verhagen & et al, 1998). Criteria selection is done through mean elimination from a group of data based on a certain cut-off point (Broomfield & Humphris, 2001).

Literature review was done to list information system factors that support university performance in nine aspects QS evaluates (Pillai, Khan, Ibrahim, & Raphael, 2012), (Nawaz & Gomes, 2014), (Khan & Zaidi, 2009), (Dukić, 2013), (Solomona, 2008), (Huang Yi, Xiaolan Ao, & Yuh-Shan Ho, 2008), (Bornmann, Schier, Marx, & Daniel, 2012), (Piwowar, Day, & Fridsma, 2007), (Rubbia, Franco, Pellizzonc, & Nannipieri, 2014), (Smith & Rankin, 2002), (Yilin Lu, Mavondo, & Qiu, 2013), (Hasan, 2013), (Gray & Fam, 2003) and (Massey & Burrow, 2012). From the extensive literature study, 67 information system factors that affect QS.

In this research, Delphi questionnaire was disseminated to seven experts from top management and administrative staffs in UI. Questionnaire consisted of 67 questions that aim to evaluate the degree of importance from each information factors to their related QS factors. Likert-5 scale was used ranging from 1 for very unimportant to 5 for very important. Geometric mean for each factor was then calculated. Cut-off point in this research is 4.5. Based on this cut-off point, *Employer Reputation* (10%) and *International Students* (2.5%) were eliminated and are excluded.

Structural Equation Modelling (SEM) is a multivariate method comprised of factor analysis and multiple regressions. SEM facilitates evaluation of relationship between measured variables/indicators and latent constructs/variables simultaneously (Dachyar & Noviannei, 2012). SEM consists of two sub models, outer model/measurement model shows relationship between latent constructs and their measured variables. Inner model/structural model shows relationship between latent constructs (Wong, 2013).

There are two types of indicators in measured model, namely formative and reflective (Wong, 2013). Reflective relationship is where indicators are segmented of their latent constructs, while formative where indicators affect their latent constructs (Freeze & Raschke, 2007). Reflective indicator is represented by one-way arrow pointing outwards from latent construct while formative indicator is represented by one-way arrow pointing from indicator to latent construct (Hair, Ringle, & Marko, 2011).

Minimum recommended sample for SEM is five times amount of measured variables (Dachyar & Hananto, 2014). Several software used for CBSEM are AMOS, EQS, LISREL, and MPlus (Wong, 2013).

PLS-SEM or partial-least square SEM (also called component-based SEM) focus on variance analysis. PLS-SEM use regression to minimalize variance within endogen latent constructs. PLS-SEM first optimizes parameters in measurement model, and then estimates the path coefficient in structural model. In PLS-SEM context, structural model is called inner model and measurement model is called outer model (Hair, Ringle, & Marko, 2011). Thus the name partial-least square is used since this method minimalizes variance partially between inner or outer model. In reflective relationship, coefficient from latent construct to measured variables is used as weight. In formative relationship, coefficient from measured variable to latent construct is used as weight. Approximation of inner and outer model will continue until the parameter estimates come together (Hsu, Chen, & Hsieh, 2006).

Based on the result from Delphi method and some previous literature (Vidaver-Cohen, 2007), (Fraj, 2014), (Memon & Rahman, 2013), (Wong, 2013) and based on interrelation factor in the literature, the proposed model of this research is displyed in Figure 1. The model consists of seven latent variables from factors used in QS University Ranking: Asia and 28 measured variables from information system factors support university in QS evaluation.

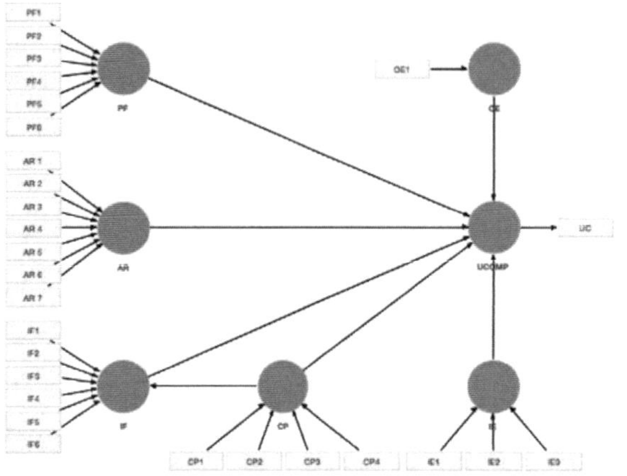

Figure 1. Proposed information system in relation to university competitiveness model

Factors in QS University Ranking: Asia report were used as latent variables and information system factors were used as measured variables. Hypothesis tested in this model are:
H1: Academic Reputation (AR) affects University Competitiveness (UCOMP)
H2: Citation per Paper (CP) affects International Faculty (IF)
H3: Citation per Paper (CP) affects University Competitiveness (UCOMP)
H4: International Exchange (IE) affects University Competitiveness (UCOMP)
H5: International Faculty (IF) affects University Competitiveness (UCOMP)
H6: Outgoing Exchange (OE) affects University Competitiveness (UCOMP)
H7: Paper per Faculty (PF) affects University Competitiveness (UCOMP)

The PLS-SEM questionnaire consisted of 28 questions corresponding to the measured variables in the model. Questions were in form of statement that performance of a certain information system component in UI is suitable, and respondents were asked to rate their agreement to the statement in Likert-5 scale. Respondents were academic and administrative staffs of UI from various faculties.

Initial evaluation was done to assess the reliability and validity of the proposed model with 25 respondents (Hair, Ringle, & Marko, 2011), (Urbach & Frederik, 2010). In this research, SmartPLS software was used.

Measurement model evaluation differs for reflective and formative latent variable (Hair, Ringle, & Marko, 2011). Reflective latent variable is evaluated based on its reliability and validity. Reliability assessment consists of composite reliability (higher than 0.7) and indicator reliability (higher than 0.7). Validity is evaluated through convergent validity (average variance extracted higher than 0.5) and cross loading (each indicator's loading should be higher for its respective latent variable). The reflective latent variable UCOMP in this research pass both the reliability and validity assessments. Formative latent variable doesn't rely on reliability and validity assessment since it is assumed that measured variables in formative latent variable have strong theoretical based. Bootstrapping was done to find significance of weight and loading in each measured variable. From initial assessment, measured variables with both insignificant weight and loading were eliminated. The α in this study is $\alpha = 0.1$ (t-value = 1.65). Variance inflation factor (VIF) evaluation was done to avoid multicollinearity within measured variables (VIF valued should be lower than 5). Measured variables that will be included further in the research after initial model evaluation are AR1, AR2, AR4, AR5, AR6, AR7, CP2, CP3, CP4, IF1, IF2, IF3, IF4, IF6, IE1, IE3, OE1, PF1, PF5, PF6, and UC.

Structural model evaluation was done with R^2 and Q^2 assessments. Endogen latent variable was assessed based on its R^2 value, where value higher than 0.67 means strong, 0.33 means moderate, and 0.19 means weak relationship. Q^2 assessment evaluate if model is predictive enough and latent variable should have Q^2 value higher than 0 (Lee & Chen , 2013).

After initial evaluation, questionnaire was disseminated again to respondents. In the end, 72 respondents were obtained. Evaluation was done again in process similar to the initial evaluation. In the final evaluation, only information system factors is included which deemed to be really important and significant compared to other factors in the same latent variable. The measured variables with significance of either weight or loading below 1.65 were eliminated. Measured variables that were included in the final model were AR5, AR6, AR7, CP3, CP4, IE3, IF1, IF6, OE, PF6, and UC which illustrate in figure 2.

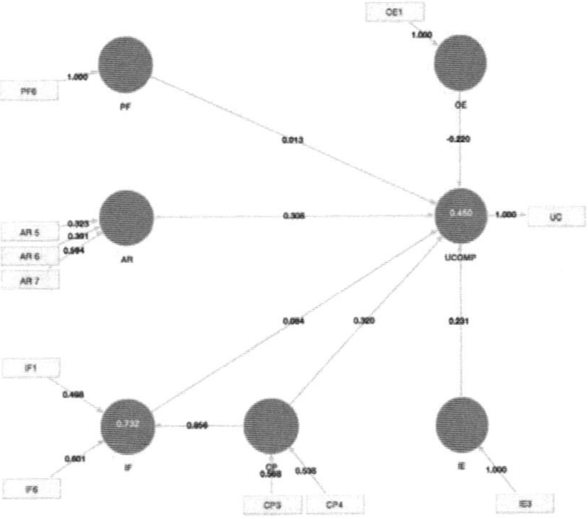

Figure 2. Final Model with Path Coefficients

Result from bootstrapping with 5000 subsamples and α = 0.1 (t-value = 1.65) from model was shown in Table 3.

Table 3. Path significance assessment result
Tabel 3. Significancy *Path*

Hypotesis	Relation	Value-t	Significant
H1	AR → UCOMP	2.0010	Yes
H2	CP → IF	22.9120	Yes
H3	CP → UCOMP	1.9933	Yes
H4	IE → UCOMP	1.5670	No
H5	IF → UCOMP	0.5945	No
H6	OE → UCOMP	1.3825	No
H7	PF → UCOMP	0.1564	No

Based on calculation with PLS-SEM method it is found out that:

1. Academic Reputation (AR) affect University Competitiveness (UCOMP) positive and significantly
2. Citation per Paper (CP) affect International Faculty (IF) positive and insignificantly
3. Citation per Paper (CP) affect University Competitiveness (UCOMP) positive and insignificantly
4. Inbound Exchange Student (IE) affect University Competitiveness (UCOMP) positive and insignificantly
5. International Faculty (IF) affect University Competitiveness (UCOMP) positive and insignificantly
6. Outbound Exchange Student (OE) affect University Competitiveness (UCOMP) negative and insignificantly
7. Paper per Faculty (PF) affect University Competitiveness (UCOMP) positive and insignificantly

Latent variable AR and CP have significant positive relation to UCOMP. Latent variable AR has three measured variables, that are AR5, AR6, and AR7. Latent variable CP has two measured variables CP3 and CP4. The five measured variables were then prioritized as alternatives to improve university competitiveness through information system.

Priority was then arranged with regard to performance score from respondents, total effect from its latent variable to university competitiveness (UCOMP), and its weight. Performance score used was 1/geometric mean from the respondent evaluation. The low score means poor performance which will be alternative to be prioritized. Weight from measured variables AR 5, AR6, AR7, CP 3, and CP4 with respect to their latent variables were also used. Priority scores were obtained by multiplying 1/performance score with Total Effect and Outer Weight. Measured variable with highest priority score will be prioritized. Performance score from respondent, Total Effect, Outer Weight, Priority Score, and Priority of Alternatives are showed in Table 4.

Table 4. Priority in Improving University Competitiveness Through Information System Management

Alternative (Variable)	Performance Score	1/Performance Score	Latent Variable Total Effect	Measured Var. Outer Weight	Priority Score	Priority
1 (AR 5)	3.52	0.28	0.31	0.32	0.03	5
2 (AR 6)	3.32	0.30	0.31	0.39	0.04	4
3 (AR 7)	3.13	0.32	0.31	0.59	0.06	3
4 (CP 3)	2.94	0.34	0.39	0.57	0.08	1
5 (CP 4)	2.88	0.35	0.39	0.54	0.07	2

From the evaluation above, the strategic priority to improve university competitiveness according to UI academic and administrative staffs is as the following:

- 1st Priority: network building in international conference
- 2nd Priority: availability of research data to public
- 3rd Priority: international conference information,
- 4th Priority: information on achievements and accreditations of each major
- 5th Priority: ease of employment for alumni

3. Conclusion

Through literature review, 67 information system factors that support university in QS University Ranking: Asia evaluation were obtained. After elimination with Delphi method by 7 experts, 28 factors were deemed important to be included further in the research. Experts find the variable Academic Reputation (AR), Citation per Paper (CP), Inbound Exchange Student (IE), International Faculty (IF), Outbound Exchange Student (OE), and Paper per Faculty (PF) are important for University Competitiveness (UCOMP).

Result from Delphi was used as base for PLS-SEM method. Assessment from 72 valid questionnaire responses through UI academic and administrative staffs resulted in two latent variables affect University Competitiveness (UCOMP) significantly, which are Academic Reputation (AR) and Citation per Paper (CP). Priorities in improving university competitiveness through information system are subsequently: (i) network building in international conference, (ii) availability of research data to public, (iii) international conference information, (iv) information on achievements and accreditations of each major, (v) ease of employment for alumni. The studies sampled on the UI, results of this study can be applied to public and private universities with same factors with this research .

Reference

(2014, 9 12). *QS University Rankings: Asia 2014.* Top Universities.

Afshari, A., Yusuff, R., & Derayatifar, A. (2012). An application of Delphi method for eliciting criteria in personnel selection problem. *Scientific Research and Essays* , 2927-2935.

Bornmann, L., Schier, H., Marx, W., & Daniel, H.-D. (2012). What factors determine citation counts of publications in chemistry besides their quality? *Journal of Informetrics*, 11-18.

Broomfield, D., & Humphris, G. (2001). Using the Delphi technique to identify the caer education requirements of general practitioners. *Medical Education*, 928-937.

Dachyar, M., & Hananto, L. (2014). Innovation and quality service factors to customer loyalty in Indonesia telecommunication company by using structural equation modeling method. *Management and Technology in Knowledge*, 41-44.

Dachyar, M., & Noviannei, M. (2012). Customer Satisfaction Index Telecommunication Industry in Indonesia. *Engineering and Technology*, 135-136.

Dukić, D. (2013). Online databases as research support and the role of librarians in their promotion: The case of Croatia. *Library Collections, Acquisitions, & Technical Services*, 56–65.

Fraj, E. (2014). Environmental strategies and organizational competitiveness in the hotel industry: The role of learning and innovation as determinants of environmental success. *Tourism Management.*

Freeze, R., & Raschke, R. (2007). AN ASSESSMENT OF FORMATIVE AND REFLECTIVE CONSTRUCTS IN IS RESEARCH. 1481-1492.

Gray, B., & Fam, K. (2003). Cross cultural values and the positioning of international education brands.

Hadaya, P., Cassivi, L., & Chalabi, C. (2012). IT project management resources and capabilities: a Delphi study. *International Journal of Managing Projects in Business*, 216-229.

Hair, J., Ringle, C., & Marko, S. (2011). PLS-SEM: INDEED A SILVER BULLET. *Journal of Marketing Theory and Practice, 19*(2), 139-151.

Hasan, R. (2013). nfluential Factors in Selecting an Overseas University for Bangladeshi Undergraduate Students.

Hsu, S.-S., Chen, W.-H., & Hsieh, M.-J. (2006). Robustness Testing of PLS, LISREL, EQS and ANN-based SEM for Measuring Customer Satisfaction. *Total Quality Management*, 355–371.

Huang Yi, Xiaolan Ao, & Yuh-Shan Ho. (2008). Use of citation per publication as an indicator to evaluate pentachlorophenol research. *Scientometrics*, 67–80.

Issa-Salwe, A., Ahmed, M., & Aloufi, K. (2010). Strategic Information Systems Alignment: Alignment of IS/IT with Business Strategy. *Journal of Information Processing Systems, Vol.6, No.1*, 121-128.

Jaf, R., Xinping, X., & Sabr Jaf , S. (2012). The Effect of the Strategic Information Systems (SIS) on the Achievement Competitive Advantage Practical in Samples of Iraqi Banks. *International Conference of Intelligent Systems Design and Engineering Application* (hal. 954 - 959). IEEE.

Jöns, H., & Hoyler , M. (2013). Global geographies of higher education: The perspective of world university rankings. *Geoforum*, 45-59.

Khan, A., & Zaidi, S. (2009). Online Databases Usage by Research Scholars of the Aligarh Muslim University. *Journal of Library & Information Technology*, 55-60.

Lane, J. (2012). *Higher Education and Economic Competitiveness.* Dipetik 7 10, 2014, dari academia.edu: https://www.academia.edu/2692830/Higher_education_and_economic_competitiveness

Louis, S., Hackney, R., & Braganza, A. (2008). Strategies, Contributions and Inhibitors of Information Systems to Organizational Competitiveness: an Empirical Analysis within the Caribbean. *Journal of International Technology and Information Management, 17.*

Massey, J., & Burrow, J. (2012). Coming to Canada to study: Factors that influence student's decisions to participate in international exchange. *Journal of Student Affairs Research and Practice*, 83–100.

Nawaz, D., & Gomes, A. (2014). Review of Knowledge Management in Higher Education Institutions. *European Journal of Business and Management*, 71-79.

Numprasertchai, S., & Yuen, P. (2006). Enhancing University Competitiveness through ICT based Knowledge Management System. *International Conference on Management of Innovation and Technology* (hal. 417-421). IEEE.

Peppard, J., Galliers, R., & Thorogood, A. (2014). Information systems strategy as practice: Micro strategy and strategizing for IS. *Journal of Strategic Information Systems*, 1-10.

Pillai, S., Khan, M., Ibrahim, I., & Raphael, S. (2012). Enhancing employability through industrial training in the Malaysian context. *Springer*, 187–204.

Piwowar, H., Day, R., & Fridsma, D. (2007). Sharing Detailed Research Data Is Associated with Increased Citation Rate. *PLOS One*, 1-5.

Rubbia, G., Franco, C., Pellizzonc, D., & Nannipieri, L. (2014). Research support services in Higher Education and Research Institutions: approaches, tools and trends. *CRIS*, 309-314.

Smith, S., & Rankin, C. (2002). *Conferences: Why to Attend and How to Benefit.* University of Texas at Austin.

Solomona, D. (2008). A longitudinal comparison of citation rates and growth among open access journals.

Turban, E., King, D., Viehland, D., & Lee, J. (2006). *Electronic Commerce 2006: A Managerial Perspective* (International ed.). NJ: Prentice Hall.

Urbach, N., & Frederik, A. (2010). Structural Equation Modeling in Information Systems Research Using Partial Least Squares. *Journal of Information Technology Theory and Application, 11*(2), 5-44.

Verhagen, A., & et al. (1998). The Delphi List: A Criteria List for Quality Assessment of Randomized Clinical Trials for Conducting Systematic Reviews Developed by Delphi Consensus. *J Clin Epidemiol*, 1235-1241.

Vidal, L.-A., Marle-Jean, F., & Bocquet, J.-C. (2010). Using a Delphi process and the Analytic Hierarchy Process (AHP) to evaluate the complexity of projects. *Expert Systems with Applications*, 5388-5405.

Vidaver-Cohen, D. (2007). Reputation Beyond the Rankings: A Conceptual Framework for Business School Research. *Palgrave Macmillan Ltd*, 278-304.

Wong, K.-K. (2013). Partial Least Squares Structural Equation Modeling (PLS-SEM) Techniques Using SmartPLS. *Marketing Bulletin*, 1-32.

Yilin Lu, Mavondo, F., & Qiu, L. (2013). Factors Influencing the Choice of Overseas Study by Undergraduate and Postgraduate Chinese Students.

IOP Conf. Series: Materials Science and Engineering **83** (2015) 012024 doi:10.1088/1757-899X/83/1/012024

A Study of Residual Amplitude Modulation Suppression in Injection Locked Quantum Cascade Lasers Based on a Simplified Rate Equation Model

J F Webb[1], K S C Yong[2] and M K Haldar[2]

[1] Department of Mechanical, Materials and Manufacturing Engineering, The University of Nottingham Malaysia Campus, Jalan Broga, 43500 Semenyih, Selangor Darul Ehsan, Malaysia
[2] Faculty of Engineering, Computing and Science, Swinburne University of Technology, Sarawak Campus, Jalan Simpang Tiga, 93350 Kuching, Sarawak, Malaysia

E-mail: `jeffwebb@physics.org`

Abstract. Using results that come out of a simplified rate equation model, the suppression of residual amplitude modulation in injection locked quantum cascade lasers with the master laser modulated by its drive current is investigated. Quasi-static and dynamic expressions for intensity modulation are used. The suppression peaks at a specific value of the injection ratio for a given detuning and linewidth enhancement factor. The intensity modulation suppression remains constant over a range of frequencies. The effects of injection ratio, detuning, coupling efficiency and linewidth enhancement factor are considered.

1. Introduction

Quantum cascade lasers (QCLs) are good candidates for wave modulation spectroscopy (WMS) because they are tunable and can operate at room temperature (unlike lead-salt based lasers which are commonly used in WMS). A simple way to achieve frequency modulation of the output from a QCL is to modulate the drive current (direct modulation)[1]. However along with the direct frequency modulation, a significant amount of amplitude modulation, known as residual amplitude modulation (RAM)[2, 3] occurs which adversely affects the sensitivity of WMS. It is possible to suppress RAM using a master-slave QCL system with direct modulation of the master QCL.

Here, results from a model based on the work of Webb and Haldar[4], extended to include injected terms are used to assess the suppression of RAM in injection locked QCLs.

2. Rate Equations and Parameters for Injection Locked QCLs

Using the simplified model of an isolated QCL[4, 5], single mode rate equations for an injection locked slave QCL, ignoring the coupling of the spontaneous emission to the lasing mode, can be written as:

$$\frac{dN_3}{dt} = \frac{I}{e} - \frac{N_3}{\tau_3} - G\left(N_3 - N_2\right)P \tag{1}$$

$$\frac{dN_2}{dt} = \frac{N_3}{\tau_{32}} - \frac{N_2}{\tau_{21}} + G\left(N_3 - N_2\right)P \tag{2}$$

$$\frac{dP}{dt} = \left[NG\left(N_3 - N_2\right) - \frac{1}{\tau_p}\right] P + 2k_c\sqrt{PP_{\text{inj}}}\cos(\phi - \phi_{\text{inj}}) \tag{3}$$

$$\frac{d\phi}{dt} = \frac{\alpha}{2}\left[NG\left(N_3 - N_2\right) - \frac{1}{\tau_p}\right] - \Delta\omega - k_c\sqrt{\frac{P_{\text{inj}}}{P}}\sin(\phi - \phi_{\text{inj}}) \tag{4}$$

where N_i is the electron number in level i; I is the injected current; G is the gain coefficient per stage, which is the product of the confinement factor and the gain due to stimulated emission; τ_{jk}, τ_{32}, and τ_{21} is the lifetime representing a transition from level j to k; τ_p is the photon lifetime; P is the photon number in the cavity and ϕ is the corresponding phase; N is the number of stages; k_c is the coupling efficiency; $\Delta\omega = \omega_{\text{master}} - \omega_{\text{slave}}$ is the detuning frequency; P_{inj} is the injected photon number; ϕ_{inj} is the injected phase; α is the linewidth enhancement factor of the slave laser and t is time. τ_3 is given by $1/\tau_3 = 1/\tau_{31} + 1/\tau_{32}$.

3. Steady state quantities and static locking range

Steady state quantities can be found by solving Eqs. (1) to (4) with the time derivatives set to zero. Such quantities will be denoted by adding a subscript of 0 (e.g. the steady state electron number in level 1 is N_{10}).

It can be shown from the above model that the static locking range is

$$-k_c\sqrt{(1+\alpha^2)R_{\text{inj}}} < \Delta\omega < k_c\sqrt{R_{\text{inj}}} \tag{5}$$

where

$$R_{\text{inj}} = \frac{P_{\text{inj0}}}{P_0} \tag{6}$$

is the injection rate under steady conditions.

4. Suppression of Intensity Modulation

Let $m = \Delta P_{\text{inj0}}/P_{\text{inj0}}$ and $M = \Delta P_0/P_0$, represent the intensity modulation (IM) indices of the master and slave laser respectively. Then an IM suppression quantity can be defined by

$$\frac{M}{m} = \left(\frac{P_{\text{inj0}}}{P_0}\right)\left(\frac{dP_0}{dP_{\text{inj0}}}\right) = R_{\text{inj}}\left(\frac{dP_0}{dP_{\text{inj0}}}\right) \tag{7}$$

By differentiating the appropriate quantities it can be shown that

$$\frac{M}{m} = R_{\text{inj}}\left[R_{\text{inj}} + \frac{I_{\text{thf}}}{I_0}\frac{\tau_p N^2 G^2 \Delta(N_{d0})^2 P_0\sqrt{(1+\alpha^2)k_c^2 R_{\text{inj}} - \Delta\omega^2}}{k_c^2 K}\right]^{-1} \tag{8}$$

with

$$P_0 = K\left(\frac{I_0}{I_{\text{thf}}}\frac{1}{\tau_p NG\Delta N_{d0}} - 1\right)$$

$$\Delta N_{d0} = N_{30} - N_{20} = \frac{I_0}{e}\frac{\tau_3\left(1 - \tau_{21}/\tau_{32}\right)}{1 + P_0/K}$$

and $K = \left[\tau_3\left(1 + \tau_{21}/\tau_{31}\right)G\right]^{-1}$.

5. Frequency Response of IM Suppression

For a sinusoidal current of angular frequency ω_m superimposed on the bias current of the master laser, a dynamic analysis leads to the frequency response of the IM suppression ratio being given by

$$\left|\frac{M}{m}\right| = \left|\frac{D_{IM}}{P_0\Lambda(j\omega_m)}\right| \tag{9}$$

where $j = \sqrt{-1}$,

$$\Lambda(s) = s^4 + d_3 s^3 + d_2 s^2 + d_1 s + d_0$$

with coefficients given by

$$\begin{aligned}
d_3 =& \frac{1}{\tau_3} + \frac{1}{\tau_{21}} + 2GP_0 - \left(NG\Delta N_{d0} - \frac{1}{\tau_p}\right)\\
d_2 =& -k_c\left(NG\Delta N_{d0} - \frac{1}{\tau_p}\right)\sqrt{R_{\text{inj}}}\cos(\Delta\phi_0) - \left(\frac{1}{\tau_3} + \frac{1}{\tau_{21}}\right)\left(NG\Delta N_{d0} - \frac{1}{\tau_p}\right)\\
&+ \frac{2GP_0}{\tau_p} + \frac{1}{\tau_3\tau_{21}} + \left(\frac{1}{\tau_{31}} + \frac{1}{\tau_{21}}\right)GP_0 - k_c^2 R_{\text{inj}}\left(\cos^2(\Delta\phi_0) - \sin^2(\Delta\phi_0)\right)\\
d_1 =& -k_c\left(\frac{1}{\tau_3} + \frac{1}{\tau_{21}}\right)\left(NG\Delta N_{d0} - \frac{1}{\tau_p}\right)\sqrt{R_{\text{inj}}}\cos(\Delta\phi_0) + \frac{2k_c GP_0\sqrt{R_{\text{inj}}}\cos(\Delta\phi_0)}{\tau_p}\\
&- 2k_c^2 GP_0 R_{\text{inj}}\left(\cos^2(\Delta\phi_0) - \sin^2(\Delta\phi_0)\right) - \frac{NG\Delta N_{d0} - 1/\tau_p}{\tau_3\tau_{21}} + \left(\frac{1}{\tau_{31}} + \frac{1}{\tau_{21}}\right)\frac{GP_0}{\tau_p}\\
&- 2\alpha k_c NG^2\Delta N_{d0}P_0\sqrt{R_{\text{inj}}}\sin(\Delta\phi_0) - \left(\frac{1}{\tau_3} + \frac{1}{\tau_{21}}\right)k_c^2 R_{\text{inj}}\left(\cos^2(\Delta\phi_0) - \sin^2(\Delta\phi_0)\right)\\
d_0 =& -\frac{k_c\left(NG\Delta N_{d0} - 1/\tau_p\right)\sqrt{R_{\text{inj}}}\cos(\Delta\phi_0)}{\tau_3\tau_{21}} + \left(\frac{1}{\tau_{31}} + \frac{1}{\tau_{21}}\right)\frac{k_c GP_0\sqrt{R_{\text{inj}}}\cos(\Delta\phi_0)}{\tau_p}\\
&- \left(\frac{1}{\tau_{31}} + \frac{1}{\tau_{21}}\right)\alpha k_c NG^2\Delta N_{d0}P_0\sqrt{R_{\text{inj}}}\sin(\Delta\phi_0) - \frac{k_c^2 R_{\text{inj}}}{\tau_3\tau_{21}}\left(\cos^2(\Delta\phi_0) - \sin^2(\Delta\phi_0)\right)\\
&- \left(\frac{1}{\tau_{31}} + \frac{1}{\tau_{21}}\right)k_c^2 GP_0 R_{\text{inj}}\left(\cos^2(\Delta\phi_0) - \sin^2(\Delta\phi_0)\right)
\end{aligned}$$

and

$$\begin{aligned}
D_{IM} =& \Big[k_c\sqrt{R_{\text{inj}}}P_0\cos(\Delta\phi_0)\left(j\omega_m + k_c\sqrt{R_{\text{inj}}}\cos(\Delta\phi_0)\right) + k_c\sqrt{R_{\text{inj}}}P_0\sin(\Delta\phi_0)\\
&\times\left(\alpha' j\omega_m + k_c\sqrt{R_{\text{inj}}}\sin(\Delta\phi_0)\right)\Big]\\
&\times\left[-\omega_m^2 + \left(\tau_3^{-1} + \tau_{21}^{-1} + 2GP_0\right)j\omega_m + \left(\tau_{31}^{-1} + \tau_{21}^{-1}\right)GP_0 + (\tau_3\tau_{21})^{-1}\right]
\end{aligned}$$

in which α' is the linewidth enhancement factor of the master QCL.

6. IM Suppression Plots and Discussion

In this section some plots of the IM suppression as measured through the suppression ratio in Eqs. (8) and (9) will be given and discussed.

The parameter values used are given in Table 1. There are three other required parameters: the coupling efficiency, k_c and the linewidth enhancement factors α and α'. Following Ref. [5] $k_c = v_g(1-R)/(2L\sqrt{R}) = 27.7\,\text{ns}^{-1}$ (using the values in Table 1). Note that the formula for k_c is derived for a plane wave in a Fabry-Perot cavity with a beam cross-section that matches the

Table 1. Typical parameters of for a slave QCL[7]

Parameter	Value
Wavelength, λ	8μm
Temperature, T	10K
Cavity width, W	14μm
Cavity length, L	2.25mm
Facet power reflection coefficient, R	0.28
Length of a single gain stage, L_p	44.3nm
Number of gain stages, N	30
Group velocity, v_g	$(3/3.28)\times10^8$ms^{-1}
Mirror loss coefficient, α_m	5.6cm^{-1}
Internal loss coefficient, α_i	24cm^{-1}
Confinement factor per gain stage, Γ	0.016
Gain coefficient per stage, G	3.4×10^3s^{-1}
τ_{31}	3.6ps
τ_{32}	3.1ps
τ_{21}	0.3ps
τ_p	3.69ps
Threshold current, I_{thf}	0.28A

active region of the slave laser. This cross-section is obtained by transforming the waist of the Gaussian beam of the master laser so that it equals that of the slave laser.

For the linewdith enhancement factors we set $\alpha = \alpha' = 0.3$[6].

The detuning frequency and injection ratio values in the work here are chosen such that the IM suppression study is carried out within the locking range given by expression (5).

In what follows the IM suppression is plotted in decibels and is defined by $-20\log(M/m)$; with this definition, large values imply better IM suppression of the slave laser light.

6.1. Quasi-static IM suppression

Figs. 1 to 4 show how IM suppression varies with injection ratio R_{inj} for different parameter values. In all cases a single maximum is evident so that there is a particular injection ratio value that optimizes the suppression. In Fig. 1 it can be seen that overall, the IM suppression curve shifts in the direction of higher injection ratio values as the coupling efficiency decreases. Fig. 2 compares IM suppression plots at different injected currents. In both figures, a suppression above 40 dB is apparent, consistent with experimental observations[6]. Figs. 3 and 4 show IM suppression versus injection ratio for different detuning frequencies and linewidth enhancement factors respectively. Suppression increases as detuning frequency decreases, and also increases with increasing linewidth enhancement. The suppression peak moves to higher injection ratios when the detuning frequency increases and moves in the same direction as the linewidth enhancement factor decreases.

At the edge of the static locking range, IM suppression, as is evident from (8), is 0 dB ($M/m = 1$). For all of the curves discussed so far, the suppression decreases monotonically after the peak value is reached—the same behaviour as is observed in injection-locked interband lasers[8].

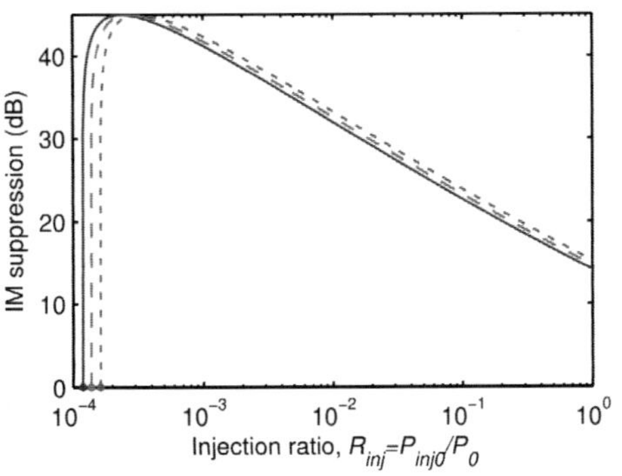

Figure 1. IM suppression against injection ratio for $k_c = 27.7\text{ns}^{-1}$ (solid line), 25.7ns^{-1} (dashed-line, long dashes) and 23.7ns^{-1} (dashed-line, short dashes). Detuning frequency, injected current and linewidth enhancement factor are set at -50MHz, $1.6I_{\text{thf}}$ and 0.3 respectively. IM suppression is given by $-20\log(M/m)$ (dB).

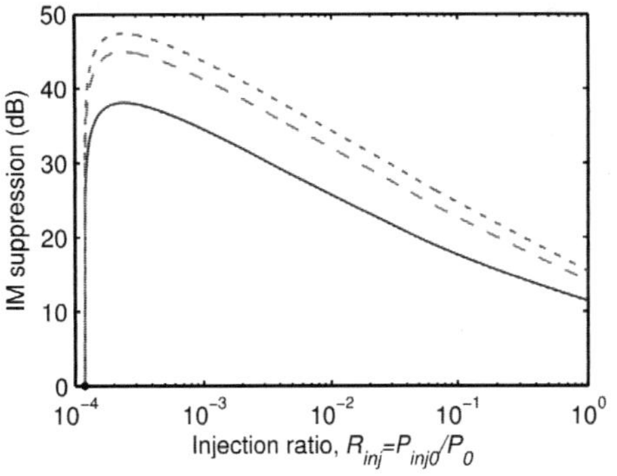

Figure 2. IM suppression against injection ratio for $I_0 = 1.2I_{\text{thf}}$ (solid line), $1.6I_{\text{thf}}$ (dashed-line, long dashes) and $2I_{\text{thf}}$ (dashed-line, short dashes). Detuning frequency and linewidth enhancement factor are set at -50MHz and 0.3 respectively.

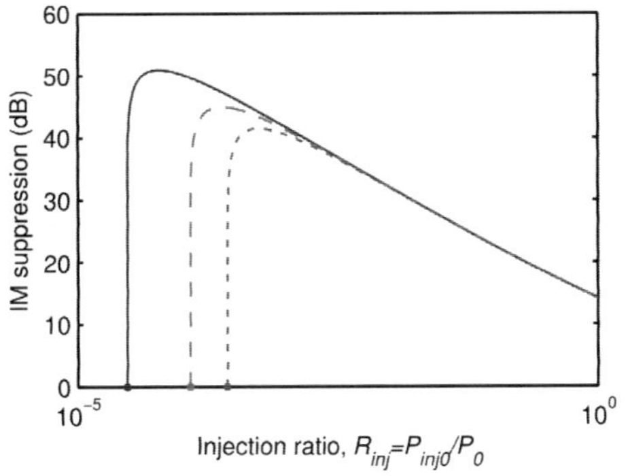

Figure 3. IM suppression against injection ratio for $\Delta f = -25$MHz (solid line), -50MHz (dashed-line, long dashes) and -75MHz (dashed-line, short dashes). Injected current and linewidth enhancement factor are set at $1.6I_{\text{thf}}$ and 0.3.

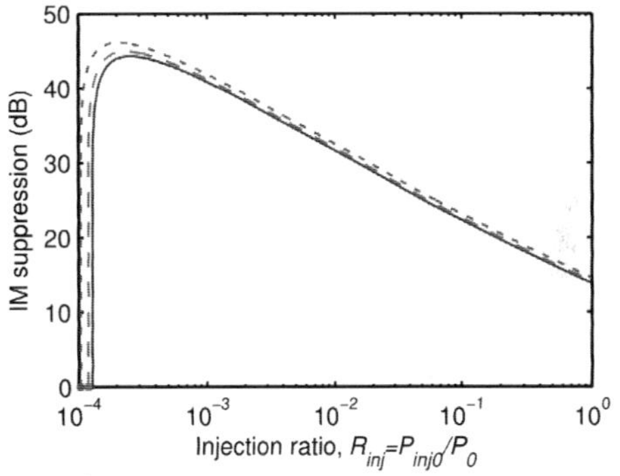

Figure 4. IM suppression against injection ratio for $\alpha = 0.1$ (solid line), 0.3 (dashed-line, long dashes) and 0.5 (dashed-line, short dashes). Detuning frequency and injected current are set at -50MHz and $1.6I_{\text{thf}}$ respectively.

6.2. Frequency response of IM suppression

The variation of IM suppression against modulating frequency is plotted in Fig. 5 for different values of injection ratio, and in 6 for different values of injection current. As the modulating frequency approaches zero, IM suppression approaches its quasi-static value. However, unlike interband lasers in which the IM suppression decreases at high modulating frequencies, the suppression observed here does not vary greatly with modulating frequency. This is because QCLs are overdamped systems, whereas interband lasers are underdamped.

IOP Conf. Series: Materials Science and Engineering **83** (2015) 012024 doi:10.1088/1757-899X/83/1/012024

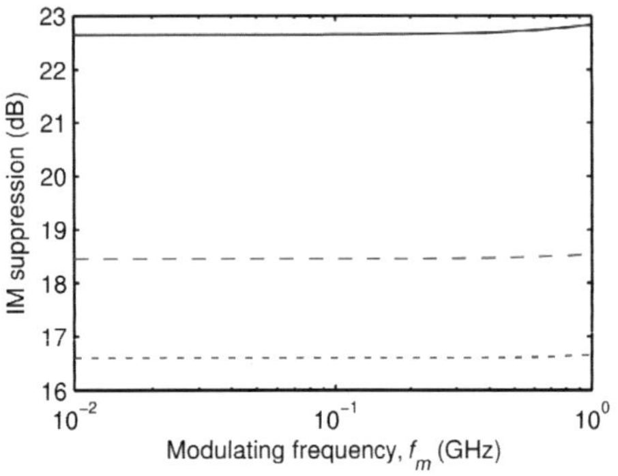

Figure 5. IM suppression against modulating frequency for injection ratio of 0.1 (solid line), 0.3 (dashed-line, long dashes) and 0.5 (dashed-line, short dashes). Detuning frequency, injected current and linewidth enhancement factor for both master and slave QCLs are set at -50MHz, $1.6I_{\text{thf}}$ and 0.3 respectively. IM suppression is given by $-20\log(M/m)$(dB).

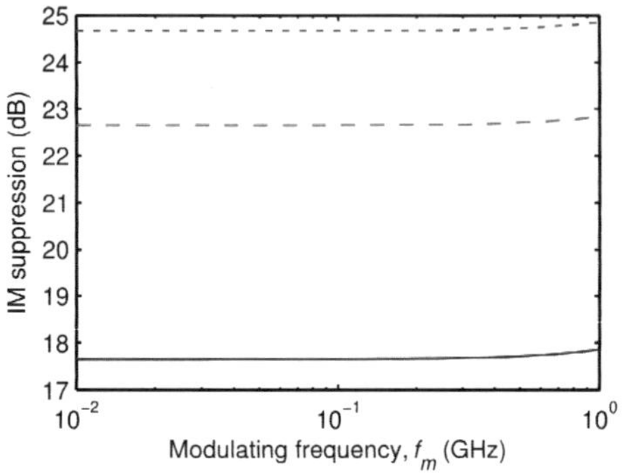

Figure 6. IM suppression against modulating frequency for $I_0 = 1.2I_{\text{thf}}$ (solid line), $1.6I_{\text{thf}}$ (dashed-line, long dashes) and $2I_{\text{thf}}$ (dashed-line, short dashes). Detuning frequency, injection ratio and linewidth enhancement factor for both master and slave QCLs are set at -50MHz, 0.1 and 0.3 respectively.

7. Conclusion

In this paper, expressions derived from a relatively simple two-level model have been used to study IM suppression for an injection locked slave QCL. A high degree of suppression is favourable for applications such as wave modulation spectroscopy. It has been demonstrated that maximum IM suppression occurs at a specific injection ratio for a given detuning frequency value and linewidth enhancement factor. Large injected currents, small detuning frequencies, and large linewidth enhancement factors increase the suppression. The IM suppression peaks at a lower injection ratios as the detuning frequency decreases and as linewidth enhancement factor increases. IM suppression remains almost constant throughout the range of modulating frequencies considered.

References

[1] G Gagliard and S B, Tamassia F, Capasso F, Gmachl C, Sivco D L, Baillargeon J N, Hutchinson A L and Cho A 2005 *Isotopes in Environmental and Health Studies* **41** 313–321
[2] Schilt S, Thevenaz L and Robert P 2003 *Appl. Opt.* **42** 6728–6738
[3] Silver A 1992 *Appl. Opt.* **31** 707–717
[4] Webb J F and Haldar M K 2012 *J. App. Phys.* **111** 093116–1–093116–6
[5] Wang C, Grillot F, Kovanis V I, Bodyfelt J D and Even J 2013 *Opt. Lett.* **38** 1975–1977
[6] Taubman M S, Myers T L, Cannon B D and Williams R M 2004 *Spectrachimica Acta Part A* **60** 3457–3468
[7] Gmachl C, Capasso F, Tredicucci A, Sivco D L, Köhler R, Hutchinson A L and Cho A Y 1999 *IEEE J. Sel. Topics Quantum Electron.* **5** 808–816
[8] Haldar M K, Coetzee J C and Gan K B 2005 *IEEE J. Quantum. Electron.* **41** 280–286

AUTHOR INDEX

Institute of Physics
Dirac House, Temple Back
Bristol BS1 6BE UK

CURRAN ASSOCIATES INC
proceedings.com